DIET AND CANCER

Molecular Mechanisms of Interactions

ADVANCES IN EXPERIMENTAL MEDICINE AND BIOLOGY

Recent Volumes in this Series

Volume 368
HEPATIC ENCEPHALOPATHY, HYPERAMMONEMIA, AND AMMONIA TOXICITY
Edited by Vicente Felipo and Santiago Grisolia

Volume 369
NUTRITION AND BIOTECHNOLOGY IN HEART DISEASE AND CANCER
Edited by John B. Longenecker, David Kritchevsky, and Marc K. Drezner

Volume 370
PURINE AND PYRIMIDINE METABOLISM IN MAN VIII
Edited by Amrik Sahota and Milton W. Taylor

Volume 371A
RECENT ADVANCES IN MUCOSAL IMMUNOLOGY, Part A: Cellular Interactions
Edited by Jiri Mestecky, Michael W. Russell, Susan Jackson, Suzanne M. Michalek, Helena Tlaskalová, and Jaroslav Sterzl

Volume 371B
RECENT ADVANCES IN MUCOSAL IMMUNOLOGY, Part B: Effector Functions
Edited by Jiri Mestecky, Michael W. Russell, Susan Jackson, Suzanne M. Michalek, Helena Tlaskalová, and Jaroslav Sterzl

Volume 372
ENZYMOLOGY AND MOLECULAR BIOLOGY OF CARBONYL METABOLISM 5
Edited by Henry Weiner, Roger S. Holmes, and Bendicht Wermuth

Volume 373
THE BRAIN IMMUNE AXIS AND SUBSTANCE ABUSE
Edited by Burt M. Sharp, Toby K. Eisenstein, John J. Madden, and Herman Friedman

Volume 374
CELL ACTIVATION AND APOPTOSIS IN HIV INFECTION: Implications for Pathogenesis and Therapy
Edited by Jean-Marie Andrieu and Wei Lu

Volume 375
DIET AND CANCER: Molecular Mechanisms of Interactions
Edited under the auspices of the American Institute for Cancer Research

---

A Continuation Order Plan is available for this series. A continuation order will bring delivery of each new volume immediately upon publication. Volumes are billed only upon actual shipment. For further information please contact the publisher.

DIET AND CANCER

Molecular Mechanisms of Interactions

Edited under the auspices of the

American Institute for Cancer Research

Washington, D.C.

PLENUM PRESS • NEW YORK AND LONDON

Library of Congress Cataloging-in-Publication Data

Diet and cancer : molecular mechanisms of interactions / edited under the auspices of the American Institute for Cancer Research.
p. cm. -- (Advances in experimental medicine and biology ; v. 375)
"Proceedings of the American Institute for Cancer Research's Fifth Annual Conference onDiet and Cancer: Molecular Mechanisms of Interactions, held September 1-2, 1994, in Washington, D.C."--t.p. verso.
Includes bibliographical references and index.
ISBN 0-306-45067-4
1. Cancer--Nutritional aspects--Congresses. 2. Cancer--Molecular aspects--Congresses. I. American Institute for Cancer Research. II. Conference on Diet and Cancer: Molecular Mechanisms of Interactions (1994 : Washington, D.C.) III. Series.
[DNLM: 1. Neoplasms--drug therapy--congresses. 2. Neoplasms--prevention & control--congresses. 3. Molecular Biology--congresses. AD559 v. 375 1995 / QZ 267 1995]
RC267.45.D53 1995
616.99'4--dc20
DNLM/DLC
for Library of Congress 95-17694
CIP

Proceedings of the American Institute for Cancer Research's Fifth Annual Conference on Diet and Cancer: Molecular Mechanisms of Interactions, held September 1–2, 1994, in Washington, D.C.

ISBN 0-306-45067-4

A Division of Plenum Publishing Corporation
233 Spring Street, New York, N. Y. 10013

10 9 8 7 6 5 4 3 2 1

Printed in the United States of America

PREFACE

The fifth of the annual research conferences of the American Institute for Cancer Research was held September 1-2, 1994, at the L'Enfant Plaza Hotel in Washington, DC. Appropriately, in view of current directions in research, the theme was "Diet and Cancer: Molecular Mechanisms of Interactions". This proceedings volume contains chapters from the platform presentations and abstracts from the poster session held on the end of the first day.

The subtopics for the three sessions held were "Retinoids, Vitamins A and D in Cancer Prevention and Therapy," "Choline and Lipids: Signal Transduction, Gene Expression and Growth Regulation," and "Dietary Factors and Regulation of Oncogenes, Growth and Differentiation." A general overview on vitamins A and D emphasized that A and D, in addition to their established roles in vision, reproduction, and bone mineral homeostasis, may play significant roles in regulating cell function. Vitamin A metabolites, *trans*-retinoic acid and 9-*cis*-retinoic acid, regulate growth and differentiation. Furthermore, vitamin A-deprived animals were more susceptible to both spontaneous and carcinogen-induced tumors. Epidemiological studies showed a correlation between low A intake and higher incidences of certain types of human cancers. Conversely, all-*trans* retinoic acid is useful in treatment and control of certain types of cancer.

Physiologically, Vitamin D is converted to the active form, 1,25-dihydroxyvitamin D_3 (VD_3). VD_3 regulates hormone production and secretion, myocardial contractility, vascular tone, and growth inhibition and differentiation. Although therapeutic use of VD_3 has been limited due to hypercalcemia and hypercalciurea, new analogs have been developed that retain the beneficial effects of VD_3 but show decreased action on calcium mobilization.

Molecular studies are demonstrating that retinoic acid (RA) functions through three receptor subtypes, alpha, beta and gamma (RARs), while the receptor proteins for 9-*cis* retinoic acid are termed RXR, also found as three subtypes. In the B16 mouse melanoma cell line, the alpha and gamma forms of RAR were expressed, but the mRNA for the beta type was induced by RA, independent of new protein synthesis. However, cyclic adenosine monophosphate can antagonize the actions of RA through inhibition of RAR expression.

Retinoids may have a beneficial action even in cancers where a viral factor plays a role. Cervical cancer is fairly common in the USA, and it represents a leading cause of cancer deaths in many countries. Human papillomavirus (HPV) has been detected in most cervical cancers and is implicated in the process of cancer development. Studies with cervical cell lines in culture led to the conclusion that cells immortalized with HPV16 had greater sensitivity to retinoids which suppressed their growth. Cytokeratin expression may serve as a useful marker to follow the course of retinoid therapy. However, retinoids did not inhibit proliferation of the HPV infected cells by acting on the viral E6 and E7 factors that interact with at least two tumor suppressor genes, namely p53 and pRB. Interferon reduced the expression of E6 and E7, while retinoids may suppress the activity of the epidermal growth factor and insulin-like growth factor signaling pathways. Thus a combination of interferon and retinoid therapy would have an enhanced beneficial action in inhibiting the progress of disease since the two agents act by distinct but reinforcing pathways. Likewise, human keratinocytes immortalized by HPV16 DNA require epidermal growth factor and bovine pituitary extract for proliferation, but after prolonged culture, some cell lines became independent. The early passages of these lines were very sensitive to control by RA, but they lost sensitivity to both RA and transforming growth factor with progress in culture. Thus retinoids may be most effective in prevention and treatment of HPV-induced lesions at early preneoplastic stages,

A double-blind, placebo-controlled six month intervention study with a combination of β-carotene and retinol in subjects at high risk of lung cancer, due to smoking, and of parenchymal fibrosis, due to exposure to asbestos, has afforded the opportunity to investigate relevant biomarkers to follow any effect. It appears that metaplasia in airway biopsies as well as examination of macrophages from bronchoalveolar lavage for inflammatory cells, combined with analyses for retinol, retinyl esters, carotenoids and vitamin E levels, may yield data on a suitable early biomarker of lung disease in humans. The analyses should also provide a basis for correlation of clinical status and vitamin levels.

Breast cancer in women and prostate cancer in men continue to increase despite programs on early detection and improved treatment. Thus the papers on the effects of Vitamin D on human breast cancer cells, specifically the MCF-7 line and on several human prostate lines were quite relevant. The multifaceted action of this vitamin, or rather the VD_3 metabolite, was shown. In MCF-7 cells, VD_3 caused the cells to undergo apoptosis (programmed cell death), not by causing DNA fragmentation but by increasing expression of proteins associated with the various phases of apoptosis, by decreasing cell numbers through G_O/G_1 arrest, or by disrupting estrogen-dependent signals. Delineation of the specific targets necessary for triggering the apoptotic process may lead to improved therapeutic agents. Receptors of VD_3 were also detected in various human prostate cancer cell lines and in primary cultures from normal, benign prostatic hyperplasia, and malignant prostate. VD_3 led to growth inhibition in these cancer cell lines, and stimulation of prostate specific antigen, thus indicating potential therapeutic utility in prostate cancer.

Turning to other dietary factors, the role of choline deficiency in promoting hepatocarcinogenesis in rats was explored. Lack of choline led to disturbing two types of molecular mechanisms. One, protein kinase C (PKC)-mediated signal transduction, involves phosphatidylcholine and a series of messengers such as 1,2-sn-diacylglycerol and certain unsaturated free fatty acids that sustain the PKC phosphorylation cascade. In turn, PKC targets many proteins involved in control of gene expression. In choline-deficient rats, these activators of PKC are increased; sustained activation may be associated with fatty liver and eventual tumor. In addition, regulation of apoptosis is also perturbed in choline deficiency. Although feeding such a deficient diet to rats for four weeks led to many changes in biochemical parameters, resumption of an adequate diet reversed many of the effects in a few weeks. However, lipotrope deficiency of longer than one month led to changes in lipid storage in the liver and loss of methylation of specific sites (CCGG sites) in c-Ha-ras, c-fos and p53 genes. Fortunately, the usual normal human diet is not likely to lead to lipotrope deficiency, comparable to that used in the animal studies. Under some specific conditions, hypomethylation may contribute to cancer in humans. Further study of lipotrope deficiency, specifically methionine deprivation, indicated that the deficiency affects start codon selection and leads to alterations of the c-myc gene. The balance of production of c-myc-l protein, which inhibits cell growth, and of c-myc- 2 protein which is involved in tumorigenesis, is perturbed in the deficient state. Normally these proteins balance each other and regulate transcription.

There continues to be controversy on the role of dietary fat in breast cancer, although there are many animal studies showing a clear correlation. For example, investigation of high-versus low-fat diets in mice bearing the mouse mammary tumor virus showed that a high fat diet accelerated hormonally controlled gene expression or transcription of the mtv-l locus in C3Hf female mice, leading to early expression of the mtv-l specific transcripts and early development of tumors after fewer litters of young. Conflicting indications came from two rat studies. In one, rats were implanted with cells from a mammary tumor line and given either fish oil or safflower oil in the diet. Tumor growth was approximately 30% slower in the fish oil group, and there was a significant prolongation of the S phase of the cell cycle, compared with the safflower oil group. However, in a trial with nitrosomethylurea-induced mammary tumors in female rats, substitution of fish oil for half the dietary corn oil had no effect on ras mutations in mammary tissue, despite the differences *in vivo*. Further investigation of the mechanisms is needed.

Likewise, the association between dietary factors and colon cancer is often obscured by confounding factors. The roles and interactions of various genetic alterations in the heterogeneity of the disease are not understood. The hereditary conditions which lead to a high frequency of colon cancer are evident at a relatively late period in life. This inherited condition led to pleiotropic and heterogeneous effects on gene expression. Nevertheless, a pattern characteristic of intestinal mucosa at risk could be distinguished. Certain short-chain fatty acids, produced by fermentation of fiber, could induce differentiation and a return of expression of mitochrondrial genes to a low-risk category.

There have been many advances in understanding the molecular mechanisms involved in carcinogenesis, which are now being applied to the problems of the diet-cancer interaction. Despite the advances, there remain many avenues to be explored. But continued effort should provide a better picture of the exact molecular events implicated in diet-cancer interactions and lead both to better means of prevention and of therapy of cancer.

CONTENTS

Chapter 4
Role of Apoptosis in the Growth Inhibitory Effects of Vitamin D in MCF-7 Cells 45
JoEllen Welsh, Maura Simboli-Campbell, Carmen J. Narvaez, and Martin Tenniswood

Chapter 5
Vitamin D and Prostate Cancer ... 53
David Feldman, Roman J. Skowronski, and Donna M. Peehl

Chapter 6
Choline and Hepatocarcinogenesis in the Rat ... 65
Steven H. Zeisel, Kerry-Ann da Costa, Craig D. Albright, and Ok-Ho Shin

1

USE OF VITAMINS A AND D IN CHEMOPREVENTION AND THERAPY OF CANCER: CONTROL OF NUCLEAR RECEPTOR EXPRESSION AND FUNCTION

Vitamins, Cancer and Receptors

R. M. Niles

Department of Biochemistry and Molecular Biology
Marshall University School of Medicine
1542 Spring Valley Drive
Huntington, WV 25755

INTRODUCTION

Vitamin A is an essential nutrient which is well known for its role in vision and reproduction [1]. Vitamin D is best known for its role in maintaining bone mineral homeostasis. However, over the last 8-10 years, research results have indicated that the active metabolites of these vitamins play a much broader role in regulating the functions of a variety of cell types[2-4]. The vitamin A metabolites, all-*trans* retinoic acid and 9-*cis* retinoic acid regulate growth and differentiation, while retinaldehyde regulates vision. Vitamin D is converted to the active form, 1,25-dihydroxyvitamin D_3 (VD_3). In addition to regulating Ca^{++} homeostasis, it regulates hormone production and secretion, myocardial contractility and vascular tone, and growth and differentiation[5-7]. Early observations that vitamin A deprived animals had a higher incidence of spontaneous and carcinogen-induced tumors suggested a link between this vitamin and cancer. This link was strenthened by epidemiological studies that found a correlation between low vitamin A intake in humans and a higher incidence of certain types of cancers[8,9]. Laboratory studies have established that all-*trans* retinoic acid can inhibit the proliferation and stimulate the differentiation of many different types of tumor cells in vitro[10]. These results have led to use of all-*trans* retinoic acid in treating certain cancers such as promyelocytic leukemia, where it induces a high rate of remissions, which are unfortunately of short duration[11]. This same vitamin A metabolite shows promise for treatment of early pre-malignant lesions of head and neck cancer[12] and in combination with γ interferon for controling cervical cancer[13]. Vitamin D_3 receptors have been found in a variety of cells not normally associated with mineral metabolism. The presence of these receptors implies that the cells are responsive to vitamin D. Indeed, vitamin D_3 has been found to inhibit human breast cancer cell growth in culture[14] and to induce the differentiation of certain leukemic

Diet and Cancer, Edited under the auspices of the American Institute of Cancer Research
Plenum Press, New York, 1995

cells[7]. Use of vitamin D_3 for cancer therapy has been limited by problems with hypercalcemia and hypercalciuria. New analogs of vitamin D which retain their ability to inhibit tumor growth or induce differentiation, but have diminished calcium moblilizing activity have recently been developed. Current research is exploring the feasibility of using these new analogs to treat breast and prostate cancer. Vitamin D_3 and retinoic acid receptors are located in the nucleus and have a structural organization similar to steroid receptors. While there is only one known receptor for vitamin D_3, there are three retinoic acid receptor (RAR) subtypes; RARα, RARβ and RARγ[15]. Each of these subtypes can generate several isotypes by using different promoters or by differential splicing of the initial transcript[16]. In addition, there is another class of receptors termed RXR, which have some homology to RAR, but specifically bind the vitamin A metabolite 9-*cis*-retinoic acid[17]. There are three subtypes of RXR (RXRα, β, and γ), each of which can form heterodimers with both the RAR, vitamin D_3 receptors, as well as other steroid receptors[18,19]. Since the RARs are expressed in low copy number (~1,000 receptors/cell), it is possibile that the control of response to retinoic acid could be governed by the availability of receptors. To examine this possibility, we tested the ability of several regulatory molecules to alter the expression of these receptors. We found that retinoic acid itself could increase the expression of RARβ and RARγ, while cyclic AMP decreased the expression of all three RAR subtypes.

MATERIALS AND METHODS

Cell Culture

B16-F1 cells were obtained from Dr. J. Fidler (M.D. Anderson Hospital and Tumor Clinic, Houston, TX). Cells under passage 25 were plated at $2x10^5$/100mm dish and cultured in Delbecco's Modified Eagle's Medium containing 10% heat-inactivated calf serum (HyClone, Logan UT), 2 mM L-glutamine, 1 mM sodium pyruvate, 50 U/ml penicilin G and 50 μg/ml streptomycin sulfate. Retinoic acid (Fluka Chemicals) at a final concentration of 10^{-5} M in DMSO and/or 8-bromo-cyclic AMP (Sigma) plus 1-methyl, 3-isobutylxanthine (IBMX) (Aldrich Chemicals) at a final concentration of 0.5 mM and 0.2 mM respectively were added to cells 24 h after seeding in order to avoid any potential effect of these agents on cell attachment. Control groups were treated with medium containing the retinoic acid solubilization vehicle (DMSO). Cells were harvested for RNA or protein extraction after various times in culture. For experiments where these agents were exposed to cells for shorter time periods (no more than 8 h), cells were plated at $5x10^5$/100 mm dish.

RNA Isolation and Northern Analysis

Total RNA was isolated by the acid guanidinium thiocyanate/phenol-chloroform single-step method[20]. Poly A^+ RNA was prepared using QuickPrep™ Micro mRNA purification kits (Pharmacia). RNA samples were electrophoretically separated on a formaldehyde agarose gel and transferred to Hybond-N (Amersham) nylon membrane as described by Sambrook et al[21]. Glyceraldehyde phosphate dehydrogenase (GAPDH) cDNA was purchased from the American Type Culture Collection. Human RARα[22], RARβ[23], and RARγ[24] cDNA fragments shown in Fig. 1 were labeled with $dCTP^{32}$ (DuPont NEN) using a random priming kit (Multi-Prime, Amersham). Hybridizations were carried out in 5X SSC, 50% formamide, 1X Denhardt's and 0.1% SDS at 42° C overnight aftr pre-hybridization with 20 μg/ml Herring sperm DNA for 2 h. Blots were sequentially washed in the following conditions until low background was achieved: 0.6X SSC, 0.1% SDS at room temperature for 2 x 15 min; 0.6X SSC, 0.1% SDS at 42° C for 2 x 15 min; 0.6X SSC, 0.1% SDS at 65C for 2 x 15 min; 0.1X SSC, 0.1% SDS at 42° C for 15 min. Blots were exposed to XAR film (Kodak) at -70° C with an enhancing screen for 2 h to 1 week. Autoradiograms were quantitated by densitometry using a Molecular Dynamics instrument with appropriate imaging and quantitation software.

Protein Extraction and Western Analysis

Whole cell extracts were prepared as described previously[25]. Briefly, a protease inhibitor cocktail was used to maintain the integrity of proteins and high salt (0.6 M KCl) to extract proteins that may be tightly associated to certain subcellular structures. Protein concentrations were determined by BCA (Pierce) protein assay. Proteins were separated by 10% SDS-polyacrylamide gel electrophoresis and then electrophoretically transferred to Hybond C-Extra (Amersham) membrane. Rainbow™ protein molecular weight markers (Amersham) were simultaneously run on each gel to determine the molecular weight of the protein bands. The transferred membranes were incubated overnight at room temperature in blocking solution: 5% nonfat dry milk in PBS with 0.1% Tween 20. Monoclonal antibody to PKCα (UBI, Saranac Lake, NY) was diluted 1:2000 in PBS and incubated with the blots for 2 h. After washing for 3 x 15 minutes in PBS with 0.1% Tween 20, the blots were incubated in 1:5000 PBS-diluted rabbit anti-mouse horseradish peroxidase-conjugated secondary antibody (Amersham), and then washed 6 to 7 x 15 min in PBS with 0.1% Tween 20. All incubations and washing were at room temperature. Bands that reacted with the primary antibody were visualized by 5 seconds to 5 minutes autoradiography following enhanced chemiluminescence (Amersham).

Isolation of Nuclei and Preparation of Nuclear Extracts

B16 cells were treated with or without 8-bromo-cyclic AMP/IBMX as described previously for 24 h. Cells were washed with ice-cold PBS, scrapped into PBS and collected by centrifugation in a microfuge. Cells were disrupted in NP-40 lysis buffer (10 mM Tris-HCl, pH 7.4, 3 mM $CaCl_2$, 2 mM $MgCl_2$ and 1% NP-40). Nuclei were recovered by centrifugation (5 min at 500xg) and washed twice with ice-cold NP-40 lysis buffer. The nuclear pellet was then resuspended in extraction buffer (0.05 M Tris-HCl, pH 7.4, 10% (v/v) glycerol, 0.01 M monothioglycerol, 0.001 M Na_2 EDTA, 0.001 M PMSF, 0.6 M KCl and 2 μg/ml each of aprotinin, leupeptin and pepstatin), sonicated 3x10 s, incubated for 2-3 h at 4° C and centrifuged at 100,000xg for 1 h to obtain a soluble nuclear extract.

Gel Mobility Shift Analysis

Two 33 base DNA oligonucleotides were annealed to form a double-stranded oligo which contained a retinoic acid response element corresponding to that found within the human RARβ gene promoter[26]. This βRARE was labeled with [^{32}P]dCTP (3000 Ci/mmol) by end-filling of the 5' protruding termini with the Klenow fragment of DNA polymerase I. The labeled βRARE was purified using a Qiagen Tip-20, precipitated with 10 μg of glycogen carrier in ethanol:acetone (1:3) at -80° C and resuspended in ddH_2O. Gel shift reactions typically contained B16 nuclear extract (25 μg) or Sf9 cell extracts from recombinant hRARα1 baculovirus-infected cells (1 μg), 10% (v/v) glycerol, 10x reaction buffer (0.2 M HEPES, pH 7.9, 0.6 M KCl, 0.01 M Na_2EDTA, 0.1 M monothioglycerol, and 0.01 M PMSF), 2.5 μg poly(dI-dC), and $2x10^4$ CPM of [^{32}P]βRARE in 20 μl total volume. The reactions were incubated for 20 min at 23-24 C° and immediately resolved at 20 mA at 23° C in a 5% non-denaturing gel (60:1 acrylamide:bis) in 0.5X TBE (0.045 M Tris borate, pH 8.0, 0.001 M Na_2EDTA). Dried gels were exposed to film between intensifying screens at -80 C°.

Transfections

B16 melanoma cells were cultured as described above and were transfected using the calcium phosphate DNA precipitation protocol [21]. βRARE oligonucleotide prepared as described for the gel shift assays was cloned into pGL2 vector (Promega) at MluI and Bgl II sites upstream of an SV40 promoter. Transfections included 5 μg of βRARE-SV40-luciferase vector and 10 μg of the pGL2 vector without insert. Each transfection also included 5 μg of pSV-β-galactosidase (Promega) in order to correct for transfection efficiency. Cells were treated with 8-bromo-cyclic AMP or with vehicle alone for 72 h and the last 24 h were treated with or without RA (10 μM). Cells were then harvested and assayed for luciferase using the Promega kit and for β-galactosidase. Luciferase assays were evaluated in the linear range and values were normalized to β-galactosidase activity. All transfections were performed in triplicate and the entire experiment repeated two additional times.

RESULTS

Expression of RARs in B16 Mouse Melanoma Cells and the Effect of Treatment with RA and/or Cyclic AMP

Specific restriction fragments of RAR α, β and γ cDNAs that are relative less GC-rich were used as probes to avoid the strong cross reaction of the whole cDNA inserts with ribosomal RNAs (Fig. 1).

An initial survey was performed to determine which RAR mRNAs were being expressed in B16 cells and whether RA had any effect on their levels. Cells were grown overnight after plating and then treated with DMSO (control), 10 μM RA, 1.0 mM 8-bromo-cyclic AMP/0.2 mM IBMX or RA plus 8-bromo-cyclic AMP/IBMX for 48 hours. At this time all treatment groups were harvested and RNA extracted. Northern analysis showed that B16-F1 cells constitutively expressed RAR α and γ mRNAs but almost undetectable levels of RAR β mRNA (Fig. 2). After 48h of 10 μM RA treatment, a point at which RA induces substantial growth inhibition in B16 cells[27], RAR β mRNA was induced at least 10 fold while RAR α and γ message levels were not dramatically altered (Fig. 2). In contrast, a 48h 8-bromo-cyclic AMP treatment dramatically decreased mRNAs of all three RARs. Simultaneous addition of RA with cyclic AMP was unable to prevent the inhibitory effect of cyclic AMP (Fig. 2). We next determined whether the expression of the RAR mRNAs changed during the growth and melanin-producing phases of culture and whether RA and cyclic AMP, both of which inhibit growth and stimulate melanin production[28], might accelerate these changes (Fig. 3). There was little change in the amount of RARα RNA in control cells except a consistent increase at 72 h in culture (time when melanin is starting to be produced). Addition of RA results in an earlier (48 h) increase in RARα mRNA, while addition of cyclic AMP decreased the amount of RARα RNA at all time periods. The inclusion of RA with cyclic AMP did not reverse the decrease in RARα RNA levels with the possible exception of the 48 h time point. The signal for RARβ RNA in control cells was too weak to quantitate by densitometry. Addition of RA induced RARβ mRNA at the earliest time measured (4 h), and continued to increase until 48 h, after which it remained at a steady state. Cyclic AMP did not prevent the early induction of RARβ mRNA (8-24 h) by RA, but it did block any further increase beyond 24 h. RARγ RNA increased between 8-24 h of culture and then remained at a steady state. Addition of RA resulted in a similar pattern of RARγ RNA expression but the absolute amounts were higher than in control cells. Cyclic AMP dramati-

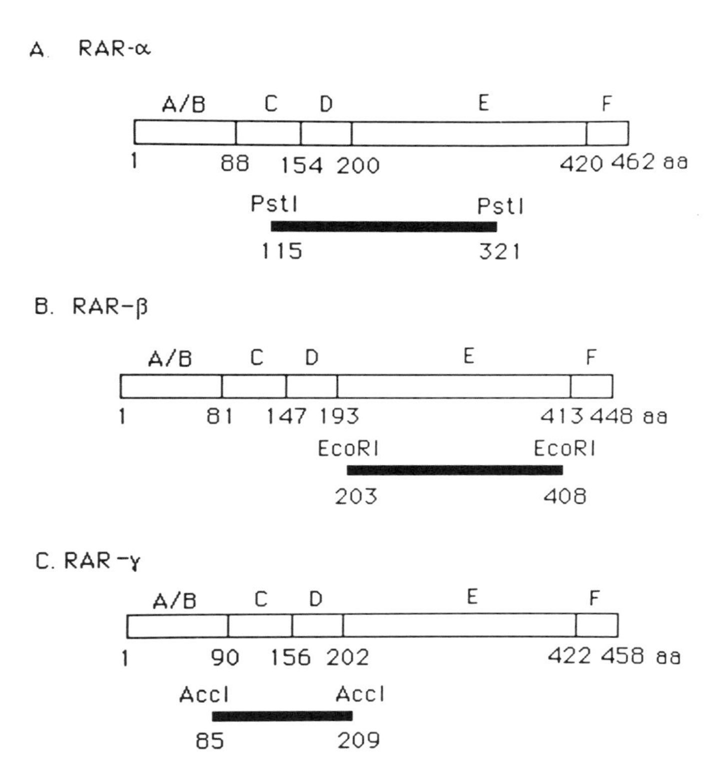

Figure 1. Structure of retinoic acid receptor cDNAs and the restriction fragments used for probing RAR mRNAs. A/B: cell and promoter specific activation region; C: DNA binding region, highly conserved; D: hinge region; E: ligand binding region; F: dimerization region.

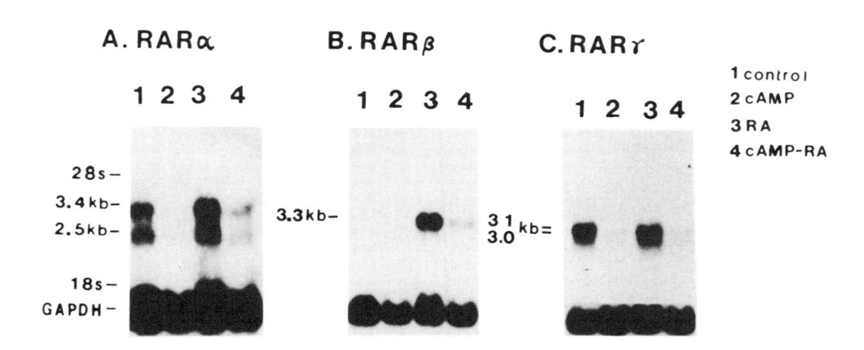

Figure 2. The effect of RA and/or cyclic AMP on RAR mRNAs levels in B16 cells. B16 cells were cultured and treated as described in methods. 2Oμg of RNA from each treatment group was analyzed by Northern blotting. Control (lane 1); treated with 1.0mM 8-bromo-cyclic AMP and 0.2mM 1-methyl 3-isobutylxanthine (MIX), a phosphodiesterase inhibitor for 48 h (lane 2); treated with 10^{-5}M RA for 48 h (lane 3); treated with RA plus 8-bromo-cyclic AMP/MIX for 48 h (lane 4). Panel A was hybridized with RARα, B with RARβ, and C with RARγ. All the blots were also hybridized with GAPDH as an internal control. Time of the blots being exposed to X-ray film: A, 2 days; B, 18 h; C, 1 day.

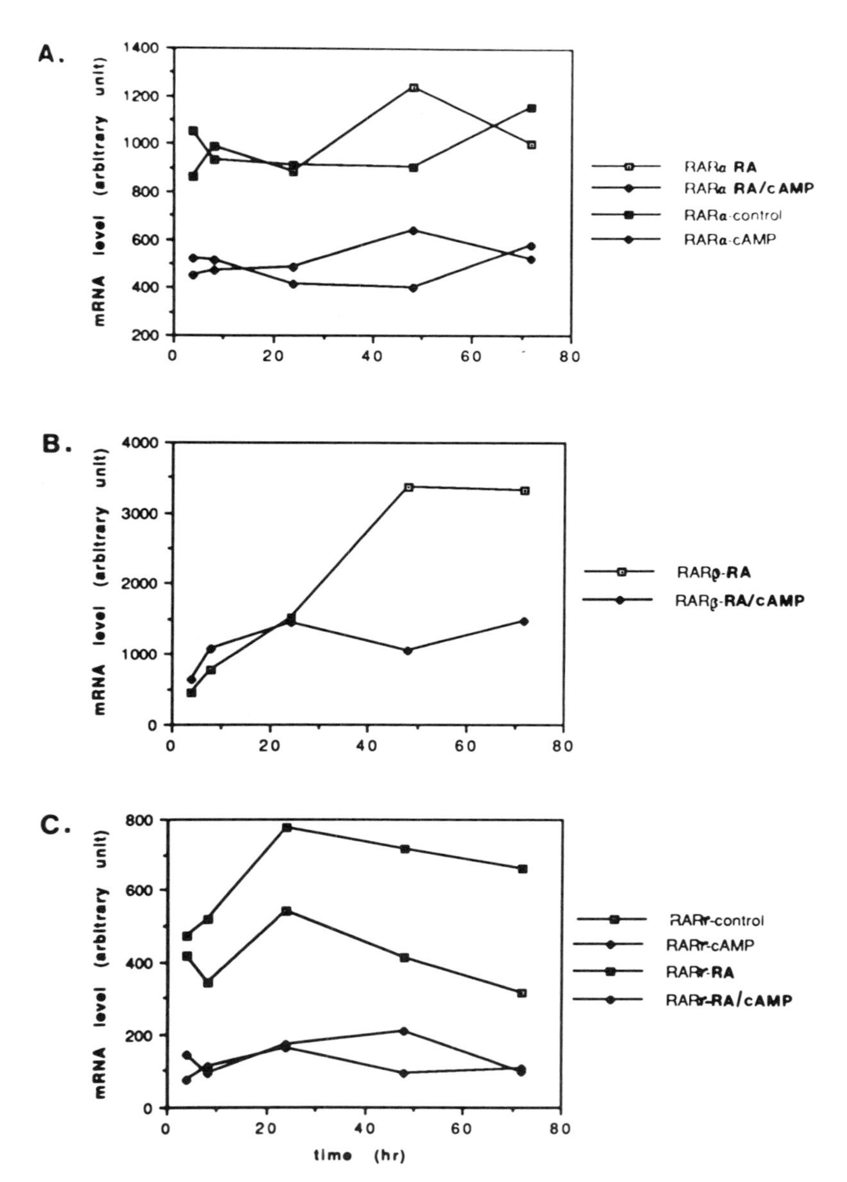

Figure 3. Time course of RAR α (panel A), β (panel B), and RARγ (panel C) expression in B16-F1 cells. Control cells or cells treated with RA, cyclic AMP/MIX or RA plus cyclic AMP/MIX for 8-72 h were harvested, RNA was extracted and 20μg/lane of total RNA was subject to Northern analysis to determine the amount of RARα, β and γ RNA. The blots were then stripped and reprobed with GAPDH as an internal control. The autoradiograms which were in the linear range were scanned using a Molecular Dynamics densitometer. All the RAR signals were normalized to the GAPDH signal. In panel B, RARβ mRNA signals in B16 cells with no treatment or treated with cyclic AMP/MIX were too weak to be detected by densitometry. The experiment was repeated several times with similar results.

cally decreased the amount of RARγ at all time points tested and the simultaneous addition of RA could not reverse this inhibition.

RA Directly Induces RARβ and Cyclic AMP Directly Represses RARγ mRNA

To determine whether the induction of RAR by RA was direct or indirect, we examined the amount of RARβ mRNA at early time points after the addition of RA. Fig. 4 shows that the induction occurred within 2 hr after adding RA and that the induction (after correction for the amount of GAPDH mRNA) was not blocked by cycloheximide, a protein synthesis inhibitor. Therefore, RA directly increased RARβ mRNA without the requirement for new protein synthesis. We chose RARγ for further study of the effect of cyclic AMP because its mRNA expression in B16 was relatively higher than RARα so that decreases in message levels were more easily detected. Fig. 5A shows that 8-bromo-cyclic AMP decreased the amount of RARγ mRNA within 1 h of treatment. At 2 h of cyclic AMP treatment the level of RARγ mRNA declined to very low levels and remained at this level through 8 h of treatment. Fig. 5B shows that the inhibitory effect of cyclic AMP on RARγ mRNA expression is direct in that the addition of the protein synthesis inhibitor cycloheximide did not reverse cyclic AMP action. To ensure that the effect obtained with the cyclic AMP analog 8-bromo-cyclic AMP was mimicking an increase in endogenous cyclic AMP, we treated cells with 0.2 μg/ml melanocyte stimulating hormone (MSH, Sigma), a polypeptide hormone which stimulates a large increase in intracellular cyclic AMP within minutes[29]. Northern analysis of RNA from control and MSH treated B16 cells revealed a large decrease in RARγ mRNA levels from MSH treated cells (data not shown).

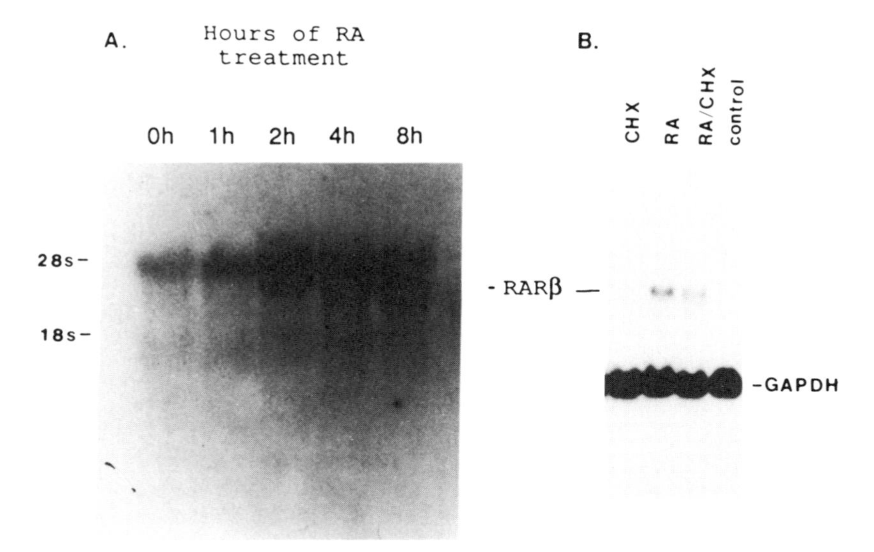

Figure 4. Time and protein synthesis dependence of RA effect on RARβ mRNA levels. A: B16 cells were treated with 1OμM RA for various periods of time as indicated, RNA was extracted from each sample and 20μg of each was analyzed by Northern blotting. The blot was hybridized with the full length RARβ cDNA insert. B: B16 cells were treated +/- 1Oμg/ml cycloheximide (CHX); 1OμM RA or RA+CHX for 4hrs and then harvested for RNA extraction. 20μg of RNA from each sample was subjected to Northern blotting and the blot was hybridized with the RARβ cDNA fragment shown in Fig. 1. Hybridization with GAPDH served as an internal control. Both blots were exposed to X-ray film overnight.

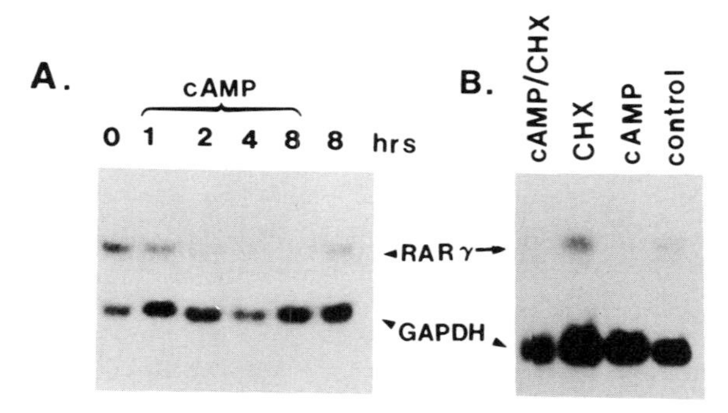

Figure 5. Time and protein dependence of cyclic AMP effect on RARγ mRNA level. A: B16 cells were treated +/- 1.OmM 8-bromocyclic AMP/0.2mM MIX for the indicated times and then the amount of RARγ mRNA determined by Northern analysis. B: B16 cells were treated +/- cycloheximide (lOμg/ml) or with cyclic AMP/MIX +/-cycloheximide (CHX) for 4h. All groups of cells were then harvested and the amount of RARγ mRNA determined by Northern analysis. For both panels, 20μg of total RNA from each sample was used in the Northern blot. Both blots were exposed to X-ray film for 18h.

Cyclic AMP Does not Alter the Stability of RARγ mRNA

The cyclic AMP-induced decrease in RARγ RNA could be due to repression of synthesis or acceleration of RNA decay. To examine the latter possibility, cells were pre-incubated +/- cyclic AMP for 3 h and then treated for short intervals with actinomycin D. The amount of RARγ RNA remaining was then determined by Northern blots. Analysis of Fig. 6 indicates that the half life of RARγ mRNA in control B16 cells is approximately 3 hours. It was extremely difficult to determine the precise half life of RARγ mRNA in cyclic AMP-treated cells, since the RNA levels were already significantly decreased. However,

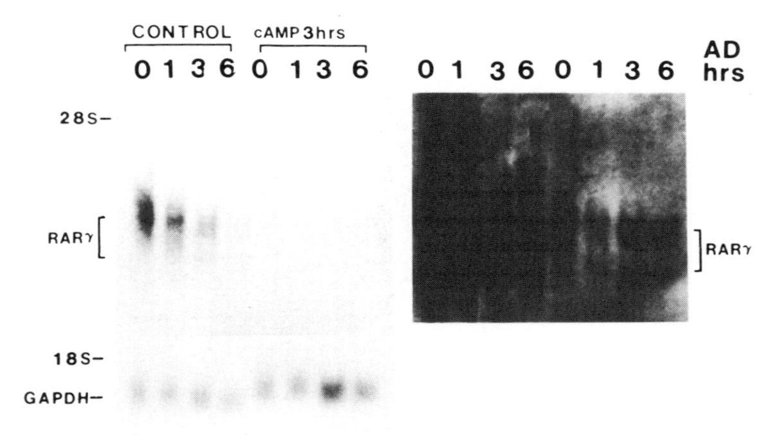

Figure 6. RARγ mRNA stability in B16 cells pre-treated with and without cyclic AMP. B16-F1 cells were pre-treated with and without 8-bromo-cyclic AMP/MIX as described in the legend for Fig. 2 for 3h. At this time cells were treated with 4μg/ml actinomycin D (a transcriptional inhibitor) for various time points. PolyA+ RNA was isolated and 5μg of each sample was used for Northern blotting. The amount of RNA was quantitated by scanning densitometry, normalizing to the GAPDH internal standard. Time of exposure to X-ray film: upper-left: 1 day: lower-left: 2h; right; 4 days.

rapid turnover of RARγ mRNA without cyclic AMP treatment makes it highly unlikely that the change in stability (if there is any) could account for the dramatic effect of cyclic AMP on RARγ mRNA expression. Since the inhibitory effect of cyclic AMP is not likely to be due to changes in the stability of RARγ mRNA, we proceeded to examine the possibility of transcriptional inhibition of RARs by cyclic AMP. However, as reported previously[30], RARγ signals were undetectable using nuclear run-on assays, possibly because the transcription rates are very low.

Effect of Cyclic AMP on RAR-DNA Binding

We attempted to determine if cyclic AMP-induced decreases in RAR mRNA levels was correlated with a decrease in the amount of the corresponding receptor protein using subtype-specific antibodies kindly provided by Dr. Chambon (Strasbourg, France). These antibodies did specifically recognize the appropriate RAR when the receptor was overexpressed in recombinant baculovirus-infected Sf9 cells. However, in our hands they were not sensitive nor specific enough to detect the endogenous receptors in B16 melanoma cells. Therefore, we used alternative methods to determine the effect of cyclic AMP on RAR protein. Gel shift analysis (Fig. 7) indicated that nuclear extracts contained a protein(s) that specifically bound a radioactively-labeled oligonucleotide corresponding to the RARE from the RARβ gene (lanes 7 and 8). The identity of this binding activity as RAR was confirmed

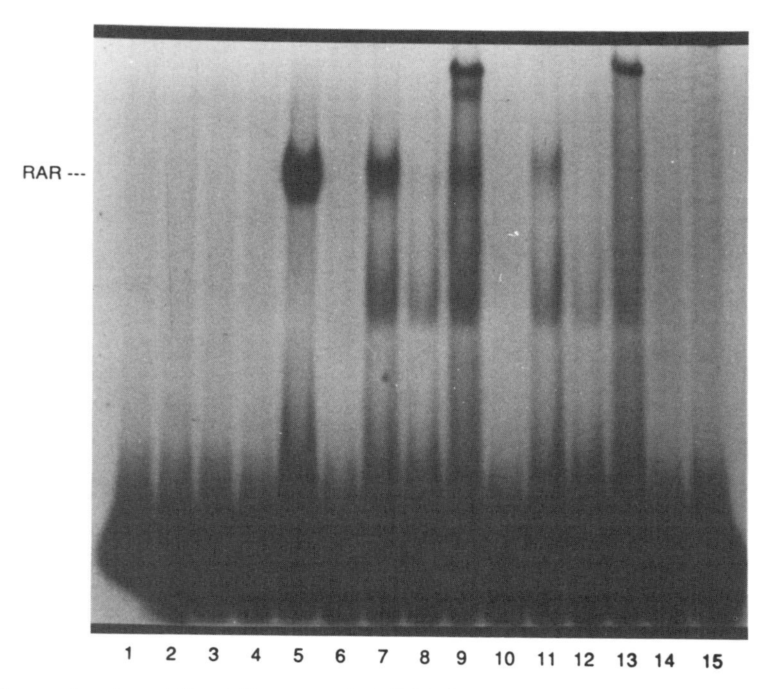

Figure 7. Electrophoretic mobility shift assay for RARE-binding proteins in nuclear extracts from control and 8-bromo-cyclic AMP-treated B16 cells. Nuclear extracts were prepared and assayed as described in Methods. Lanes 1-4, [^{32}P]-βRARE without nuclear extract. Lane 5, 1μg of protein from Sf9 insect cells infected with recombinant hRARα1 baculovirus. Lane 6, oligonucleotide probe without any nuclear extract. Lane 7, 25μg of nuclear extract from control B16 cells. Lane 8, same as lane 7 plus a 100fold excess of unlabeled βRARE oligonucleotide. Lane 9, same as lane 7 plus 5μl of RAR γIIIB antiserum. Lane 10, oligonucleotide probe without nuclear extract. Lanes 11-13, same as lanes 7-9, except 25μg of nuclear extract from 8-bromo-cyclic AMP/MIX-treated cells. Lanes 14 and 15, oligonucleotide probe without nuclear extract.

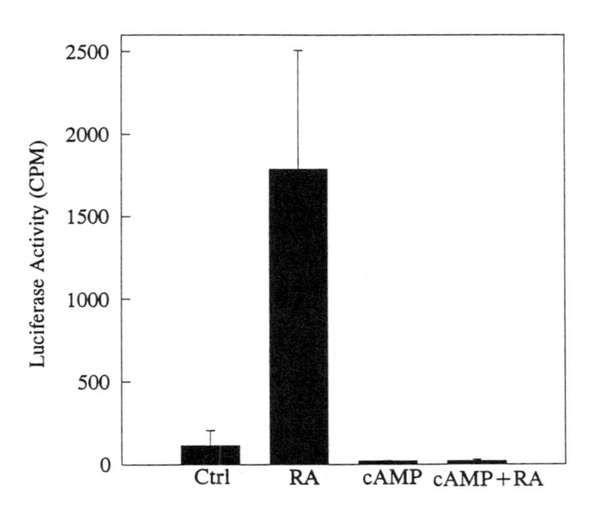

Figure 8. The effect of 8-bromo-cyclic AMP on RA-stimulated reporter gene activity. B16 cells were transfected with βRARE-SV-Luciferase and SV-β-gal plasmids as described in Methods. Cells were treated with or without 8-bromo-cyclic AMP/MIX (1.0 and 0.2mM respectively) for 72h. The last 24h, where indicated, they were also treated with 10μM RA. Cells were then harvested, extracted and assayed for luciferase activity using the Promega (Madison, WI) kit. Luciferase activity values were corrected for the amount of β-galactosidase activity in each sample. The data is presented as the mean +/- SEM (error bars) from triplicate dishes of transfected cells. The entire experiment was repeated several times with similar results.

by the resultant "supershift" (lane 9) when the assay was performed in the presence of antiserum that recognizes all RAR subtypes[31]. Nuclear extracts from B16 cells treated for 24 h with 8-bromo-cyclic AMP had a 75% reduction in binding of RAR to the RARE compared to nuclear extracts from control cells (compare lanes 7 vs 11).

Cyclic AMP Inhibits RA-Induced Reporter Gene Expression

Two copies of the β-RARE was inserted 5' to an SV-40 promoter - luciferase construct obtained from Promega (Madison, WI). This plasmid was cotransfected with a plasmid encoding β-galactosidase (to correct for transfection efficiency) into B16 cells. Cells were then subdivided and treated with DMSO (control), RA, 8-bromo-cyclic AMP, or RA + 8-bromo-cyclic AMP. Fig. 8 shows that RA increased luciferase activity by almost 4-fold. Cyclic AMP alone had a small inhibition of luciferase activity. However, when cyclic AMP was presented together with RA, luciferase activity was inhibited by approximately 65%. This finding together with the decrease in RAR-RARE binding suggest that the cyclic AMP-induced decrease in RAR mRNA was accompanied by a decrease in the amount of functional RAR protein.

Cyclic AMP Inhibits RA-Induced PKCα Expression

Retinoic acid stimulates B16 melanoma differentiation. An early event in this pathway is the induction of PKCα[32]. Overexpression of PKCα can mimic the phenotype of retinoic acid-treated cells[25], suggesting that this enzyme plays an important role in differentiation. Since cyclic AMP can decrease the amount of functional RAR, we determined whether it would have any effect on RA-induced differentiation of B16 cells as measured by induction of PKCα. A 48 h treatment of B16 cells with 10 μM

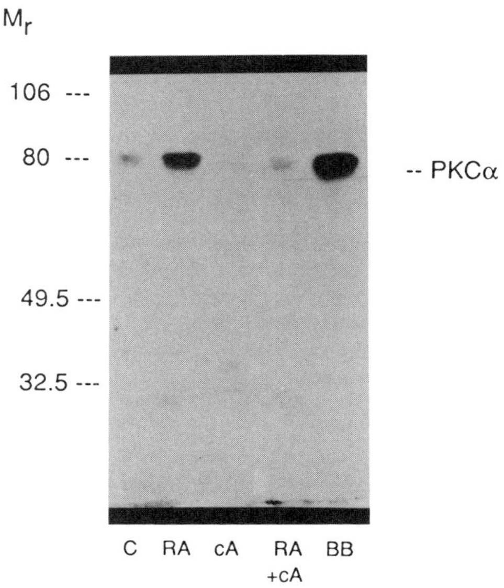

Figure 9. Effect of 8-bromo-cyclic AMP on RA-induction of PKCα. B16 cells were treated with or without 1mM 8-bromo-cyclic AMP + 0.2mM MIX for 24h before the addition of 10μM RA or vehicle (DMSO) for an additional 48h. Cells were then harvested, extracted and 50μg of total cell protein from each sample assayed for the level of PKCa by Western blotting as described in Methods. C=control, RA=10μM retinoic acid, cA=8-bromo-cyclic AMP/MIX, BB=50μg of protein from bovine brain used as a positive control for the detection of PKCα. The entire experiment was repeated several times with similar results.

retinoic acid increased PKCα levels by 6-fold (Fig 9). A 24 h treatment with cyclic AMP decreased baseline levels of PKCα by 50%, while a 24 h pre-treatment of the cells with cyclic AMP before adding retinoic acid for 48 h completely blocked the retinoic acid induction of PKCα.

DISCUSSION

In this study we have shown that RARα and RARγ mRNAs were constitutively expressed in B16 cells, while RARβ mRNA was induced by retinoic acid. Furthermore, this induction was direct in that it occurred very early after addition of RA and was not blocked by cycloheximide, indicating that new protein synthesis was not required. Induction of RARβ mRNA levels by RA has been reported to occur in many cell culture systems that are responsive to RA [33,34]. Indeed the first retinoic acid responsive element (RARE) was characterized in the 5' flanking sequence of the RARβ gene[26]. We also found that RA increased RARγ mRNA, although this increase was not as large as found for RA-induction of RARβ mRNA. Clifford et al.[35] also found that RA induced a 1.5-2.0 fold increase of RARγ in S91-C2 melanoma cells. Lehmann et al.[36] have shown that the RARγ2 promoter contains an RARE which can trans-activate a reporter gene. Thus, the increase in RARγ mRNA in our cells may have been due to RARγ2 which would not be distinguished from RARγ1 by the probes we used. It would be of interest to determine if the RA-induced increase in RARβ or γ might be linked to some of the biochemical changes induced by RA in B16 cells such as induction of protein kinase C[32] and increase in melanin production[25]. A significant finding of these studies was the ability of cyclic AMP to drastically decrease the level of all three RAR mRNAs. Two other studies, both using F9 teratocarcinoma cells, observed similar effects of cyclic AMP[30,34]. However, in contrast to the findings of one of these studies[30], the ability of cyclic AMP to decrease RAR mRNA levels in B16 cells did not require the simultaneous presence of RA. Similar results were found whether cyclic AMP was added alone or in combination with RA. Modulation of RAR by cyclic AMP is also supported by a study with PC12 cells, which generally are resistant to retinoic acid and express low levels of RAR mRNA. Cyclic AMP-dependent protein kinase deficient cells, derived from the parental line were found to differentiate upon retinoic acid treatment and had much higher RAR mRNA levels[37]. The ability of cyclic AMP to decrease RAR mRNA levels in B16 cells was rapid and independent of protein synthesis. The half-life that we calculated for RARγ mRNA in control B16 cells was similar to that reported by Hu and Gudas for RARγ mRNA in F9 teratocarcinoma cells[30]. Rapid and large decreases in RARγ mRNA induced by cyclic AMP prevented us from obtaining an accurate half-life under these conditions. However, after long-term exposure of the autoradiograms, it was concluded that the stability of RARγ mRNA did not appear to be altered by cyclic AMP. A similar finding was reported by Hu and Gudas [30] who observed similar half-lives for all RAR mRNA's in control and RA + cyclic AMP treated cells. We conclude from both our own data and that of Hu and Gudas that the effect of cyclic AMP on suppressing RAR mRNA levels is probably at the transcriptional level. The 5' upstream region of several RAR has been sequenced[36,38]. Thus far no CRE-like motifs have been reported to occur within the sequenced regions which could account for a transcriptional regulation by cyclic AMP[39]. In contrast to these conclusions, Martin et al.[34] reported that cyclic AMP had no effect on the transcriptional induction of RARβ by RA in F9 teratocarcinoma cells. Further investigation will be needed to understand the mechanism by which cyclic AMP can directly decrease the mRNA levels for all RAR. We were unable to determine if the cyclic AMP-induced decrease in RAR mRNAs correlated with a corresponding decrease in RAR protein due to a lack of sensitivity of the subtype-specific antibodies. Therefore, we examined the functional properties of the receptors. Both receptor-RARE binding activity and transcriptional activation of a reporter gene were drastically reduced in cyclic AMP-treated B16 cells. These data suggest that the decrease in RAR mRNA levels resulted in a similar decrease in RAR protein levels. In contrast to our results, Huggenvik et al.,[40] found that co-transfection of expression plasmids for the catalytic subunits of cyclic AMP-dependent protein kinase with RARE-CAT increased both ligand independent and ligand dependent activation of CAT expression in CV-1 and HeLa cells. The protocol in these studies was different from ours in that we pre-treated the B16 cells with cyclic AMP for 24 h prior to transfection with the reporter plasmid. Thus it is possible that cyclic AMP may have dual and opposing effects on RAR. By activating cyclic AMP-dependent protein kinase the RAR may be phosphorylated which increases their transcriptional activation activity. In addition, through phosphorylation of CREB or other related transcription factors, it may also inhibit transcription of the RARs. We found that pre-treatment of B16 cells with cyclic AMP blocked the RA induction of PKCα, an early marker of differentiation that plays an important role in the differentiation pathway. Although we can not rule out the possibility that cyclic AMP may directly inhibit the expression of PKCα, the data presented here suggest that the mechanism is through depletion of RARs. It is conceivable that hormones which elevate intracellular cyclic AMP levels could modulate the physiological actions of retinoids. Our laboratory also has evidence that protein kinase C can regulate RAR function. Depletion of protein kinase C in B16 melanoma cells by chronic treatment with phorbol dibutyrate leads to impairment of RAR binding to β-RARE and to diminished ability to transactivate a luciferase reporter gene in response to retinoic acid treatment. However, there was no change in the level of RAR mRNA (unpublished data). One possible explanation of this data is that the RAR must be phosphorylated either directly or indirectly by protein kinase C in order to be fully functional. From the studies presented here, it is clear that the expression and function of RAR can be affected by other regulatory agents. This adds an additional level of complexity in determining how cells, animal models, or patients will response to retinoic acid treatment.

SUMMARY

Vitamin A is metabolized to several biologically active compounds, the best known of which is retinoic acid. This compound has been shown to inhibit the growth of a variety of tumor cells and to induce a more differentiated phenotype in several tumor types. Vitamin D is metabolized to the active compound 1,25-dihydroxyvitamin D_3. This vitamin is well-known for its role in maintaining calcium homeostasis in the body. Recently it has been shown that vitamin D_3 can also inhibit tumor cell replication and stimulate differentiation of selected tumor types. Retinoic acid is being used clinically to treat promyelocytic leukemia, head and neck tumors as well as cervical dysplasia. Use of vitamin D_3 clincally has been restricted by its affect on calcium metabolism. Recently, however, new analogs of vitamin D_3 have been developed which have much less calcium mobilizing activity, yet still retain their tumor inhibitory properties. The action of both of these vitamins is mediated by nuclear receptors which have the same structure as steroid receptors. There are three nuclear retinoic acid receptors (RARα, β, and γ), but only one vitamin D_3 nuclear receptor. These receptors are expressed in very small amounts. Since the ligand should be in vast excess of receptor (ie not limiting), we explored the possibility that response to vitamin A might be mediated by control of RAR expression. Using B16 mouse melanoma cells as a model system, we found that RARα and γ mRNAs were constitutively expressed. RARβ mRNA was induced by treatment of the cells with RA. Induction of RARβ mRNA occurred within 1h and was not inhibited by cycloheximide. The mRNA for all three RARs was dramatically decreased with 8-bromo-cyclic AMP treatment and could not be rescued by addition of RA. Analysis of RARγ revealed that this decrease occurred within 1h of exposure to 8-bromo-cyclic AMP and was not blocked by simultaneous treatment with cycloheximide. Nuclear extracts from cyclic AMP-treated cells showed a large decrease in protein binding to a retinoic acid response element (RARE) oligonucleotide compared to control cells. This correlated with a marked reduction of RA-stimulated RARE-reporter gene activity in transfected cells which were treated with cyclic AMP. Pre-treatment of B16 cells with cyclic AMP prior to RA addition dramatically reduced induction of PKCα, an early marker of RA-induced cell differentiation. Thus, cyclic AMP can antagonize the physiological actions of RA via its ability to inhibit RAR expression.

ACKNOWLEDGMENTS

This work was supported in part by grant # 92A58 from the AICR. The author would like to thank Yonghong Xiao, Dr. Dinakar Desai, Dr. Tim Quick, and Faustina Fenton, who were involved in various aspects of these studies.

REFERENCES

1. S.B. Wolbach, P.R. Howe, Tissue changes following deprivation of fat soluble vitamin A *J Exp Med* 62: 753 (1925).
2. H.A.P. Pols, J.C. Birkenhager, J.A. Foekens, J.P.T.M. Van Leeuwen, Vitamin D: A modulator of cell proliferation and differentiation *J Steroid Biochem* 37: 873 (1990).
3. H. Reichel, H.P. Koeffler, A.W. Norman, The role of the vitamin D endocrine system in health and disease *New Engl J Med* 320: 980 (1989).
4. T. Suda, T. Shinki, N. Takahashi, The role of vitamin D in bone and intestinal cell differentiation *Ann Rev Nutr* 10: 195 (1990).
5. K. Colston, M.J. Colston, D. Feldman, 1,25 dihydroxyvitamin D_3 and malignant melanoma: The presence of receptors and inhibition of cell growth in culture *Endocrinology* 108: 1083 (1981).

6. J.A. Eisman, D.H. Barkla, P.J.M. Tutton, Suppression of in vivo growth of human cancer solid tumor xenografts by 1,25- $(OH)_2D_3$ *Cancer Res* 47: 21 (1987).
7. G.P. Studzinski, A.K. Bhandal, Z.S. Brelvi, Cell cycle sensitivity of HL-60 cells to the differentiation-inducing effects of 1- alpha, 25-dihydroxyvitamin D_3 *Cancer Res* 45: 3898 (1985).
8. T. Kummet, T.E. Moon, F.L. Meyskens, Vitamin A: evidence for its preventive role in human cancer *Nutr Cancer* 5: 96 (1983).
9. P. Nettesheim, C. Snyder, J.C.S. Kim, Vitamin A and the susceptibility of respiratory tract tissues to carcinogenic insult *Environ Health Perspect* 29: 89 (1979).
10. M.B. Sporn, A.B. Roberts, Role of retinoids in differentiation and carcinogenesis *Cancer Res* 43: 3034 (1983).
11. S.R. Frankel, A. Eardley, G. Lauwers, M. Weiss, R.P. Warrel, The retinoic acid syndrome in acute promyelocytic leukemia *Ann Intern Med* 117: 292, 1992.
12. W.K. Hong, S.M. Lippman, L. Itri, D.D. Karp, J.S. Lee, R.M. Byers, S.S. Schantz, A.M. Kramer, R. Lotan, L.L. Peters, I.W. Dimery, B.W. Brown, H. Goepfert, Prevention of second primary tumors with isotretinoin in squamous cell carcinoma of the head and neck *N Engl J Med* 323: 795 (1990).
13. S.M. Lippman, J.J. Kavanagh, M. Paredes-Espinoza, F. Delgadillo- Madrueno, P. Paredes-Casillas, W.K. Hong, E. Holdener, I. H. Krakoff, 13-cis-retinoic plus interferon α-2a: highly active systemic therapy for squamous cell carcinoma of the cervix *J Natl Cancer Inst* 84: 241 (1992).
14. J.A. Eisman, R.L. Sutherland, M. Lynne McMenemy, J-C. Fragonas, E.A. Musgrove, G.Y.N. Pang, Effects of 1,25-dihydroxyvitamin D_3 on cell-cycle kinetics of T 47D human breast cancer cells. *J Cell Physiol* 138: 611 (1989).
15. A. Zelent, A. Krust, M. Petkovich, P. Kastner, P. Chambon, Cloning of murine retinoic acid receptor α and β cDNAs and of a novel third receptor predominantly expressed in skin. Nature 339:714 (1989).
16. P. Kastner, A. Krust, C. Menselsohn, et. al., Murine isoforms of retinoic acid receptor γ with specific patterns of expression *Proc Natl Acad Sci USA* 87: 2700 (1990).
17. R.A. Heyman, D.J. Mangelsdorf, J.A. Dyck, R.M. Stein, G. Eichele, R.M. Evans, C. Thaller 9-cis retinoic acid is a high affinity ligand for the retinoid X receptor *Cell* 68: 397 (1992).
18. X-K. Zhang, B. Hoffmann, P.B-V. Tran, G. Graupner, M. Pfahl, Retinoid X receptor is an auxiliary protein for thyroid hormone and retinoic acid receptors *Nature* 355: 441 (1992).
19. S.A. Kliewer, K. Umesono, D.J. Mangelsdorf, R.B. Stein, R.M. Evans, Retinoic X receptor interacts with nuclear receptors in retinoic acid, thyroid hormone and vitamin D_3 signaling. *Nature* 355: 446 (1992).
20. P. Chomczynski, N. Sacchi, Single-step method of RNA isolation by acid guanidinium thiocyanate-phenol-chloroform extraction. *Anal Biochem* 162:156 (1987).
21. J. Sambrook, E.F. Fritsch, T.Maniatis, Molecular cloning-a laboratory manual. 2nd ed. CSH lab (1989).
22. M. Petkovich, N.J. Brand, A. Krust, P.Chambon, A human retinoic acid receptor which belongs to the family of nuclear receptors. *Nature* 330:444 (1987).
23. N.M., Brand, M. Petkovich, A. Krust, P. Chambon, H. de The, A. Marchio, P. Tiollais, A. Dejean, Identification of a second human retinoic acid receptor. *Nature* 332:850 (1988).
24. A. Krust, P. Kastner, M. Petkovich, A. Zelent, P. Chambon, A third human retinoic acid receptor, hRARγ *Proc. Natl. Acad. Sci. USA* 86:5310 (1989).
25. J.R. Gruber, S. Ohno, R.M. Niles, Increased expression of protein kinase Cα plays a key role in retinoic acid-induced melanoma differentiation. *J Biol Chem* 267: 13356 (1992).
26. H.M. Sucov, K.K. Murakami, R.M. Evans, Characterization of an autoregulated response element in the mouse retinoic acid receptor type β gene. *Proc Natl Acad Sci USA* 87: 5392 (1990).
27. R.M. Niles, Retinoic acid-induced growth arrest of mouse melanoma cells in Gl without inhibition of protein synthesis. *In Vitro* 23:803 (1987).
28. R.M. Niles, B. Loewy, B16 mouse melanoma cells selected for resistance to cyclic AMP-mediated growth inhibition are cross-resistant to retinoic acid-induced growth inhibition. *J Cell Physiol* 147:176 (1991).
29. R.M. Niles, J.S. Makarski, Hormonal activation of adenylate cyclase in mouse melanoma variants. *J Cell Physiol.* 96:355 (1978).
30. L. Hu, L.J. Gudas, Cyclic AMP analogs and retinoic acid influences the expression of retinoic acid receptor α, β and γ mRNAs in F9 teratocarcinoma cells, *Mol Cell Biol* 10:391 (1990).
31. T.C. Quick, A.M. Traish, A.M. R.M. Niles, Characterization of human retinoic acid receptor α1 expressed in recombinant baculovirus-infected Sf9 insect cells. *Receptor* 4: 65 (1994).
32. S.E. Rosenbaum, R.M. Niles, Regulation of protein kinase C gene expression by retinoic acid in B16 mouse melanoma cells. *Arch Biochem Biophys* 294: 123 (1992).
33. H. de The, A. Marchio, P. Tiollais, A. Dejean, Differential expression and ligand regulation of the retinoic acid receptor α and β genes. *EMBO J* 8:429 (1989).

34. C.A. Martin, L.M. Ziegler, J.L. Napoli, Retinoic acid, dibutyryl-cyclic AMP, and differentiation affect the expression of retinoic acid receptors in F9 cells. *Proc Natl Acad Sci USA* 87:4804 (1990).
35. J.L., Clifford, M. Petkovich, P. Chambon, R. Lotan, Modulation by retinoids of mRNA levels for nuclear retinoic acid receptors in murine melanoma cells. *Mol Endocrinol* 4:1546 (1990).
36. J.M., Lehmann, X-K. Zhang, X-K., M. Pfahl, RARγ2 expression is regulated through a retinoic acid response element embedded in Sp1 sites. *Mol Cell Biol* 12: 2976 (1992).
37. R.J. Scheibe, D.D. Ginty, J.A. Wagner, Retinoic acid stimulates the differentiation of PC12 cells that are deficient in cyclic AMP-dependent protein kinase. *J Cell Biol* 113: 1173 (1991).
38. P. Leroy, H. Nakshatri, P. Chambon, Mouse retinoic acid receptor α is transcribed from a promoter that contains a retinoic acid responsive element. *Proc Natl Acad Sci USA* 88:10138 (1991).
39. J.L. Meinkoth, M.R. Montminy, J.S. Fink, J.R. Feramisco, Induction of a cyclic AMP responsive gene in living cells requires the nuclear factor CREB. *Mol Cell Biol* 11:1759 (1991).
40. J.I. Huggenvik, M.W. Collard, Y-W. Kim, R.P. Sharma, Modification of the retinoic acid signaling pathway by the catalytic subunit of protein kinase-A. *Mol Endocrinol* 7: 543 (1993).

2

VITAMIN A CHEMOPREVENTION OF LUNG CANCER

A Short-Term Biomarker Study

Carrie A. Redlich,[1] Ariette M. Van Bennekum,[2] Joel A. Wirth,[4] William S. Blaner,[2] Darryl Carter,[3] Lynn T. Tanoue,[4] Carole T. Holm,[4] and Mark R. Cullen[1]

[1] Occupational and Environmental Medicine Program
Department of Medicine
Yale University School of Medicine
New Haven, Connecticut

[2] Institute of Human Nutrition
Columbia University
New York, New York

[3] Department of Pathology
Yale University School of Medicine
New Haven, Connecticut

[4] Pulmonary and Critical Care Section
Department of Medicine
Yale University School of Medicine
New Haven, Connecticut

I. INTRODUCTION

This article describes an ongoing short-term biomarker study of Vitamin A in subjects at high risk for lung cancer. Workers exposed to asbestos and cigarettes have a markedly increased risk of developing lung cancer and parenchymal fibrosis. Epidemiologic and experimental studies have shown that dietary vitamin A has significant anticancer and immunomodulatory effects (1-6). However, whether intervention with supplemental vitamin A can reduce the mortality or morbidity from either disease and the mechanisms involved remains unclear. Airway metaplasia on bronchial biopsy and inflammation on bronchoalveolar lavage (BAL) are considered potential markers for the development of lung cancer and parenchymal fibrosis respectively. Our prior clinical studies (see below) have shown an high incidence of both of these lesions in asbestos-exposed subjects, findings consistent with the idea that the processes of inflammation and carcinogenesis are linked. We have hypothesized that 1) the vitamin A intervention may reduce both bronchial metaplasia and lung inflammation, 2) the mechanism of this effect may be through modulation of relevant pulmonary

Diet and Cancer, Edited under the auspices of the American Institute of Cancer Research
Plenum Press, New York, 1995

cytokines, growth factors, and/or oncogenes, and 3) local lung vitamin A status may be a key modifiable host determinant or biomarker of susceptibility. We are performing a double-blind placebo controlled 6 month trial of combination β-carotene and retinol in 50 subjects at high risk for both lung cancer and parenchymal fibrosis to address these hypotheses. This study should provide valuable data on whether vitamin A can modify potential early markers of lung cancer and fibrosis, the mechanisms involved, and the role of local vitamin A. The findings may lead to effective preventive strategies for lung cancer.

A. Lung Cancer

Lung cancer is the leading cause of cancer death in the United States, accounting for almost 1/3 of all cancer deaths. Unfortunately screening of persons at high risk for lung cancer has not been shown to reduce cancer mortality and the five year survival for lung cancer remains poor. Thus there is a critical need to develop preventive strategies for those at high risk. Dietary modification such as vitamin A supplementation may play an important role in the prevention of lung cancer.

Both asbestos and cigarette exposure are associated with a significantly increased risk of lung cancer (7). Genetic alterations in oncogenes such as p53 and cJun/cFos are felt to play an important role in the multistep process of lung carcinogenesis, and may be potential biomarkers of early disease and/or markers of host susceptibility (8,9). The processes of lung inflammation and carcinogenesis are likely linked (10,11). Those with an inflammation-related lung lesion (fibrosis or obstructive airways) have been found to be at significantly increased risk of lung cancer (12,13). The finding that asbestos-related lung cancers usually develop in areas of parenchymal fibrosis supports this hypothesis, as do our prior findings of a strong association between the presence of bronchial metaplasia and inflammatory lavage cells in asbestos-exposed subjects (13,14). Several mediators classically associated with inflammation and/or cell growth such as transforming growth factor-β (TGF-β), interleukin-6 (IL-6), platelet-derived growth factor (PDGF), and scatter factor (SF) may also modulate tumor development and growth (11,15,16).

B. Effects of Vitamin A on Cancer and Inflammation

It has been estimated that up to 70% of all cancers may be attributed to dietary factors, of which vitamin A is considered one of the most important (7). Vitamin A has been defined as retinol plus that part of the provitamin β-carotene which is converted into retinol. Extrahepatic vitamin A is felt to play an important role in local vitamin A effects (17). Because plasma retinol levels are maintained within a narrow range despite large fluctuations in dietary vitamin A intake, serum retinol levels are not good indicators of total body vitamin A stores, making the evaluation of vitamin A status in humans difficult (1). It has been estimated that 20-40% of the United States population may have marginal vitamin A nutritional status (18). We recently have been able to detect retinol and β-carotene in BAL macrophages (see below). Such levels may potentially be used as an accessible biomarker for lung vitamin A.

Epidemiologic, animal and *in vitro* studies have shown vitamin A to have potent anticancer effects (1-6). An inverse relationship between dietary vitamin A intake and lung cancer risk has been found in a number of studies (6), although it remains unclear whether retinol, β-carotene, or some other dietary factor is the protective agent. Animal studies have shown that vitamin A deficiency is associated with an increased incidence of cancer (19). *In vitro* studies have also clearly shown retinoids, potent regulators of epithelial proliferation and differentiation, to have potent antineoplastic effects (4,20).

Vitamin A has also been shown to have significant immunomodulatory effects (5). Vitamin A treatment has been shown to reduce morbidity and mortality from infectious diseases (21,22) and reduced vitamin A intake has been associated with reduced lung function (23).

Recent studies suggest that the pleomorphic effects of retinoids are mediated through specific retinoic acid nuclear receptors (RARs and RXRs) (1). There is also evidence that retinoic acid may regulate mediators such as TGF-β (24). We have recently demonstrated that retinoids inhibit IL-6 and IL-8 production *in vitro* (25,26). Retinoids may also protect the lung by promoting cell differentiation and mucosal integrity (4). β-carotene, a naturally occurring carotenoid and antioxidant, has been shown to inhibit tumor growth, but its anticancer effects are relatively weak compared with the effects of retinoids (27).

Despite the significant literature documenting anticancer effects of vitamin A, human data on whether dietary supplementation with vitamin A can reduce the risk of lung cancer is quite limited. In several clinical trials retinoids have been shown to significantly reduce the incidence of second primaries in patients with lung cancer and head and neck cancer (2,28). In an uncontrolled trial, reduced airway metaplasia was seen after a six month trial of etretinate (29). However, a more recent randomized controlled trial of isotretinoin showed no effect on squamous metaplasia (30). A recent large intervention trial of β-carotene and vitamin E, but no retinoids, showed no reduction in the incidence of lung cancer (31). Several other large intervention trials are in progress (32), but results will not be available for a number of years. We are currently involved in the CARET project, a multi-center long-term vitamin A (retinol and β-carotene) chemoprevention trial on a population at increased risk for lung cancer (32).

C. Biomarkers as Intermediate Endpoints in Future Chemo-Preventive Trials

These clinical and epidemiologic studies have been hampered by the very large number of subjects and long follow-up time needed to detect a change in mortality or morbidity following the intervention. The identification and validation of early markers of lung cancer and biomarkers of host susceptibility will strengthen epidemiologic research by elucidating the relationships between exposure and disease and identifying those at risk for disease. Such markers should greatly facilitate chemo-preventive studies by reducing the number of subjects and length of follow-up needed, and the costs involved.

Bronchial metaplasia is considered a potential early marker for lung cancer. Metaplasia can be seen throughout the bronchial tree following exposure to cigarette smoke and other respiratory toxins (33) and may progress to lung cancer (34). Bronchial metaplasia can serve as a marker for mucosal injury and lung cancer risk. Moreover, metaplasia is reversible and could be modified by a dietary intervention such as vitamin A.

The presence of parenchymal alveolitis with increased inflammatory cells on BAL is considered a potential marker for pulmonary fibrosis and inflammation (35,36). Our prior studies have also shown a strong association between the presence of inflammatory cells on BAL and airway metaplasia (14). This alveolitis is believed to result from the release of several mediators such as TFG-β, IL-6, IL-8, PDGF, and IL-11 (35 - 39). Inhibition of these mediators may modulate the processes of alveolitis and metaplasia, and potentially prevent lung cancer.

II. MATERIAL AND METHODS

A. Study Population

Asbestos-exposed subjects with a high likelihood of having bronchial metaplasia on airway biopsy and/or lung inflammation on BAL are being recruited, such as smokers with radiographically apparent asbestosis. Subjects also meet the CARET eligibility criteria including heavy asbestos exposure, current or prior smoking, age between 45 and 70, absence of severe liver disease, and not currently taking supplemental vitamin A (32). All studies have been approved by the Yale Committee on Human Investigation and informed consent is obtained on all subjects.

B. Study Design

The study is designed as a randomized two arm, placebo controlled, double-blinded trial. Each subject is being used as his own control. Baseline studies including chest xray, pulmonary function testing (PFTs) and methacholine challenge testing to assess airway hyper-responsiveness, liver function tests, and serum nutrient levels are performed on each eligible subject to ensure that the patient is not medically disqualified, and to establish baseline characteristics. Intake questionnaires are administered to provide data on occupational exposures, cigarette smoking, medications, and diet, as well as comparability to the large CARET study population. Subjects then undergo bronchoscopy with multiple airway biopsies and BAL. Assays to be performed on this material are described below. Each subject is then randomly assigned to either the placebo or the study arm. The study intervention is the same as that used in the CARET trial, β-carotene 30 mg and vitamin A 25,000 IU (32). At six months the full protocol, including repeat bronchoscopy and PFTs, is performed again on each subject. Subjects are contacted monthly to assess compliance and any potential side effects. Risks are minimal, and the CARET strategy for drug side effect monitoring is employed (32).

Sample size estimates were difficult, as the incidence and test-re-test variability for metaplasia or any of the other endpoints, as well as the response to treatment are not known. Loss to follow-up and noncompliance are expected to be low, based on our prior experience (14,32). A sample size of 50 subjects, 25 randomized to each treatment arm was decided upon.

C. Pulmonary Function Tests

Complete pulmonary function testing including standard spirometry, static lung volumes by helium dilution, and single breath diffusion capacity are performed on all subjects pre and post the intervention using the same equipment and according to ATS recommendations. Airway hyper-responsiveness is assessed using standard methacholine challenge testing with increasing doses of methacholine.

D. Fiberoptic Bronchoscopy and BAL

Fiberoptic bronchoscopy is performed as described in prior publications (14,40). Briefly, the upper airways are anesthetized with topical xylocaine and a fiberoptic bronchoscope inserted. If unexpected abnormalities are noted, appropriate evaluation is performed. At least eight mucosal biopsies are obtained from branch points in the right and left lungs and from the main carina. Cytology brushings of the right upper lobe are also obtained. BAL is performed from subsegments of the lingula by alternately instilling and aspirating five 50 ml aliquots of sterile saline (total 250 ml per BAL) and carried to the laboratory for immediate processing. Cells from the first aliquot are processed separately whereas cells from the subsequent aliquots are pooled for further analysis. All subjects are biopsied and lavaged at identical sites to the extent possible.

E. Processing of Biopsy Specimens

Biopsies are processed as follows: A minimum of 6 biopsies are fixed in glutaraldehyde- paraformaldehyde fixative, imbedded in paraffin, and four micron sections cut and stained with hematoxylin-eosin. Biospsies are also frozen fresh in liquid nitrogen, and embedded and frozen in O.C.T. Compound (Miles) for further studies. Metaplasia is determined as in our prior studies (14,40). The presence of keratinizing squamous metaplasia is assessed in at least 6 bronchial biopsy specimens and cytology brushes. Slides from the cytology brushes are placed in alcohol and processed promptly. All biopsies and cytology are scored in a blinded fashion by one of us (DC) Each individual specimen is scored with respect to the worst metaplastic lesion present (14,40). Lesions are scored as normal, basal cell hyperplasia, full-thickness hyperplasia or squamous metaplasia (Figure 1).

F. Analysis of BAL Fluid

BAL fluid is processed as previously described (14). Briefly, BAL fluid and cells from aliquots one and two though five are pooled,filtered and centrifuged (500 x g) to pellet cellular elements. The cells are resuspended in a balanced salt solution and a cell count performed. Concentrated smears are stained with modified Wright's Giemsa. Differential counts are performed by counting the percentage from 200 stained cells. The remaining supernatant fluid and cells are frozen (-70 °C) for subsequent use.

G. HPLC Nutrient Determinations

Macrophage and serum concentrations of retinol, retinyl esters, carotenoids and vitamin E are measured by reverse phase high pressure liquid chromatography (HPLC) using procedures described previously (41) by one of us (WB). Briefly, pelleted macrophages, frozen at -70°C prior to analysis, are resuspended in phosphate buffered saline. Absolute ethanol, containing internal standards to correct for retinoid and carotenoid recoveries, are added. The retinoids and carotenoids are extracted into hexane and evaporated and resuspended in benzene for injection onto the HPLC. The reverse phase HPLC analysis is carried out on a Beckmann Ultrasphere C_{18} column. The carotenoid echinenone and retinyl acetate are employed as internal standards to assess for recovery. Our low limits of detection for β-carotene, α-carotene, lycopene, cryptoxanthin, zeaxanthin, retinol, retinyl esters, and α-tocopherol are respectively, 1 ng, 1 ng, 0.5 ng, 0.5 ng, 0.5 ng, 0.5 ng, 1 ng, and 0.03 ug.

H. Assessment of Lung Cytokines and Oncogenes

The expression of relevant cytokines, growth factors, and oncoproteins (Table 1) before and after the intervention are being assessed using several methods, including ELISA and bioassay techniques. Data is expressed as pg/ml of BAL fluid. If adequate numbers of cells are obtained, short term BAL cell cultures are performed and the supernatants saved at 4 and 16 hours for cytokine determination.

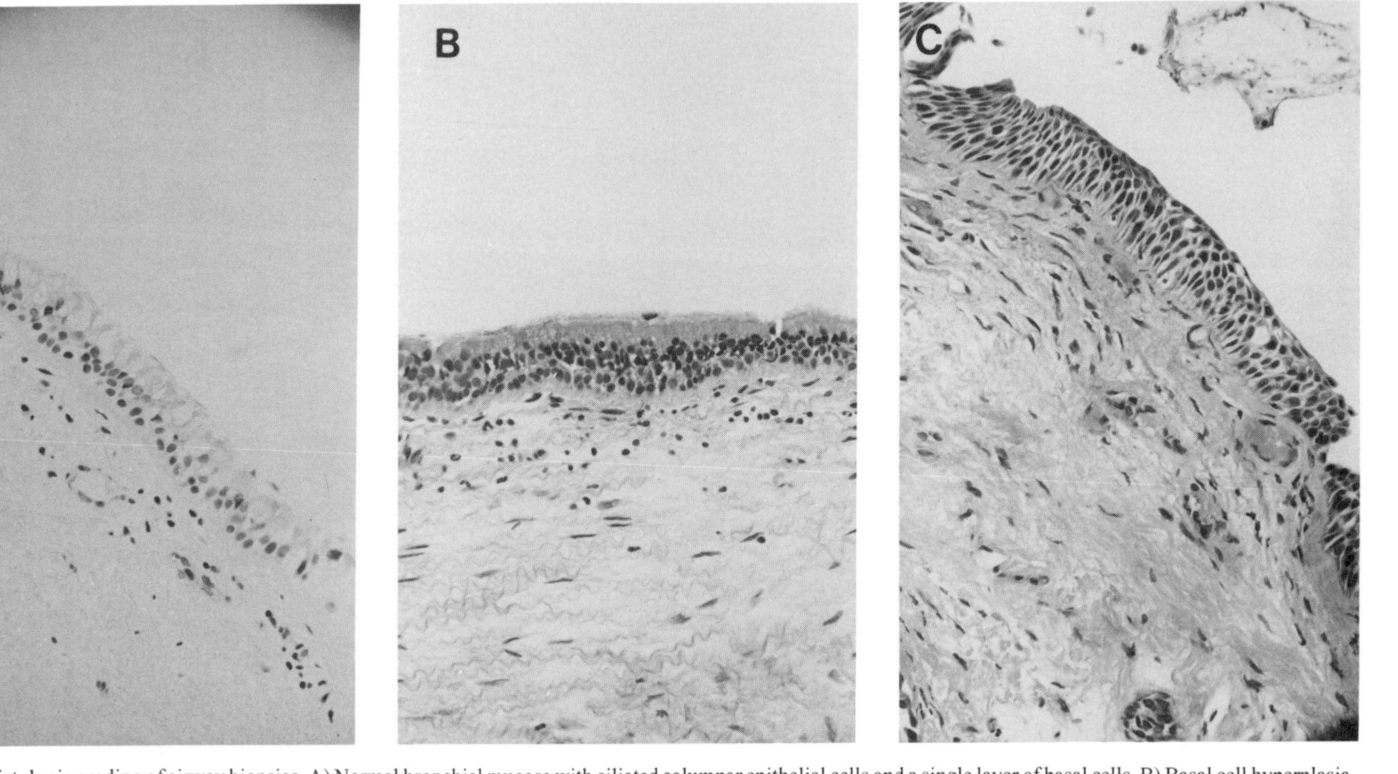

Figure 1. Histologic grading of airway biopsies. A) Normal bronchial mucosa with ciliated columnar epithelial cells and a single layer of basal cells. B) Basal cell hyperplasia. The basal cells are increased to 3 or more cell layers and columnar epithelial cells are still present. C) Squamous metaplasia with full thickness basal cell hyperplasia and keratinization resembling a stratefied squamous epithelium.

Table 1. Study Endpoints—Biomarkers

Airway histology
Metaplasia
Bronchial biopsies
Cytology brushings
Lung inflammation \ fibrosis
BAL alveolitis
BAL cell differentials
Lung function
PFTs
Airway hyperreactivity - methacholine test
Nutritional assessment
Dietary questionnaire
Serum nutrient levels
BAL nutrient levels
Cellular / Molecular Markers
Cytokines - Growth factors
IL-6, IL-8, IL-11, TGF-β, SF, PDGF
Genetic alterations in bronchial epithelium
p53

Northern blot analysis of bronchial epithelial and macrophage gene expression is performed if adequate quantities of cellular mRNA are obtained. Total cellular mRNA is obtained from the bronchial epithelial biopsies and lavage cells (42). Our preliminary studies demonstrate adequate recovery of intact mRNA for these studies. If mRNA expression levels are too low to detect with standard Northern analysis, semi-quantitative reverse transcriptase polymerase chain reaction (RT-PCR) amplification will be used (43). Immunohistochemical staining for the mediators or oncoproteins of interest is planned.

I. Dietary Assessment

Dietary information is collected by direct interview with a dietician at baseline and at 6 months using the modified Gladys Block Questionnaire employed in the CARET study, and analyzed by computer at the CARET Coordinating Center. The relationships between serum, BAL, and dietary vitamin levels are being determined.

J. Data Management & Analysis

All potential continuous variables of interest, including demographics, clinical status and measured markers at first evaluation are analyzed for normality and comparability of treatment groups. Categorical variables such as smoking status are similarly assessed for distribution by treatment groups to assure comparability. The relationships between metaplasia, lung inflammation, vitamin A levels, and other potential markers are explored using logistic regression based on the data from pre-treatment evaluation. The degree of test/re-test variability among placebo treated subjects for each outcome variable of interest will be determined. The impact of the intervention on airway metaplasia, lung inflammation and other outcome measures will be tested by comparison of proportions for categorical variables and paired T tests for continuous markers. Potential confounding factors such as preexistent lung disease and smoking status will be adjusted for using multivariate techniques or stratification as appropriate.

III. RESULTS

A. Preliminary Studies

Fifty subjects with heavy prior exposure to asbestos have previously been evaluated to identify risk factors and early biological markers for lung cancer and progressive fibrosis (14,40). The relationships between metaplasia, dietary vitamin A intake, BAL fluid cellularity, and pulmonary function were characterized. Biopsies were evaluated from 46/50 of the study subjects. Metaplasia with keratinization, the most severe pathologic change, was detected in 1/3 of subjects biopsied (14,40). Detailed dietary histories were obtained on the asbestos workers. Total vitamin A intake was lower among those with metaplasia (mean log intake 10.80) than in those with normal airways (mean log intake 11.21 $p < 0.02$) (44). A strong association between neutrophil concentration and airway metaplasia was found (odds ratio 9.93; C.I. 1.62-61.0) (14).

B. Demographic and Exposure Variables

To date, 18 subjects have been recruited, completed the baseline evaluation, and been randomized to the study intervention or placebo. Compliance and follow-up have both been excellent. There have been no side effects related to the intervention and no one has discontinued his medication or withdrawn from the study.

Baseline pre intervention data on the first 14 subjects is summarized in Tables 2 and 3. Thirteen out of 14 (93%) of the subjects have radiographic changes (plaques alone ± increased markings) consistent with asbestos-related pulmonary disease. None have significant ongoing asbestos exposure. Thirteen of 14 subjects are current or former smokers, one a lifetime nonsmoker.

The subjects had been exposed to asbestos from work in the construction trades and asbestos products manufacturing. Asbestos exposure averaged 15 insulator-year equivalents (range 5 to 36) with a mean latency between beginning exposure and entry into the study of 35 years (Table 2).

Table 2. Baseline Characteristics (n = 14)

	Mean or #	Range
Demographics		
Age	65 yrs	43–72
Race		
White	12 (85%)	
Black or Hispanic	2 (15%)	
Smoking		
current or past	13 (93 %)	
exposure (pack years)	29	0.0–67.5
Chest Xray		
Plaques	12 (86%)	
ILO > 1/0	3 (21%)	
Asbestos exposure		
Dose (insulator years)	15 yrs	5–36
Latency	30 yrs	15–40

Table 3. Baseline Endpoints, Markers (n = 14)

	Mean	Range
Pulmonary Function		
FEV_1 (% predicted)	80.3	50–108
FVC (% predicted)	82.7	44–103
DLCO (% predicted)	77.6	37–111
PD 15 FEV_1 (mg methacholine)	9.16	0.20–29.3
Bronchoalveolar Lavage		
Cell differentials		
Eos/PMNs/Lyctes(%total)	7.0	0.0–19.0
Macrophages (%total)	93.0	82.0–99.0
Cell RNA yield (ug)	21.3	2.9–52
Bronchial biopsies		
Basal cell hyperplasia	13/13	
Metaplasia	8/13	
Nutritional Assessment		
Dietary questionnaire		
retinol (ug/day)	478	26–1069
β-carotene (ug/day)	2750	975–5167
Serum		
retinol (ug/dl)	34.67	10.5–51.8
beta-carotene (ug/dl)	10.32	0.7–27.1
Lavage macrophage		
retinol (ng/10^6 cells)	0.90	0.02–1.7
β-carotene (ng/10^6 cells)	3.78	0.8–7.4

C. Airway Histology

Bronchial biopsies were scored on a scale from normal to squamous metaplasia with keratinization (14). The histologic scoring system is shown in Figure 1. For each subject each biopsy was graded separately. Abnormal bronchial airway histology was very common. All subjects had basal cell hyperplasia on at least one biopsy (Figure 1A). Eight of 13 (62%) subjects had squamous metaplasia with keratinization, the most severe pathologic change on at least 1 biopsy (Figure 1C, Table 3). These findings are consistent with our earlier studies, and suggest that airway metaplasia may be an appropriate biomarker of airway injury and / or increased risk for lung cancer in these subjects.

D. Markers of Nutrient Status

Baseline serum levels of retinol (mean 34.67 ug/dl) and β-carotene (mean 10.3 ug/dl) were determined as described above (Table 3). Retinol, β-carotene, and α-tocopherol intake, as assessed by a modified Gladys Block questionnaire administered by a trained dietician revealed dietary intake of retinol and β-carotene of 478 ug/day and 2750 ug/day, respectively. These serum and dietary values are comparable to those found in the larger CARET population and other studies. Using HPLC methodology we have been able to detect and quantitate BAL macrophage retinol and β-carotene levels. Preliminary data (not shown) suggests that these levels may reflect dietary intake and potentially be a better biomarker of lung vitamin A status than serum levels.

E. Markers of Lung Function

Bronchoalveolar lavage demonstrated a potentially mild alveolitis with 5.3% lymphocytes, 1.2 % eosinophils and 1.5% neutrophils, similar to our prior studies (Table 3). Of interest, methacholine challenge studies demonstrated greater bronchial hyper- responsiveness than expected in a non asthmatic population, consistent with the findings that asbestos can cause airways disease (Table 3).

F. Additional Studies

RNA has been processed from lavage macrophages, with a mean yield of 21.3 ug per subject, sufficient RNA for Northern blot and RT-PCR assays (Table 3). Cytokine assays will be performed as a group to minimize variability. Adequate quantities of mRNA (> 10 ug) have also been obtained from airway biopsy samples, demonstrating the feasibility of obtaining adequate quantities of mRNA from lavage macrophages and bronchial biopsies for the studies proposed.

IV. DISCUSSION AND CONCLUSIONS

In summary, this article has summarized the rationale, study design, methodology, preliminary studies and initial baseline results of an ongoing placebo controlled short-term biomarker study of retinol and β-carotene in subjects at high risk for lung cancer. We are investigating biomarkers of early disease and host susceptibility in these subjects and determining the effects of retinol and β-carotene on these biomarkers for several reasons. Despite extensive epidemiologic, animal and *in vitro* studies supporting a beneficial effect of such an intervention on both lung cancer and inflammatory lung diseases, there is a paucity of clinical data addressing this important question. Further, the mechanisms of such an effect are poorly understood. Well validated early biomarkers of disease or the effect of a chemopreventive intervention, and biomarkers of host susceptibility are greatly needed to facilitate such studies.

Our prior studies in asbestos-exposed workers have demonstrated an high frequency of squamous metaplasia and alveolitis, potential early biomarkers of lung cancer and progressive fibrosis, respectively, and a strong association between low vitamin A intake and metaplasia. Additional preliminary studies (data not shown) have documented low lung tissue vitamin A levels in patients with lung cancer, detectable macrophage retinoid and carotenoid levels, and a possible correlation between tissue and BAL macrophage retinol levels (data not shown), suggesting that lung macrophages may serve as a biomarker for local lung nutrient status. Our *in vitro* studies have demonstrated that retinoids are potent inhibitors of two pro-inflammatory cytokines, IL-6 and IL-8 (25,26).

Baseline data on our first 14 subjects pre-intervention demonstrate a high frequency of airway metaplasia, and serum and dietary vitamin A levels within the normal range. Follow-up and compliance have been excellent, and there have been no complications. Together these data demonstrate the validity and feasibility of this ongoing chemoprevention trial to determine: 1) the effects of vitamin A on airway metaplasia and lung inflammation, 2) whether these effects may be mediated through altered cytokine and growth factor production, and 3) whether BAL retinol and β-carotene levels are a marker of susceptibility. In addition to the two major clinical endpoints we are examining, metaplasia and presence of inflammatory cells on BAL, a number of other biochemical, cellular and molecular biomarkers are also being investigated. Because all subjects, including controls will be studied at two time points, these studies will provide much needed information on the incidence, reproducibility, and variability of these potential biomarkers. Asbestos-exposed smokers, because of their previously shown high likelihood of having squamous metaplasia on airway biopsy and lung inflammation on BAL, are being used. However, the findings may be applicable to others at high risk for lung cancer.

Several potential problems in this ongoing study should be addressed: Because of the limited data available to determine our sample size estimates, our proposed sample size of 50 subjects may be inadequate to detect an important benefit from the intervention. Significant inter-individual differences in the degree of airway injury, inflammatory cells, lung function, or other parameters may make it difficult to detect an effect of the vitamin A supplementation. However, the use of each subject as his own control should reduce this problem. We can expand the sample size should the preliminary data warrant further investigation.

The two major clinical endpoints we are examining, metaplasia and presence of inflammatory cells on BAL, are promising and feasible early markers for lung cancer and ongoing inflammation. However, the significance of these findings in this population, although indicative of airway injury and lung inflammation, is unclear and needs further evaluation. More long term follow-up of study subjects is planned. Additional early markers of disease activity, response to therapy, or host susceptibility which may be valuable for future studies may be identified.

There are a number of other potential mediators and genetic alterations, such as the expression of other cytokines or oncogenes, which may be involved and be relevant markers in this study. We have chosen to study those parameters which appear most promising and feasible, but will save samples of all biologic specimens for potential additional studies in the future.

The use of both dietary retinol and β-carotene together as the chemopreventive intervention will make it difficult to determine which agent is responsible for any effects noted. In designing the study it was decided that, since it is unclear which agent may be beneficial, both agents would be used to maximize the chance of finding a significant response. Both agents are also being used together in the CARET study, and the results of this study may be applicable to the much larger CARET population. Should a significant reduction in bronchial metaplasia and/or lung inflammation be detected further studies are warranted to determine which agent is responsible. It is possible that retinol or β-carotene could have an effect at either an earlier or later stage in the process of lung carcinogenesis than that being studied. Thus negative findings will not exclude the possibility of an effect of the intervention at a different stage.

Macrophage and epithelial mRNA expression may be difficult to detect using standard northern analysis. We are prepared to use PCR to increase the sensitivity and immunohistochemistry to localize protein expression. These studies are by necessity limited by the small quantities of human tissue and cells obtainable.

The proposed study will address many of these issues and the results obtained here should provide important information on the effects of supplemental vitamin A administration, and also facilitate further studies with vitamin A and other chemo-preventive agents.

ACKNOWLEDGMENTS

This work was supported in part by grant 92A69 from the American Institute for Cancer Research and the Clinical Research Center at Yale University School of Medicine.

REFERENCES

1. Blomhoff R, Green MH, Norum KR. Vitamin A: Physiological and biochemical processing, *Annu Rev Nutr* 12:37-57 (1992).
2. Brenner SE, Lippman SM, Hong WK. Chemopreventive strategies for lung and upper aerodigestive tract cancer, *Cancer Research* (suppl), 52:2758s-2761s (1992).
3. Mayne ST. β-Carotene and cancer prevention: What Is the evidence? *Conn Med* 54:547-551 (1990).
4. Chytil F. The lungs and vitamin A, *Am J Physiol* 262:L51 (1992).
5. Ross CA, Hammerling UG. Retinoids and the immune system, *in* "The Retinoids: Biology, Chemistry, and Medicine", Sporn MB, Roberts AB, eds. 1994.
6. Hong WK, Itri LM. Retinods and human cancer, *in* "The Retinoids: Biology, Chemistry, and Medicine, Sporn MB, Roberts AB, eds. 1994.
7. Doll R, Peto R. The causes of cancer: Quantitative estimates of avoidable risks of cancer in the United States today, *J Natl Cancer Inst* 66:1192-1308 (1981).
8. Gazdar AF. The molecular and cellular basis of human lung cancer, *Anticancer Res* 13:261-268 (1994).
9. Minna JD. The molecular biology of lung cancer pathogenesis, *Chest* 103:445S-56S (1993).
10. Ames BN, Gold LS. Too many rodent carcinogens: Mitogenesis increases mutagenesis, *Science* 249:970-12 (1990).
11. Sporn MB, Roberts AB. Peptide growth factors and inflammation, *JClin Invest* 78:329-32 (1986).
12. Tockman Ms, Antonisen NR, Wright EC, Donithan MG. Airways obstruction and the risk for lung cancer, *Ann Int Med* 106:512-8 (1987).
13. Kipen HM, Lilis R, Suzuki Y, Valcuikas JA, Selikoff IJ. Pulmonary fibrosis in asbestos workers with lung cancer: A radiologic and histopathologic evaluation, *Brit J Ind Med* 44:96-100 (1987).
14. Cullen MR, Merrill WW. The relationship between acute inflammatory cells in lavage fluid and bronchial metaplasia, *Chest* 102:688-93 (1992).
15. Cross M, Dexter M. Growth factors in development, transformation,and tumorigenesis, *Cell* 64:271-80 (1991).
16. Tsao MS, Zhu H, Giaid A, Viallet J, Nakamura T, Park M. Hepatocyte growth factor/scatter factor is an autocrine factor for human normal bronchial epithelial and lung carcinoma cells. *Cell Growth and Diff* 4:571-9 (1993).
17. Blaner WS, Olson JA. Retinol and retinoic acid metabolism, *in*: "The Retinoids: Biology, Chemistry, and Medicine", Sporn MB, Roberts AB, eds. 1994.
18. Underwood BA, Siegel H, Weisell RC, Dolinski M. Liver stores of Vitamin A in a normal population dying suddenly or rapidly from unnatural causes, *Am J Clin Nutri* 23:1037 (1970).
19. Moon RC, Mehta RG, Rao KVN. Retinoids and cancer in experimental animals, *in* "The Retinoids: Biology, Chemistry, and Medicine", Sporn MB, Roberts AB, eds. 1994.
20. Jetten AM, Vollberg TM, Nervi C and George MD. Positive and negative regulation of proliferation and differentiation in tracheobronchial epithelial cells, *Am Rev Resp Dis* 142:S36-9 (1990).
21. Rahmathullah L, Underwood BA, Thulasiraj RD, Milron RC, Ramawamy K, Rahmathullah R, Babu G. Reduced mortality among children in Southern India receiving a small weekly dose of Vitamin A, *New Eng J Med* 323:929-35 (1990).
22. Hussey GD, Klein M. A Randomized, Controlled trial of Vitamin A in children with severe measles, *New Eng J Med* 323:160-4 (1990).
23. Morabia A, Menkes MJS, Comstock GW, Tockman MS. Serum Retinol and airway obstruction, *Am J Epid* 132:77-82 (1990).
24. Glick AB, Flanders KC, Danielpour D, Yuspa SH, and Sporn MB. Retinoic Acid induces transforming growth factor-β_2 in cultured keratinocytes and mouse epidermis, *Cell Reg* 1:87-97 (1989).
25. Redlich CA, Sikora AG, Zitnik RJ, Elias JA. Retinoid regulation of Interleukin-1-induced Interleukin-8 production by human lung fibroblasts, *Am Rev Resp Dis* 147:A752 (1993).
26. Zitnik RJ, Kotloff RM, Zheng T, Whiting NL, Schwalb J, and Elias JA. Retinoic acid inhibition of Interleukin-1-induced Interleukin-6 production by human lung fibroblasts, *J Immunol* 152:1419-27 (1994).
27. Devet, HCW. The puzzling role of Vitamin A in cancer prevention. *Anticancer Res* 9:145-152 (1989).
28. Hong WK, Lippman SM, Itri LM, Karp DD, Lee JS, Byers RM, Schantz SP, Kramer AM, Lotan R, Peters LJ, Dimery IW, Brown BW, Goepfert H. Prevention of second primary tumors with isotretinoin in squamous cell carcinoma of the head and neck, *New Eng J Med* 323: 795-801 (1990).
29. Gouveia J, Hercend T, Lemaigre G et al. Degree of bronchial metaplasia in heavy smokers and its regression after treatment with a retinoid, *Lancet* 1:710-12 (1982).

30. Lee JS, Lippman SM, Benner SE, Lee JJ, Ro JY, Lukeman JM, Morice RC, Peters EJ, Pang AC, Fritsche HA Jr. Randomized placebo-controlled trial of isotretinoin in chemoprevention of bronchial squamous metaplasia, *J Clin Onc* 12(5):937-45 (1994).
31. The Alpha-Tocopherol, Beta Carotene Cancer Prevention Study Group. The Effect of Vitamin E and beta carotene on the incidence of lung cancer and other cancers in male smokers, *New Eng J Med* 330:1029-35 (1994).
32. Omenn GS, Goodman G, Thornquist M, Grizzle J, Rosenstock L, Barnhart S, Balmes J, Cherniak M, Cone J, Cullen M, Glass A, Keogh J, Meyskens F, Valanis B, Williams, J. The β-carotene and retinol efficacy trial (CARET) for chemoprevention of lung cancer in high-risk populations: smokers and asbestos-exposed workers, *Cancer Res* 54:20385-20435 (1994).
33. Slack JMW. Epithelial metaplasia and the second anatomy, *Lancet* 2:268-71 (1986).
34. Auerbach O, Hammond EC, Garfinkle L. Changes in bronchial epithelium in relation to smoking, *New Eng J Med* 300:381-6 (1979).
35. Rom WN, Travis WD, Brody AR. Cellular and molecular basis of the asbestos-related diseases, *Am Rev Resp Dis* 143:408-422 (1991).
36. Rochester CL, Elias JA. Cytokines and cytokine networking in interstitial and fibrotic lung disorders, *Semim in Resp Med* 14:389-416 (1993).
37. Khalil N, O'Connor RN, Unruh H, Warren PW, Flanders KC, Kemp A, Bereznay H, Greenberg AH. Increased production and immunohistochemical localization of transforming growth factor β - in idiopathic pulmonary fibrosis, *Am J Respir Cell Mol Biol* 5:155-162 (1991).
38. Martinet Y, Rom WN, Grotendorst GR, Margin GR, Crystal RG. Exaggerated spontaneous release of platelet-derived growth factor by alveolar macrophages from patients with idiopathic pulmonary fibrosis, *New EngJ Med* 317:202-9 (1987).
39. Elias JA, Zheng T, Whiting NL, Trow TK, Merrill WW, Zitnik R, Ray P, Alderman EM. IL-1 and transforming growth factor-Beta regulation of fibroblast-derived IL-11, *J Immunol* 152:2421-29 (1994).
40. Merrill WW, Carter D, Cullen MR. The relationship between bronchial inflammatory cells and large airway metaplasia, *Chest* 100:131-135 (1991).
41. Yamada M, Blaner WS, Soprano WS, Dixon JL, Kjeldbye HM, and Goodman, DS. Biochemical characteristics of isolated rat liver stellate cells, *Hepatology* 1224-1229 (1987).
42. Chomczyski P, and Sacchi N. Single-step method of RNA isolation by acid guanidinium thiocyanate-phenol-chloroform extraction, *Anal Biochem* 162:156-61 (1987).
43. Omiecinski CJ, Redlich CA, Costa P. Induction and developmental expression of cytochrome P450IAI mRNA in rat and human tissues: Detection by the polymerase chain reaction. *Ca Research* 50:4315- 4321 (1990).
44. Mohr SN, Redlich CA, , Cullen MR, Eckhoff C, Mayne ST. Beta-carotene and retinoids as predictors of bronchial metaplasia among asbestos workers, submitted for publication.

3

HUMAN CERVICAL CANCER

Retinoids, Interferon and Human Papillomavirus

Richard L. Eckert,* Chapla Agarwal, Joan R. Hembree, Chee K. Choo, Nywana Sizemore, Sheila Andreatta-van Leyen and Ellen A. Rorke

Departments of Physiology and Biophysics, Dermatology, Reproductive Biology, Biochemistry, and Environmental Health Sciences
Case Western Reserve University School of Medicine
2109 Adelbert Road
Cleveland, Ohio 44106-4970

I. INTRODUCTION

Cervical cancer is the third most common type of cancer in women in the United States and is the leading cause of cancer deaths in women in third world countries (**1,2**). Although the pathogenesis of the disease is incompletely understood, human papillomavirus (HPV) is present in over 90% of all high grade cervical lesions and is strongly implicated in the process of cancer development. A common feature of these tumors is expression of the E6 and E7 viral reading frames (**3,4,5**). These viral genes encode factors, E6 and E7, that interact with the p53 and pRB tumor suppressor genes (**6,7,8,9,10**). E6 promotes the degradation of p53 via a ubiquitin-dependent pathway (**10,11**) and E7 interferes with pRB function (**8,9,10**). Expression of E6 and E7 is sufficient for immortalization of cultured epidermal keratinocytes and cultured cervical epithelial cells (**12,13,14,15**).

Human papillomaviruses comprise a family of related viruses, each of which displays a specific tissue tropism (**3,4**). Several members of this family, including HPV types 16, 18, 31 and 33, are classified as "high risk" viruses and are strongly implicated as causative agents in the genesis of cervical cancer. HPV16 is the type most commonly associated with cervical tumors. The genomic organization of HPV16 is shown in Fig. 1. The viral genome is an 8 kilobase, closed circular, double stranded DNA genome which encodes several early (E1, E2, etc.) and late (L1 and L2) gene products from multiple reading frames. The P_{97} promoter is the major transcription initiation site within the viral genome.

In most high grade cervical tumors, HPV DNA is found integrated into the host cell DNA, a process which frequently disrupts the HPV genome within the E1 and E2 reading frames (**16**). This integration event results in a loss of expression of E2, the viral protein that

* Correspondence: Richard L. Eckert, Ph.D., Department of Physiology/Biophysics, Rm E532, Case Western Reserve University School of Medicine, 2109 Adelbert Road, Cleveland, Ohio 44106-4970. Ph: 216-368-5530; Fx: 216-368-5586.

Diet and Cancer, Edited under the auspices of the American Institute of Cancer Research
Plenum Press, New York, 1995

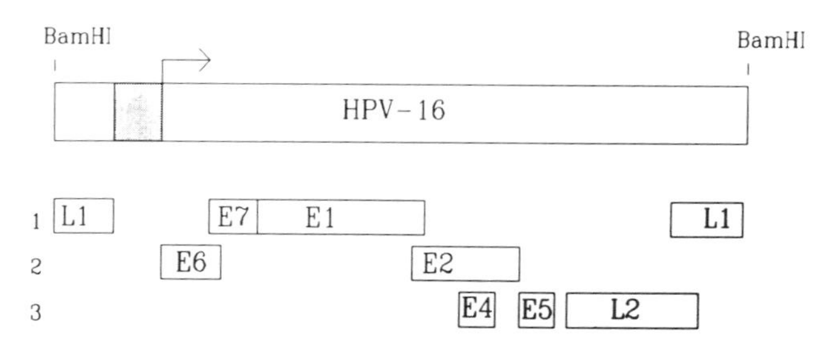

Figure 1. Structure of the HPV16 genome. HPV16 is a circular genome of approximately 8 kb in length that encodes proteins produced from reading frames 1, 2 and 3. The genome is shown linearized at the unique BamHI site. The early (E) and late (L) gene products are shown as is the upstream regulatory region (cross-hatched) located between the L1 and E6 reading frames. The arrow shows the site of transcription initiation of the P_{97} promoter.

binds to specific cis-acting DNA sequence elements within the HPV promoter (i.e., the P_{97} promoter in HPV16) and regulates the production of HPV mRNA (**17,18**). The loss of E2 expression is thought to permit increased production of E6 and E7 (**19,20,21**).

Vitamin A and Cervical Cancer

The prolonged time between HPV infection and development of cancer suggests that cervical cancer progression requires multiple events. In this regard, cigarette smoking, poor hygiene and nutritional deficiencies, particularly deficiencies of dietary vitamin A, are risk factors for cervical cancer (**22,23,24**), suggesting that vitamin A may be protective against cervical cancer. Additional studies show that retinoid therapy is effective for the treatment of cervical cancer (**25,26,27**). Our recent studies show that an HPV16-immortalized cervical cell line, ECE16-1, displays an enhanced sensitivity to retinoids (**28**). These cells are more readily growth suppressed by retinoids and regulation of cell differentiation is more retinoid-sensitive compared to normal cells (**28,29**). Similar results have been noted in HPV-immortalized epidermal foreskin keratinocytes (**30**).

In the present manuscript we review our work on the retinoid effects on cell differentiation (**28,29,31,32,33,34,35,36,37**), HPV16 oncogene expression (**29**) and insulin-like growth factor-1 (IGF1)(**38,39**) and epidermal growth factor (EGF)(**40**) signalling in HPV16-immortalized cervical epithelial cells. Our results indicate that HPV-immortalized cells are more sensitive to the effects of retinoids, that retinoids inhibition of growth is not always correlated with reduced E6/E7 transcript levels and that retinoids may, at least in part, inhibit cervical tumor cell growth by reducing the activity of the IGF1 and EGF signalling pathways.

II. MATERIALS AND METHODS

A. Cell Culture

Normal ectocervical epithelial cells and HPV16-positive cervical cell lines (ECE16-1, ECE16-D1, ECE16-D2 and CaSki) were maintained in growth medium as previously described (**28,29,40**) and were transferred to experimental medium prior to experiments. ECE16-1, ECE16-D1 and ECE16-D2 were produced by transfection of normal cells with the HPV16 genome (**28,41**). These cells are immortal, but do not form tumors in nude mice or grow on soft agar. The CaSki cell line is a fully transformed HPV16-positive cell line that is immortal and tumorigenic (**42**). In general, growth conditions, etc. are described for each experiment; however, additional details can be found in the manuscript(s) referenced in each figure legend.

B. Detection of Cytokeratins

To evaluate the level and types of cytokeratin protein present, cells were incubated in medium containing ^{35}S-methionine (**28**). After labeling for 20 h, the cytokeratins were extracted and characterized by two-dimensional gel electrophoresis as previously described (**28**).

C. Detection of IGFBP-3

The level of IGFBP-3 in conditioned medium was detected by ligand blot (**38,39**). Conditioned medium was prepared in reducing agent-free sample buffer, electrophoresed on a 10% acrylamide gel and transblotted to nitrocellulose. After transfer, the membranes were incubated with ^{125}I-IGF1, washed and dried, and ^{125}I-IGF1 binding to membrane-immobilized IGF binding protein was visualized by exposing the membrane to x-ray film (**38,39**). For direct detection of IGFBP-3 protein, cell culture medium was electrophoresed on 10% acrylamide gels, transferred to nitrocellulose and incubated with an antibody specific for IGFBP-3 (**38,39**). The relative level of expression was determined by densitometry.

D. Nucleic Acids Methods

Total RNA was isolated using the guanidine/CsCl gradient method and poly(A)+ RNA was prepared by oligo(dT) cellulose chromatography. Denaturing RNA gel electrophoresis and transfer blotting were performed exactly as previously described (**43,44**). Plasmids encoding the cytokeratins (**43,44**) and IGFBP-3 (**38,39**) and polymerase chain reaction products encoding specific regions of the HPV16 genome were radiolabeled by random priming in the presence of ^{32}P-dCTP and hybridized to the blots. After hybridization, the blots were washed under stringent conditions to remove non-specific binding and exposed on x-ray film (**43,44**).

E. Measurement of EGF-Receptor Levels

EGF receptor (EGF-R) levels were quantitated by immunoprecipitation of EGF-R protein from metabolically labeled cells or by ligand binding assay using ^{125}I-EGF (**40**).

III. RESULTS

Retinoid Effects on Normal and HPV16-Immortalized Ectocervical Epithelial Cell Differentiation

To develop *in vitro* model systems for the study of HPV16-involved cervical cancer, we transfected normal cervical epithelial cells with the molecularly cloned HPV16 genome and selected cell lines that displayed unlimited growth potential. This process resulted in the production of the ECE16-1, ECE16-D1 and ECE16-D2 cell lines (**28,41**). Each of these lines express HPV16-specific transcripts (**28,41**), produce HPV16 E7 protein (**Choo and Eckert,**

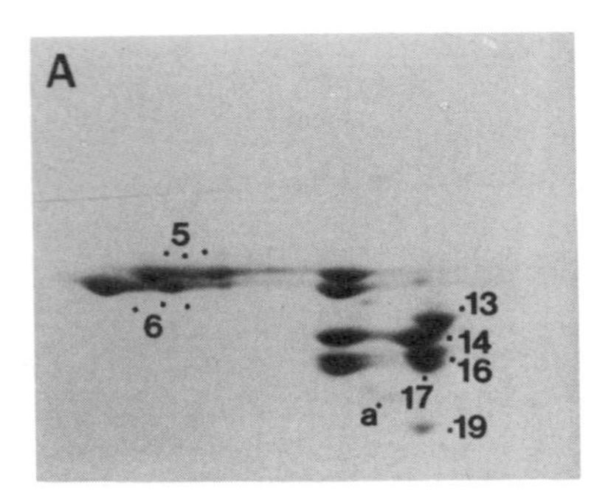

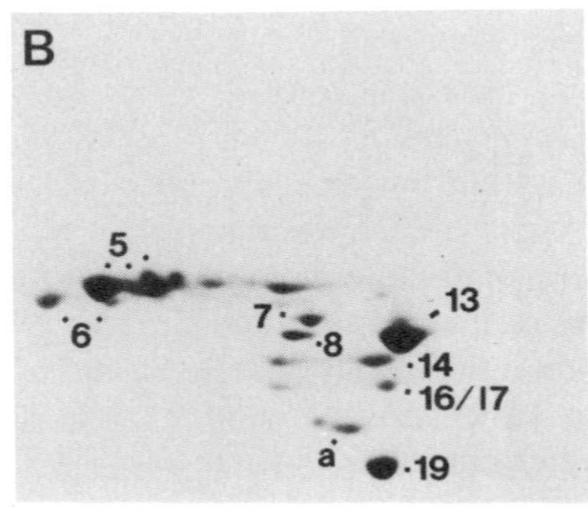

Figure 2. Comparison of cytokeratin expression in normal and HPV16-immortalized cells grown under identical conditions. Cultures of normal cervical cells (A) or ECE16-1 cells (B) were grown in medium containing normal fetal calf serum and then labeled with ^{35}S-methionine prior to extraction of cytokeratins (**28**). The extracts were electrophoresed on two dimensional gels, fluorographed, dried and exposed on x-ray film. The cytokeratins are numbered according to Moll et al., (**52**) and a indicates actin which is a variable contaminant in cytokeratin preparations.

unpublished) and display unlimited growth potential. We have used these cell lines to study the effects of retinoids on HPV16-immmortalized cervical cell function.

Dietary vitamin A has been shown to be important as a protective agent against a variety of cancers (**45**). Moreover, a recent study has demonstrated that cervical tumors are responsive to therapy consisting of combined treatment with a vitamin A metabolite (13-cis-retinoic acid) and interferon-α (IFNα)(**25,26,27**). One of our major goals is to elucidate the mechanism(s) underlying the beneficial effects of retinoid and interferon therapy. We therefore evaluated the effects of retinoids on cervical cell differentiation. As a measure of the differentiation status of the cells, we measured the effects of retinoids on cytokeratin gene expression (**28,36,41**). Cytokeratins consist of a family of over twenty proteins that form the intermediate filaments in epithelial cells (**46,47**). The pattern of keratin expression is tissue- and differentiation-specific (**47**) and keratin gene expression is regulated by agents that influence epithelial cell differentiation (**47**). Moreover, our studies in the cervical cell system indicate that the pattern of cytokeratin expression is an excellent indicator of the differentiation status of the cells, both *in vivo* and *in vitro* (**28,32**).

We initially observed that the pattern of keratin gene expression was shifted in the HPV16-immortalized ECE16-1 cells compared to normal cervical cells (**28**). The level of cytokeratins K5, K6, K14, K16 and K17 decreased while the level of cytokeratins K7, K8 and K19 increased (**28**)(Fig. 2). At first, the reason for this shift was not clear; however, we realized that the serum in our culture medium contained retinoids. In normal cervical cells, cytokeratin expression is retinoid responsive; however, the retinoid levels present in the culture medium are not high enough to influence keratin gene expression. The dramatic shift in the pattern of keratin gene expression in the HPV16-immortalized cells could be explained if the cells had become sensitized to the effects of retinoids. To test this hypothesis, we grew ECE16-1 cells in retinoid-free medium or in medium supplemented with various natural or synthetic retinoids and monitored the effects on keratin gene expression (**28**). As shown in Fig. 3A, growing ECE16-1 cells in retinoid-free medium shifts the pattern of cytokeratin

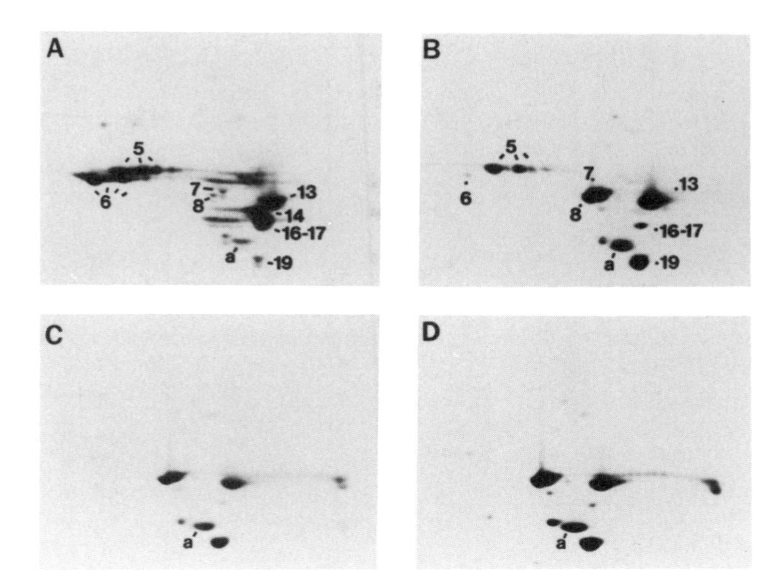

Figure 3. Retinoid regulation of cytokeratin levels in HPV16-immortalized cervical epithelial cells. ECE16-1 cells were maintained for six days in retinoid-free medium (A) or medium containing 100 nM of RA (B), 100 nM Ro 13-6298 (C) or 100 nM Ro 13-7410 (D). The cultures were labeled with ^{35}S-methionine for 24 h prior to extraction of cytokeratins. The extracts were electrophoresed, fluorographed and the photograph is labeled as indicated in Fig. 2 (**28**).

expression back to that observed in normal cells (i.e., increased K5, K6, K14, K16 and K17; decreased K7, K8 and K19). Addition of all-trans-retinoic acid (RA)(Fig. 3B), or the synthetic retinoids Ro 13-6298 (Fig. 3C) or Ro 13-7410 (Fig. 3D) produces a dedifferentiated pattern of keratin expression.

Retinoid Regulation of HPV16 Transcription

The HPV16 E6 and E7 reading frame is frequently expressed in high grade HPV-positive cervical tumors (**3,4,5**). This has led to the suggestion that elevated E6/E7 levels may be correlated with increased proliferation and the corollary suggestion that agents that reduce E6/E7 should reduce growth rate. To investigate this hypothesis in more detail, we investigated the relationship between cell proliferation rate and the level of transcript encoding the HPV16 E6/E7 reading frame.

Proliferation studies indicate that both interferon-γ and RA acid are efficient inhibitors of ECE16-1 cell growth and combined treatment with both agents suppresses growth further (Fig. 4). In a parallel study, we examined the effects of these agents on HPV16 transcription. ECE16-1 cells were maintained in retinoid-free medium (-) or in medium containing 1 uM RA (+) and the effects of treatment on actin (A), cytokeratin K5 and K19 (K5, K19) and HPV16 E6/E7 reading frame expression (E6/E7) were monitored (Fig. 5). K5 and K19 are retinoid-responsive markers in cervical cells (**28**). Actin is not retinoid regulated in these cells (**28**). As shown in Fig. 5, K5 levels decrease, K19 levels increase, actin mRNA level is not regulated and the level of transcript encoding the HPV16 E6/E7 reading frame is not regulated. Fig. 6 shows the effects of IFNγ. IFNγ does not regulate K5 or actin mRNA level; however, the transcripts encoding the E6/E7 reading frame are suppressed by >90%.

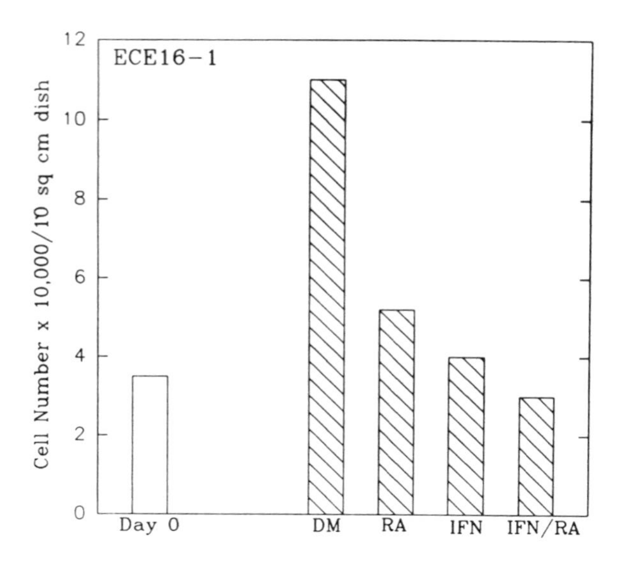

Figure 4. Regulation of ECE16-1 cell growth by all-trans-retinoic acid and IFNγ. ECE16-1 cells were plated in complete medium containing 20 ng/ml EGF and incubated for 3 days in the presence of 1 uM RA, 200 IU/ml IFNγ or combined treatment (**29**). At the end of 3 days of treatment, the cells were harvested and counted. Day 0 indicates the cell count at the start of treatment and the closed bars indicate cell number after three days of treatment.

Retinoid Regulation of the IGF1 Signalling System

The results outlined above suggest that retinoids do not always act to suppress cell growth by suppressing the level of RNA encoding the HPV16 E6/E7 oncogenes. We, therefore, began to examine whether retinoids regulate the activity of any of the signal transduction pathways. The IGF signalling pathway is a major mitogenic pathway in epithelial cell types (**48,49,50**). Our recent results indicate that normal and HPV16-

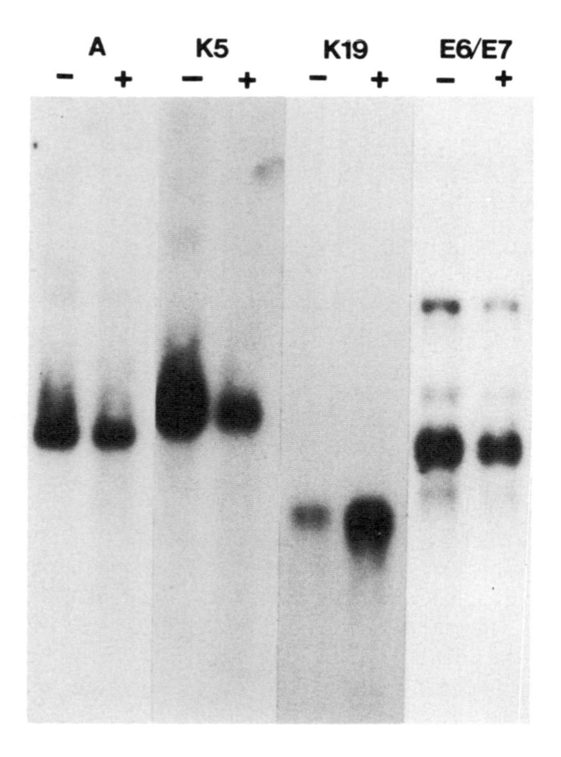

Figure 5. Retinoid effects on HPV16 E6/E7 transcript level in ECE16-1 cells. Cells were grown for three days in the presence (+) or absence (-) of 1 uM RA. Identical paired sets of lanes were hybridized with cDNAs encoding cytokeratins K5 (K5) and K19 (K19), actin (A) and the HPV16 E6/E7 coding region (E6/E7)(**29**). The blots were washed and exposed on x-ray film.

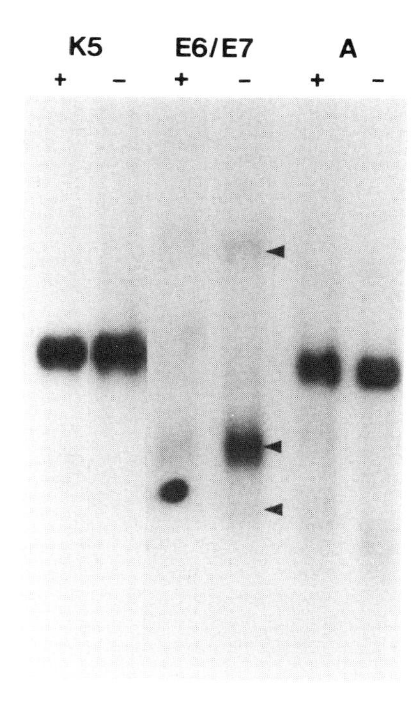

Figure 6. IFNγ effects on HPV16 E6/E7 transcript production in ECE16-1 cells. Poly(A)+ RNA from ECE16-1 cells treated for three days in the presence (+) or absence (-) of 200 IU IFNγ was hybridized with cDNAs encoding cytokeratin K5 (K5), actin (A) or the HPV16 E6/E7 (E6/E7) coding region. The blots were then washed and exposed on x-ray film (**29**).

immortalized cervical epithelial cells produce and release large amounts of insulin-like growth factor binding protein-3 (IGFBP-3)(**38,39**). IGF1 is an autocrine factor, produced and released by mesenchymal cells, that stimulates epithelial cell proliferation via binding to a cell surface receptor (**48,49**). IGFBP-3 is a small, secreted, extracellular protein that binds to IGF1 and alters the ability of IGF1 to associate with the IGF cell surface receptors. In systems where the IGFBP-3/IGF1 association decreases responsiveness to IGF1, it is proposed that the association reduces the concentration of free IGF1 in the extracellular environment and, as a result reduces the amount of IGF1 available to bind to receptors on the cell surface. This reduces the mitogenic stimulus.

Our recent experiments indicate that epidermal growth factor reduces the level of IGFBP-3 that is produced and released by HPV16-immortalized cervical epithelial cells (**39**). This is correlated with an enhanced responsiveness of the cells to IGF1 (**39**), suggesting a model where EGF increases free IGF1 concentration by inhibiting IGFBP-3 production, resulting in a higher level of IGF1 available to stimulate mitosis. This suggests that *in vivo*, EGF-like factors may enhance proliferation by increasing the amount of free IGFs (**39**). Based on these results, we were curious whether RA could act to increase IGFBP-3 levels and reduce the mitogenic activity of IGF1. To evaluate this possibility, we grew ECE16-1 cells in medium containing 20 ng/ml EGF in the presence of 0 - 1000 nM RA and monitored the level of IGFBP-3 mRNA produced per cell and the level of IGFBP-3 protein released into the culture medium at each retinoid concentration. As shown in Fig. 7A, RA produced a concentration-dependent increase in IGFBP-3 protein which was half-maximal at 1 nm and maximal a 10 nM RA. As shown in Fig. 7B, the level of IGFBP-3 mRNA increased in parallel (**38**). The increased IGFBP-3 level was correlated with reduced cell proliferation (not shown)(**38**).

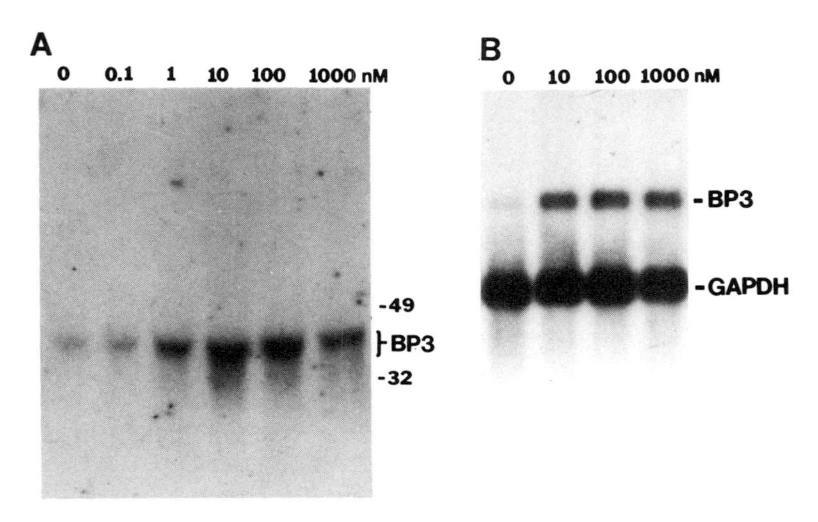

Figure 7. Regulation of IGFBP-3 protein and RNA levels by RA. ECE16-1 cells were grown for three days in the presence or absence of 0 to 1000 nM RA. Conditioned medium from the last 24 h of treatment was removed and assayed for IGFBP-3 by ligand blot (A). BP3 indicates the migration of IGFBP-3 and the migration of the 32 and 49 kDa markers is shown. In panel B, poly(A)+ RNA was isolated from cells grown for 3 days in medium containing 20 ng/ml EGF with 0 to 1000 nM RA. The mRNA was electrophoresed, transferred to membrane and hybridized with ^{32}P-labeled cDNAs encoding IGFBP-3 (BP3) and glyceraldehyde-3-phosphate dehydrogenase (GAPDH)(**38**).

Retinoid Regulation of the EGF Signalling System

A second major signalling system in cervical epithelial cells, is the EGF signalling system (**39,40**). In addition to the effects described above of EGF to reduce IGFBP-3 levels and potentiate the effects of IGF1 (**38,39**), EGF is also a direct stimulator of cervical cell proliferation (**38,39,40**). We have recently noted that increased EGF-R levels are associated with HPV16 immortalization (**40 and references therein**). As shown in Fig. 8, as measured by immunoprecipitation or binding assay, the level of EGF-R is approximately 2 times higher in ECE16-1 cells compared to normal cervical epithelial cells. Moreover, incubation of ECE16-1 cells with 100 nM RA (crosshatched bars) reduced EGF-R levels by approximately 50% compared to untreated (open bars) cultures. The reduction in EGF-R level by RA is concentration-dependent with half-maximal suppression at around 1 nM RA (Fig. 9). This reduction in EGF-R level is correlated with a parallel RA-dependent reduction in proliferation (**40**)(Fig. 10A), suggesting that RA may act to inhibit the growth of some HPV-involved tumors via suppression of the EGF signalling system. In contrast, treatment of normal cervical cells with RA did not result in a reduction in EGF-R level (Fig. 8) or proliferation rate (Fig. 10B).

IV. DISCUSSION

Retinoids and Cervical Cell Differentiation

Retinoids are important dietary agents that are known to be essential for the normal functioning of stratifying epithelial cell types (**47**). Retinoid deficiency results in extensive

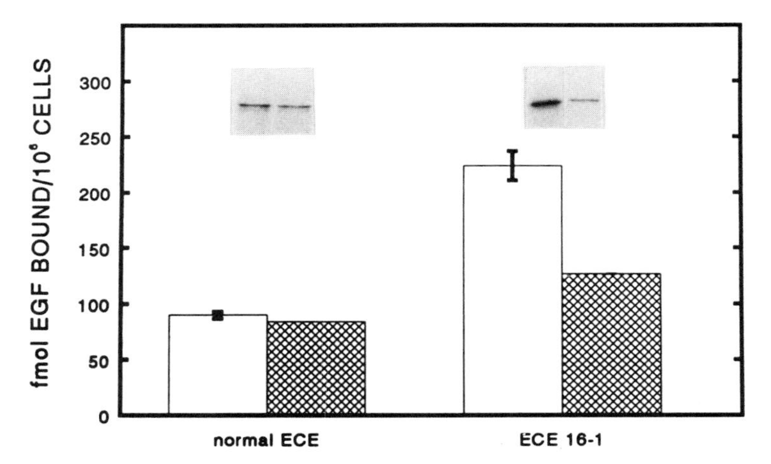

Figure 8. Retinoid regulation of EGF-receptor level in normal and HPV16-immortalized cervical epithelial cells. Cell were treated with (crosshatched bars) or without (open bars) 100 nM RA for 2 days. To assay for EGF binding, cells were incubated with ^{125}I-EGF ± 1000-fold excess of radioinert EGF for 3 h at 4 C. Specific EGF binding was evaluated as previously described (**40**). For immunoprecipitation, cells were prelabeled with ^{35}S-cysteine, solubilized and the EGF-R was immunoprecipitated, electrophoresed and autoradiographed (**40**).

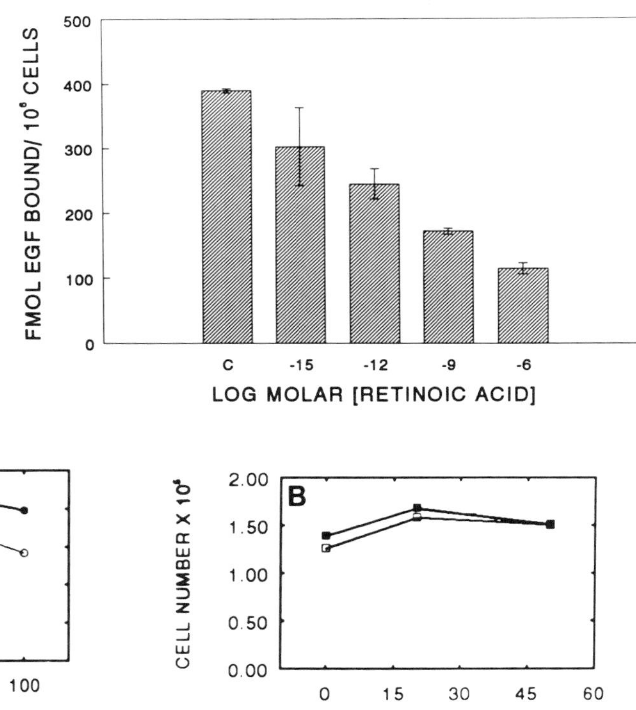

Figure 9. Concentration-dependent suppression of EGF-R levels by RA. ECE16-1 cells were incubated with 0 to 1 uM RA for two days and EGF-R levels were measured by binding assay (**40**). The values shown are the mean ± SE.

Figure 10. Retinoid suppression of EGF-stimulated growth in ECE16-1 cells. ECE16-1 cells (A) or normal cervical epithelial cells (B) were incubated with 0 to 100 ng/ml EGF in the presence (open symbols) or absence (closed symbols) of 100 nM RA. Cells were treated for seven days, harvested and counted using a coulter counter (**40**). Fresh growth factor, medium and retinoid were added every 48 h.

epithelial stratification, which is normalized by retinoid readministration (**51**). Retinoids are also useful chemopreventive agents and in the case of cervical cancer, have been successfully used therapeutically (**25,26,27**).

We and others have shown that retinoids regulate normal (**31,33,46**) and HPV16-immortalized cervical epithelial cell growth and differentiation (**28,30,41**). Our results, as reviewed above, indicate that human papillomavirus-immortalized cervical epithelial cells are more sensitive to the effects of retinoids than normal cervical epithelial cells, an effect that has also been observed in HPV-immortalized epidermal keratinocytes (**30**). This enhanced responsiveness is manifest as an increased sensitivity of the cells to retinoids as measured by ability to regulate biochemical markers of differentiation and cell proliferation (**28,38**). Cytokeratins comprise a family of structural proteins that are useful indicators of the differentiation status of the cell (**47**). Our results show that cytokeratin gene expression is markedly more sensitive to retinoid regulation in HPV16-immortalized cells than in normal cells (**28**). Moreover, the pattern of cytokeratin expression observed in cultured HPV16-immortalized cells growing in the presence of retinoids is similar to the pattern observed in cervical tumors *in vivo* (**32**). This suggests that the cell culture model is reproducing *in vivo*-like events and that retinoids are likely to be influencing the phenotype of the tumor cells *in vivo*. The finding that HPV16-immortalized cells, compared to normal cervical cells, are more sensitive to retinoid regulation, may explain why retinoids are effective therapeutic agents for the treatment of cervical cancer. That is, the tumor cells may respond to the presence of the retinoids, but the surrounding normal cells are less responsive or do not respond. In addition, these results suggest that the keratin genes may be useful markers for the study of the effects of retinoid therapy *in vivo*. For example, a biopsy of tumor tissue showing that the pattern of keratin gene expression has changed following retinoid therapy in a manner consistent with the studies shown above may be useful as an indicator that the retinoids are obtaining access to and regulating the tissue.

Retinoids, Interferon, HPV16 E6/E7 Expression and Cell Proliferation

A prevalent and reasonable hypothesis in the literature suggests that reduced HPV E6/E7 expression should be associated with decreased cell proliferation rate and that agents that suppress E6/E7 should also suppress growth. We evaluated this hypothesis for retinoids and interferon on HPV16-immortalized cervical cells. Our results, presented above, indicate that E6/E7 suppression is associated with reduced proliferation following administration of IFNγ. However, no change in E6/E7 RNA level is associated with retinoid-dependent inhibition of growth (**29**). These results suggest that IFNγ may act via suppression of E6/E7 levels (i.e., suppression of viral transcription), but that retinoids can suppress growth without affecting papillomavirus transcript level. This result suggests that retinoids may be acting via other mechanisms to suppress HPV16-immortalized cell growth that are independent of regulating viral transcription. The lack of retinoid effects on HPV16 E6/E7 mRNA level was verified for CaSki cells, an HPV16-immortalized transformed cell line (**29**), where retinoids actually increase proliferation, but do not regulate E6/E7 mRNA levels (**29**). Parallel studies with CaSki cells show that IFNγ suppresses both proliferation and E6/E7 RNA levels (**29**). The results with CaSki cells demonstrate that the absence of retinoid regulation of E6/E7 transcript levels is not a peculiarity of the ECE16-1 cell line.

Retinoid Effects on the IGF Signalling Pathway

Our previous results indicate that retinoids do not necessarily act by regulating the level of HPV16 E6/E7 transcript produced (**29**). We, therefore, have examined the effects of retinoids on other signalling systems. The insulin-like growth factors are strong mitogens for epithelial cells (**38,39**). IGF1 is produced by mesenchymal cell types and released into the extracellular environment where it then interacts with cell surface IGF receptors on epithelial cells to stimulate proliferation. However, most of the IGF1 produced becomes bound to members of a family of proteins called the insulin-like growth factor binding proteins. The IGFBPs are a family of proteins that sequester IGFs (**50**) and enhance or suppress their ability to stimulate cell proliferation. Our results show that cervical epithelial cells produce large quantities of IGFBP-3, which in most systems acts to inhibit the activity of IGF1. In ECE16-1 cells, epidermal growth factor, and presumably other EGF-like activities, reduces the level of this binding protein and potentiates the activity of IGF1 (**39**).

Treatment with retinoids, as shown in Fig. 7, increases the IGFBP-3 level in HPV16-immortalized cervical cells treated with EGF. Although our understanding of the system is not yet complete, our results are consistent with the idea that elevating the IGFBP-3 level inhibits the activity of IGF1 by sequestering IGF1 (**39**). In addition, our results suggest that IGFBP-3 may have some direct inhibitory effect on cervical cell proliferation independent of the presence of IGF1 (**38**). It is likely that tumors are exposed to a variety of paracrine factors that may reduce the level of IGFBP-3 and facilitate IGF1 growth stimulatory activity. Our results indicate that retinoids may act to counter this effect by increasing IGFBP-3 levels. In this regard, it is important to note that the retinoid-dependent increase in IGFBP-3 is dominant over the EGF-dependent suppression (**38,39**). This suggests that any therapeutic effects of RA resulting from elevation of IGFBP-3 levels should be achievable *in vivo*, even in the presence of growth factors.

Retinoid Effects on the EGF-R Signalling Pathway

As mentioned above, EGF is a potent mitogen for cervical epithelial cells. We have examined the effects of retinoids on the activity of this important signalling system. Our observation is that the level of EGF-R is increased 2-fold compared to normal cells, in a number of HPV16-immortalized cell lines (**40**). Interestingly, treatment with RA reduces the level of EGF-R back to normal levels. This appears to be an effect strictly on receptor number and not on receptor affinity for EGF, receptor internalization, or receptor stability. The retinoid-dependent reduction in EGF-R level correlates with a reduction in retinoid-dependent cell proliferation (**40**). This suggests that retinoids may act *in vivo* to inhibit tumor cell proliferation by reducing EGF-R levels.

V. SUMMARY

Our studies highlight the importance of dietary vitamin A (retinol) and other retinoids in maintaining normal cervical cell function and in inhibiting the growth of cervical tumors. Based on our results we conclude that 1) HPV16-immortalization enhances cervical cell sensitivity to retinoids, 2) cytokeratin expression may be useful as a marker for evaluating the success of retinoid therapy *in vivo*, 3) retinoids do not necessarily act to inhibit proliferation of HPV-immortalized cervical cells via effects on HPV E6 and E7 RNA levels and 4) retinoids may act to inhibit cervical cell proliferation by "suppressing" the activity of the EGF and IGF signalling pathways. Based on these and other results, it is worth considering the possibility that vitamin A or related retinoids could be administered therapeutically, early in the neoplastic process (either systemically or locally), to inhibit the progress of the disease. These results also suggest that combined interferon/retinoid therapy may provide an enhanced beneficial effect to reduce cervical tumor size due to the fact that each agent is inhibiting cervical cell proliferation via distinct, but reinforcing, pathways (i.e.,

IFNγ reduces E6/E7 expression, RA inhibits the function of the EGF and IGF1 signalling pathways).

ACKNOWLEDGEMENTS

This work was supported by grants from the American Institute for Cancer Research (RLE) and utilized the facilities of the Skin Diseases Research Center of Northeast Ohio (NIH, AR39750) and the Case Western Reserve University Ireland Cancer Center.

REFERENCES

1. P.J. Disaia, and W.T. Creasman, Preinvasive disease of the cervix, *in*: "Clinical Gynecologic Oncology," P.J. Disaia, and W.T. Creasman, eds. Mosby Year Book, St Louis (1993) pp. 1-36.
2. American Cancer Society: Cancer Facts and Figures. American Cancer Society, Atlanta (1992).
3. T.R. Broker, Structure and genetic expression of papillomaviruses, *Obstet. Gynecol. Clin. North. Am.*, 14: 329 (1987).
4. H. zur Hausen, Human papillomaviruses and their possible role in squamous cell carcinomas, *Curr. Top. Microbiol. Immunol.*, 78: 1 (1977).
5. L. Gissmann, M. Boshart, M. Durst, H. Ikenberg, D. Wagner, and H. zur Hausen, Presence of human papillomavirus in genital tumors, *J. Invest. Dermatol.*, 83: 26s (1984).
6. M.S. Lechner, D.H. Mack, A.B. Finicle, T. Crook, K.H. Vousden, and L.A. Laimins, Human papillomavirus E6 proteins bind p53 in vivo and abrogate p53-mediated repression of transcription, *EMBO J.*, 11: 3045 (1992).
7. V. Band, J.A. De Caprio, L. Delmolino, V. Kulesa, and R. Sager, Loss of p53 protein in human papillomavirus type 16 E6- immortalized human mammary epithelial cells, *J. Virol.*, 65: 6671 (1991).
8. P.S. Huang, D.R. Patrick, G. Edwards, P.J. Goodhart, H.E. Huber, L. Miles, V.M. Garsky, A. Oliff, and D.C. Heimbrook, Protein domains governing interactions between E2F, the retinoblastoma gene product, and human papillomavirus type 16 E7 protein, *Mol. Cell. Biol.* 13: 953 (1993).
9. D.V. Heck, C.L. Yee, P.M. Howley, and K. Munger, Efficiency of binding the retinoblastoma protein correlates with the transforming capacity of the E7 oncoproteins of the human papillomaviruses, *Proc. Natl. Acad. Sci. U S A.* 89: 4442 (1992).
10. K. Munger, M. Scheffner, J.M. Huibregtse, and P.M. Howley, Interactions of HPV E6 and E7 oncoproteins with tumour suppressor gene products. *Cancer. Surv.*, 12: 197 (1992).
11. J.M. Huibregtse, M. Scheffner, and P.M. Howley, A cellular protein mediates association of p53 with the E6 oncoprotein of human papillomavirus types 16 or 18, *EMBO J.*, 10: 4129 (1991).
12. P. Hawley-Nelson, K.H. Vousden, N.L. Hubbert, D.R. Lowy, and J.T. Schiller, HPV16 E6 and E7 proteins cooperate to immortalize human foreskin keratinocytes, *EMBO J.*, 8: 3905 (1989).
13. L. Pirisi, S. Yasumoto, M. Feller, J. Doniger, and J.A. DiPaolo, Transformation of human fibroblasts and keratinocytes with human papillomavirus type 16 DNA, *J. Virol.*, 61: 1061 (1987).
14. M.S. Barbosa, and R. Schlegel, The E6 and E7 genes of HPV-18 are sufficient for inducing two- stage in vitro transformation of human keratinocytes, *Oncogene*, 4: 1529 (1989).
15. C.D. Woodworth, J. Doniger, and J.A. DiPaolo, Immortalization of human foreskin keratinocytes by various human papillomavirus DNAs corresponds to their association with cervical carcinoma, *J. Virol.*,. 63:159 (1989).
16. A.P. Cullen, R. Reid, M. Campion, and A.T. Lorincz, Analysis of the physical state of different human papillomavirus DNAs in intraepithelial and invasive cervical neoplasm, *J. Virol.*, 65: 606 (1991).
17. C.M. Chiang, G. Dong, T.R. Broker, and L.T. Chow, Control of human papillomavirus type 11 origin of replication by the E2 family of transcription regulatory proteins, *J. Virol.*, 66: 5224 (1992).
18. C.M. Chiang, M. Ustav, A. Stenlund, T.F. Ho, T.R. Broker, and L.T. Chow, Viral E1 and E2 proteins support replication of homologous and heterologous papillomaviral origins, *Proc. Natl. Acad. Sci. U S A.*, 89: 5799 (1992).
19. H. Romanczuk, and P.M. Howley, Disruption of either the E1 or the E2 regulatory gene of human papillomavirus type 16 increases viral immortalization capacity, *Proc. Natl. Acad. Sci. U S A.*, 89: 3159 (1992).

20. B.C. Sang, and M.S. Barbosa, Increased E6/E7 transcription in HPV 18-immortalized human keratinocytes results from inactivation of E2 and additional cellular events, *Virology*, 189: 448 (1992).
21. R. Sousa, N. Dostatni, and M. Yaniv, Control of papillomavirus gene expression, *Biochim. Biophys. Acta*, 1032:19 (1990).
22. S.L. Romnay, P.R. Palan, and C. Duttagupta, S. Wassertheil-Smolter, J. Wylie, G. Miller, N.S. Slagle, and D. Lucido, Retinoids and the prevention of cervical dysplasia, *Am. J. Obst. Gyn.*, 141:890 (1981).
23. W.J. Winkelstein, E.J. Shillitoe, R. Brand, and K.K. Johnson, Further comments on cancer of the uterine cervix, smoking, and herpesvirus infection, *Am. J. Epidemiol.*, 119:1 (1984).
24. H.Y. Ngan, R.J. Collins, K.Y. Wong, A. Cheung, C.F. Lai, and Y.T. Liu, Cervical human papilloma virus infection of women attending social hygiene clinics in Hong Kong, *Int. J. Gynaecol. Obstet.*, 41:75 (1993).
25. S.M. Lippman, B.S. Glisson, J.J. Kavanagh, R. Lotan, W.K. Hong, M. Paredes Espinoza, W.N. Hittelman, E.E. Holdener, and I.H. Krakoff, Retinoic acid and interferon combination studies in human cancer, *Eur. J. Cancer*, 29A Suppl. 5: S9 (1993).
26. S.M. Lippman, J.J. Kavanagh, M. Paredes Espinoza, F. Delgadillo Madrueno, P. Paredes Casillas, W.K. Hong, G. Massimini, E.E. Holdener, and I.H. Krakoff, 13-cis-retinoic acid plus interferon-alpha 2a in locally advanced squamous cell carcinoma of the cervix, *J. Natl. Cancer Inst.*, 85: 499 (1993).
27. S.M. Lippman, and W.K. Hong, 13-cis-retinoic acid and cancer chemoprevention, *Monogr. Natl. Cancer Inst.*, 13:111 (1992).
28. C. Agarwal, E.A. Rorke, J.C. Irwin, and R.L. Eckert, Immortalization by human papillomavirus type 16 alters retinoid regulation of human ectocervical epithelial cell differentiation, *Cancer Res.*, 51: 3982 (1991).
29. C. Agarwal, J.R. Hembree, E.A. Rorke, and R.L. Eckert, Interferon and retinoic acid suppress the growth of human papillomavirus type 16 immortalized cervical epithelial cells, but only interferon suppresses the level of the human papillomavirus transforming oncogenes, *Cancer Res.*, 54:2108 (1994).
30. L. Pirisi, A. Batova, G.R. Jenkins, J.R. Hodam, and K.E. Creek, Increased sensitivity of human keratinocytes immortalized by human papillomavirus type 16 DNA to growth control by retinoids, *Cancer Res.*, 52:187 (1992).
31. G.I. Gorodeski, R.L. Eckert, W.H. Utian, L. Sheean, and E.A. Rorke, Retinoids, sex steroids and glucocorticoids regulate ectocervical cell envelope formation but not the level of the envelope precursor, involucrin, *Differentiation* 42:75 (1989).
32. G.I. Gorodeski, R.L. Eckert, W.H. Utian and E.A. Rorke, Maintenance of in vivo-like keratin expression, sex steroid responsiveness and estrogen receptor expression in cultured human ectocervical epithelial cells, *Endocrinology* 126:399 (1990).
33. G.I. Gorodeski, R.L. Eckert, W.H. Utian, L. Sheean and E.A. Rorke, Cultured human ectocervical epithelial cell differentiation is regulated by the combined direct actions of sex steroids, glucocorticoids and retinoids, *J. Clin. Endocrinol. Metab.* 70:1624 (1990).
34. C.K. Choo, E.A. Rorke and R.L. Eckert, Calcium regulates the differentiation of human papillomavirus type 16 immortalized ectocervical epithelial cells, but not the expression of the papillomavirus E6 and E7 oncogenes, *Exper. Cell Res.* 208:161 (1993).
35. C.K. Choo, E.A. Rorke, and R.L. Eckert, Differentiation-dependent constitutive expression of the human papillomavirus type 16 E6 and E7 oncogenes in CaSki cervical tumor cells, *J. Gen. Virol.* 75:1139 (1994).
36. N. Sizemore, and E.A. Rorke, Retinoid regulation of human ectocervical epithelial cell transglutaminase activity and keratin gene expression, *Differentiation* 54: 219 (1993).
37. L. Kasturi, N. Sizemore, R.L. Eckert, K. Martin, and E.A. Rorke, Calcium modulates cornified envelope formation, involucrin content, and transglutaminase activity in cultured human ectocervical epithelial cells, *Exp. Cell. Res.* 205: 84 (1993).
38. S. Andreatta-van Leyen, J.R. Hembree, and R.L. Eckert, Regulation of IGF-I binding protein 3 levels by epidermal growth factor and retinoic acid in cervical epithelial cells, *J. Cell. Physiol.* 160:265 (1994).
39. J. Hembree, C. Agarwal and R.L. Eckert, Epidermal growth factor suppresses insulin-like growth factor binding protein-3 levels in human papillomavirus type 16-immortalized cervical epithelial cells and thereby potentiates the effects of IGF-I, *Cancer Res.* 54:3160 (1994).
40. N. Sizemore, and E.A. Rorke, Human papillomavirus 16 immortalization of normal human ectocervical epithelial cells alters retinoic acid regulation of cell growth and epidermal growth factor receptor expression, *Cancer Res.* 53:4511 (1993).
41. C.K. Choo, E.A. Rorke, and R.L. Eckert, Human papillomavirus E6 and E7 oncogene expression is not correlated with cell proliferation or c-myc RNA expression in HPV16-immortalized human cervical epithelial cells treated with retinoids, *Cancer Res.* (submitted).

42. R.A. Pattillo, R.O. Hussa, M.T. Story, A.C. Ruckert, M.R. Shalaby, and R.F. Mattingly, Tumor antigen and human chorionic gonadotropin in CaSki cells: a new epidermoid cervical cancer cell line, *Science* 196: 1456 (1977).
43. B.M. Gilfix, and R.L. Eckert, Coordinate control by vitamin A of keratin gene expression in human keratinocytes, *J. Biol. Chem.* 260: 14026 (1985).
44. R.L. Eckert and H. Green, Cloning of cDNAs specifying vitamin A-responsive human keratins, *Proc. Natl. Acad. Sci. USA* 81: 4321 (1984).
45. M. Athar, R. Agarwal, Z.Y. Wang, J.R. Lloyd, D.R. Bickers and H. Mukhtar, All-trans-retinoic acid protects against conversion of chemically induced and ultraviolet B radiation induced skin papillomas to carcinomas, *Carcinogenesis* 12:2325 (1991).
46. N. Sizemore, L. Kasturi, G. Gorodeski R.L. Eckert, A.M. Jetten and E.A. Rorke, Retinoids regulate human ectocervical epithelial cell transglutaminase activity and keratin gene expression, *Differentiation* 54:219 (1993).
47. R.L. Eckert, Structure, function, and differentiation of the keratinocyte, *Physiol. Rev.*, 69:1316 (1989).
48. W.S. Cohick and D.R. Clemmons, Regulation of insulin-like growth factor binding protein synthesis and secretion in a bovine epithelial cell line, *Endocrinology*, 129:1347 (1991).
49. D.P. DeLeon, B. Bakker, D.M. Wilson, R.L. Hintz, and R.G. Rosenfeld, Demonstration of insulin-like growth factor (IGF-I and -II) receptors and binding proteins in human breast cancer cell lines, *Biochem. Biophys. Res. Comun.*, 152:398 (1988).
50. R.C. Baxter, and J.L. Martin, Binding proteins for insulin-like growth factors: structure, regulation and function, *Prog. Growth Factor Res.* 1:49 (1989).
51. N. Darwiche, G. Celli, L. Sly, F. Lancillotti, and L.M. De Luca, Retinoid status controls the appearance of reserve cells and keratin expression in mouse cervical epithelium, *Cancer Res.*, 53:2287 (1993).
52. R. Moll, W.W. Franke, D.L. Schiller, B. Geiger, and R. Krepler, The catalog of human cytokeratins: patterns of expression in normal epithelia, tumors and cultured cells, *Cell* 31:11 (1982).

4

ROLE OF APOPTOSIS IN THE GROWTH INHIBITORY EFFECTS OF VITAMIN D IN MCF-7 CELLS

JoEllen Welsh, Maura Simboli-Campbell, Carmen J. Narvaez, and Martin Tenniswood

W. Alton Jones Cell Science Center
10 Old Barn Rd
Lake Placid, New York 12946

I. INTRODUCTION

1,25-dihydroxycholecalciferol ($1,25(OH)_2D_3$) is the biologically active form of vitamin D, a fat soluble vitamin originally identified as an anti-rachitic factor in the early 1920s. Since that discovery, the molecular pathways by which $1,25(OH)_2D_3$ functions to maintain calcium homeostasis and bone turnover have been intensely investigated. $1,25(OH)_2D_3$ interacts with the vitamin D receptor, which is a member of the steroid/thyroid/retinoic acid superfamily of nuclear receptors, and induces or represses expression of specific genes whose protein products mediate the biological responses attributed to vitamin D. Vitamin D modulated proteins include calcium binding proteins (calbindins D9K, D28K), bone matrix proteins (osteocalcin, osteopontin), digestive enzymes (alkaline phosphatase) and vitamin D metabolizing enzymes (24-hydroxylase). The expression of the vitamin D receptor in tissues not normally involved in calcium homeostasis (such as mammary gland, pancreas and cells of the immune system) suggested additional roles for this hormone, and led to the identification of $1,25(OH)_2D_3$ as a potent regulator of cell differentiation and proliferation. Studies from a number of laboratories have clearly demonstrated that $1,25(OH)_2D_3$ inhibits growth of breast cancer cells *in vitro*, and causes regression of tumors growing *in vivo* (1-3), however, the underlying mechanism of the effect of vitamin D has yet to be clarified.

The uncontrolled growth characteristic of both benign and metastatic tumors may result from either an increase in the rate of cell proliferation or a decrease in the rate of cell death, or both (Figure 1). Furthermore, it is now recognized that therapeutic agents which cause tumor regression can exert their effects by inhibition of proliferation or by activation of cell death pathways (4). Cell death can be classified as either necrosis, the result of tissue insult or injury; or apoptosis, an active process of cellular self-destruction. In our lab, we have been testing the hypothesis that modulation of breast cancer cell growth by $1,25(OH)_2D_3$ involves activation of apoptosis.

Diet and Cancer, Edited under the auspices of the American Institute of Cancer Research
Plenum Press, New York, 1995

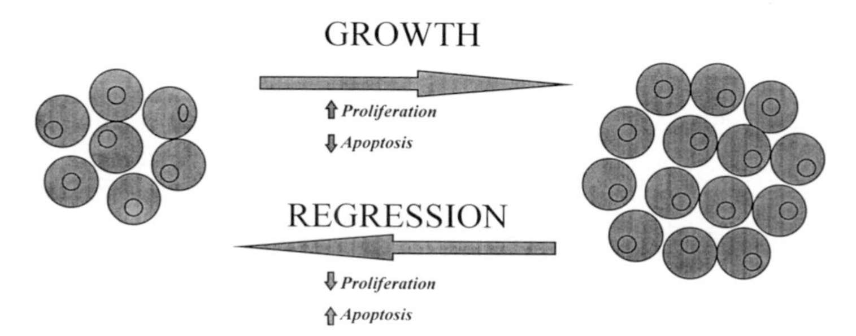

Figure 1. Interplay of apoptosis and proliferation in tumor growth and regression. Expansion of a tumor may result from either an increase in the net rate of cell proliferation and/or a decrease in the net rate of cell death (apoptosis). Tumor regression may result from a decrease in the net rate of proliferation, and/or an increase in the net rate of apoptosis. See text for details.

II. OVERVIEW OF APOPTOSIS

1. Morphology

Apoptosis is an asynchronous process comprised of a series of distinct steps which are outlined in Table 1. The resting cell in homeostasis expresses a complement of genes which promote survival and suppress apoptosis. Following receipt of a signal which initiates apoptosis, the cell enters the precondensation stage. In this stage, the expression of specific genes involved in apoptosis is activated, yet no morphological changes can be detected relative to the cell in homeostasis. The first distinguishable morphological changes mark the condensation phase of the apoptosis process, which involves both cytoplasm and nucleus. Disruption of the cytoskeleton and condensation of intermediate filaments around the nucleus results in cell shrinkage and subsequently loss of cell-cell contacts. *In vitro*, adherent cells are released from the monolayer and can be recovered in the media. Nuclear events include chromatin condensation, which results in the characteristic pyknotic nuclei of apoptotic cells. Condensation of chromatin can be visualized with Hoescht 33258 or other dyes which intercalate with DNA (5). Condensation of nuclear matrix proteins also occurs

Table 1. Stages and events during apoptosis of epithelial cells

Stage	Events	Morphology
Homeostasis	Survival	Normal
Precondensation	Altered gene expression -TRPM-2/clusterin -TGFβ -Cathepsin B	Normal
Condensation		
-cytoplasmic	Cytoskeletal breakdown Loss of cell-cell contacts Degradation of ECM	Cell shrinkage, release of cell from monolayer
-nuclear	Chromatin condensation Nuclear matrix reorganization Endonuclease activation	Pyknotic nuclei DNA fragmentation
Fragmentation	Cell fragments into membrane bound bodies containing condensed nuclei and organelles	Apoptotic bodies, phagocytosis

and can be observed by immunofluorescence using antibodies for specific nuclear matrix proteins such as the nuclear mitotic apparatus protein, NuMA (5-7).

The condensation phase is followed by the fragmentation stage, in which the dying cell fragments into several membrane bound apoptotic bodies. This process is accomplished without the release of intracellular contents and necessitates remodelling of the plasma membrane. The re-organization of chromatin and the degradation of nuclear matrix proteins are thought to render DNA susceptible to attack by one or more unidentified nucleases, which first degrade DNA into 50-300 kbp fragments (8). These large fragments can be detected by pulsed field gel electrophoresis. Further degradation results in generation of 200bp fragments which can be visualized as a nucleosome "ladder" using conventional agarose gel electrophoresis. Although often used as a hallmark of apoptosis, recent evidence indicates that nucleosome ladders are not universally detected (9), even when other biochemical and morphological markers of apoptosis are observed. These observations support the idea that the final fragmentation of the DNA into 200bp fragments may be a relatively late event and thus not always evident in apoptotic cells.

2. Biochemical and Molecular Markers

As discussed above, apoptosis is an active process which involves alterations in gene expression during the pre-condensation phase. Induction of several genes, including TRPM-2, TGFβ and cathepsin B (Table 1) has been linked to apoptosis (10,11), however, as yet the majority of the genes involved in the process remain to be identified. Although the mRNAs for all of these genes are up regulated during the pre-condensation phase, their protein products appear to act during later stages to trigger or complete the apoptotic process. For example, the expression of the extracellular protease cathepsin B may be necessary for the release of dying cells from the basement membrane (one of the earliest morphological signs of an apoptotic cell), thereby initiating the condensation phase. The TRPM-2 gene codes for a secreted, plasma membrane-associated glycoprotein (clusterin) which may be involved in the extensive membrane remodelling required during the fragmentation phase. Of the known apoptosis related genes, TRPM-2 is the most widely used molecular marker for apoptosis, as its expression is up-regulated during regression of prostate and mammary gland *in vivo* (10-12), as well as many other tissues.

3. Induction of Apoptosis in Breast Cancer Cells

It is now clear that many chemotherapeutic agents target the apoptotic pathway. Table 2 summarizes data from studies documenting the induction of apoptosis in human breast cancer cells in response to a variety of manipulations. Apoptosis is induced in estrogen dependent MCF-7 human breast cancer cells following estrogen deprivation or anti-estrogen treatment (13-15). In these cells, apoptosis is associated with arrest of cells in G_0/G_1, DNA fragmentation and up regulation of TRPM-2 and TGFβ gene expression. These studies have clearly demonstrated that tamoxifen, an anti-estrogen commonly used in human breast cancer therapy, activates the apoptotic pathway in hormone dependent tumor cells. Although estrogen independent cell lines do not undergo apoptosis following anti-estrogen treatment, these cells retain the ability to activate apoptosis in response to other agents. For example estrogen receptor negative BT-20 cells undergo apoptosis which is associated with a nuclear calcium surge and DNA fragmentation in response to the cytokine TNFα (16). The chemotherapeutic agents 5-fluoro-2'-deoxyuridine or trifluorothymidine induce apoptosis (characteristic morphology, DNA fragmentation and enhanced expression of TGFβ) in MDA-MB-468 cells (17), another estrogen receptor negative cell line. These findings are significant since anti-estrogen treatment of human patients is commonly associated with emergence of hormone resistant cells in which the apoptotic pathway is no longer coupled to hormone dependent pathways. Agents which induce apoptosis in such estrogen

Table 2. Summary of studies assessing the role of apoptosis in breast cancer cell growth inhibition

Agent/cells	Markers assessed	Reference
Estrogen withdrawal or anti-estrogen treatment *in vivo* (MCF-7 cells)	Tumor regression DNA fragmentation ↑TRPM-2 mRNA ↑TGFβ mRNA	Kyprianou et al, 1991 (13) Warri et al, 1993 (14)
Anti-estrogen or anti-progestin treatment *in vitro* (MCF-7, T47D cells)	↓Cell number G_o/G_1 arrest Morphology DNA fragmentation ↑TRPM-2 mRNA ↑TGFβ mRNA	Bardon et al, 1987 (15) Warri et al, 1993 (14)
Chemotherapeutic agents *in vitro* (MDA-MB-468 cells)	↓Cell number Morphology DNA fragmentation ↑TGFβ mRNA TRPM-2 mRNA unchanged	Armstrong et al, 1992 (17)
TNFα *in vitro* (BT-20 cells)	Morphology DNA fragmentation ↑intracellular Calcium	Bellomo et al, 1992 (16)
Vitamin D compounds *in vitro* (MCF-7 cells)	↓Cell number G_o/G_1 arrest Morphology ↓estrogen receptor ↑TRPM-2 mRNA, protein ↑Cathepsin B mRNA, protein ↓bcl-2 protein	Simboli-Campbell and Welsh, 1994 (18, 20) Welsh et al, 1994 (19)

independent cells may represent important adjuncts to existing anti-hormonal therapies for breast cancer patients.

Previous work has clearly demonstrated that $1,25(OH)_2D_3$ reduces cell numbers of both estrogen dependent and estrogen independent breast cancer cells (1-3). These studies have demonstrated that long term (6-10 days) treatment with $1,25(OH)_2D_3$ reduces breast cancer cell proliferation. However, the relative contributions of proliferation and apoptosis to the reduction in cell number, particularly in the initial response to $1,25(OH)_2D_3$ treatment, has yet to be clarified. In our lab, we have been characterizing the effects of $1,25(OH)_2D_3$ on proliferation and apoptosis of estrogen dependent MCF-7 human breast cancer cells *in vitro*. These studies have been specifically designed to investigate the effects of $1,25(OH)_2D_3$ on apoptosis in breast cancer cells by monitoring morphological, biochemical and molecular markers associated with the active cell deqath pathway.

III. INDUCTION OF APOPTOSIS BY VITAMIN D COMPOUNDS

1. Morphology

We have extensively characterized the effects of $1,25(OH)_2D_3$ on MCF-7 cell morphology (18-20). In control MCF-7 cultures, cells exhibit typical epithelial morphology and numerous proliferating cells are evident under phase contrast microscopy. The mitotic spindles of these proliferating cells are clearly identified with the DNA dye, Hoescht 33258. In contrast, cells treated for 48 hours with $1,25(OH)_2D_3$ are rounded up (above the adherent cells of the monolayer) and contain condensed cytoplasm and frequent intracellular vesicles. In these cultures, Hoescht 33258 staining clearly identifies the hyperchromatic, pyknotic nuclei resulting from chromatin condensation typical of apoptotic cells. After 48 hours treatment with $1,25(OH)_2D_3$, apoptotic cells display enhanced nuclear fluorescence when stained with an anti-NuMA antibody due to the re-organization of nuclear matrix proteins which occurs during nuclear condensation. This reorganization of NuMA parallels the chromatin condensation visualized with Hoescht 33258. Despite clear evidence of nuclear condensation in $1,25(OH)_2D_3$ treated cells, however, no DNA fragmentation could be detected on agarose gels. This finding is consistent with reports indicating no DNA fragmentation in MCF-7 cells undergoing ACD in response to serum deprivation or anti-estrogens (8, 14).

2. TRPM-2 Expression

The dramatic induction of TRPM-2 in mammary gland regression following weaning (12) and MCF-7 tumors following estrogen deprivation (13), suggests an important role for this protein during apoptosis of breast epithelial cells. In support of this concept, our data indicate increases in TRPM-2 mRNA levels in MCF-7 cells within 48 hours of $1,25(OH)_2D_3$ treatment (20). Western blotting indicates that the increase in TRPM-2 gene expression correlates with an increase in the expression of the clusterin protein (18,19). On immunofluorescence, control MCF-7 cells exhibit a low level of clusterin expression, whereas dramatic increases in clusterin intensity are evident in apoptotic cells in cultures treated with $1,25(OH)_2D_3$. Thus, all three techniques indicate that more clusterin is synthesized in $1,25(OH)_2D_3$ treated MCF-7 cells undergoing apoptosis. The enhanced expression of clusterin, a plasma membrane associated glycoprotein, with dying cells is consistent with a role in membrane remodelling during the fragmentation phase of apoptosis (10, 25).

3. Cell Growth and Cell Cycle

Cell growth indices were assessed in relation to the onset of morphological and biochemical evidence of apoptosis. Cell number is significantly lower in $1,25(OH)_2D_3$ treated cultures within 48 hours. Surprisingly, ^{3}H-thymidine incorporation data indicate minor effects of $1,25(OH)_2D_3$ on cell proliferation during the first 72 hours after treatment. These effects of $1,25(OH)_2D_3$ on cell proliferation are too small to account for the large differences in cell number. Flow cytometric analysis confirmed that the major effect of $1,25(OH)_2D_3$ after 48 hours is to induce accumulation of cells in G_o/G_1 (21). $1,25(OH)_2D_3$ treated cells exhibit a decrease in the ratio of phosphorylated to dephosphorylated Rb protein, a finding consistent with G_o/G_1 arrest. $1,25(OH)_2D_3$ also increases the percentage of cells detected as a discrete population in the $< G_o$ peak, which represents apoptotic cells. Recent evidence indicates that early G_1 may be the point from which switching between cell cycle progression and induction of ACD occurs (22), suggesting that $1,25(OH)_2D_3$ may be an important regulator of this switch.

4. Other Effects of $1,25(OH)_2D_3$ on MCF-7 Cells

The expression of cathepsin B, a lysosomal protease, is enhanced during regression of normal mammary gland following weaning, and during regression of mammary tumors following estrogen ablation (10, 11, 23, 24). We have observed up regulation of cathepsin B mRNA and protein in $1,25(OH)_2D_3$ treated cells (Narvaez and Welsh, unpublished data).

These data support the suggestion that regression/apoptosis in mammary cells involves activation of the autophagic lysosomal pathway. According to this model, cathepsin B and other lysosomal enzymes, packaged in autophagic lysosomes, are secreted and participate in degradation of the basement membrane during the condensation phase of apoptosis. We hypothesize that secreted proteases, such as cathepsin B, are also involved in disruption of growth factor signalling via their effects on growth factor binding proteins in the extracellular milieu (25).

We have also found that treatment of MCF-7 cells with $1,25(OH)_2D_3$ for 48 hours reduces the level of bcl-2, a 26kDa protein which has been implicated in the protection of cells from apoptotic cell death (26). Further studies will be necessary to clarify the significance of this change in bcl-2, particularly in relation to the newly discovered heterodimer partners of bcl-2 such as bax and bcl-x (26, 27).

Within 24 hours of $1,25(OH)_2D_3$ treatment, there is a 30% reduction in estrogen receptor level as assessed by estradiol binding, and down regulation of the amount of estrogen receptor protein on western blots. Since estrogen withdrawal and anti-estrogen treatments induce apoptosis in MCF-7 cells (Table 2), vitamin D may induce apoptosis via disruption of estrogen dependent signals. This possibility is discussed further in section IV, below.

5. Effect of Vitamin D Analogs on MCF-7 Cells

Therapeutic use of $1,25(OH)_2D_3$ is extremely limited due to its inherent calcemic activity. A wide variety of vitamin D analogs with growth modulatory properties but without calcemic activity *in vivo* have been developed by several pharmaceutical firms. EB1089 is a non-hypercalcemic vitamin D analog developed by LEO Pharmaceutical Products (Ballerup, Denmark) which induces regression of chemically induced breast tumors in rats (28). In parallel studies, we have observed that EB1089 also induces apoptosis in MCF-7 cells. EB1089, which is 50-100 fold more potent than $1,25(OH)_2D_3$, elicits effects which are identical to those of $1,25(OH)_2D_3$ on all of the morphological and biochemical indices discussed above (21). These findings indicate that EB1089, a representative of the novel class of non hypercalcemic vitamin D compounds, also target the apoptotic pathway, and we are currently testing other non hypercalcemic vitamin D analogs for their efficacy in inducing apoptosis in our sytem. Further, it is now imperative to clarify the role of apoptosis during tumor regression in response to these novel vitamin D compounds *in vivo* (28-30).

IV. COMPARISON OF $1,25(OH)_2D_3$ WITH OTHER INDUCERS OF APOPTOSIS IN BREAST CANCER CELLS

To date, only five studies have specifically addressed the issue of apoptosis in breast cancer cell growth regulation (Table 2). These have included the effects of anti-hormonal therapies, the cytokine TNFα, and DNA active chemotherapeutic agents. Our lab also has preliminary data indicating that TGFβ induces apoptosis in MCF-7 cells. The diversity of agents which trigger the cell death pathway suggests that no single known signal transduction pathway is utilized. However, the similarity in morphology and biochemistry indicate that a common target for these diverse agents may exist.

The effects of $1,25(OH)_2D_3$ in MCF-7 cells which have been characterized in this lab are also summarized in Table 2. Under the conditions used in our studies, $1,25(OH)_2D_3$ treated cells exhibit cytoplasmic and chromatin condensation, NuMA reorganization and enhanced expression of both TRPM-2 and cathepsin B; two proteins which have been implicated in apoptosis. Flow cytometric data indicate that $1{,}25(OH)_2D_3$ treated cells arrest in G_0/G_1. These observations are very similar to those reported for anti-estrogen treatment of MCF-7 cells (13, 14). Since we observed down regulation of estrogen receptors prior to the reduction in cell number in $1{,}25(OH)_2D_3$ and EB1089 treated cultures, vitamin D compounds may induce apoptosis in estrogen sensitive cells via interference with estrogen signalling. However, since vitamin D compounds inhibit growth of estrogen insensitive cell lines (1-3), and in our system, anti-estrogens and anti-progestins potentiate the actions of vitamin D compounds (Narvaez and Welsh, unpublished data) we believe that the effects of vitamin D include both estrogen dependent and estrogen independent pathways. To clarify this issue, we are currently characterizing the effects of $1{,}25(OH)_2D_3$ and EB1089 on apoptosis in MDA-MB -231 cells, which lack functional estrogen receptors.

Further work will be necessary to determine whether the effects of anti-estrogens and vitamin D compounds are distinct from non-steroidal agents such as TNFα or chemotherapeutic agents which also induce apoptosis in breast cancer cells. Detailed, direct comparison of the effects of a variety of agents on apoptotic markers in distinct breast cancer cell lines will be necessary to identify specific targets which are critical for triggering the apoptotic machinery. Identification of such targets would offer insight into designing novel therapeutics which activate the cell death pathway. These therapies would likely be useful adjunctive therapies to current anti-hormonal approaches as well as more novel agents under investigation, such as retinoids (31).

V. ACKNOWLEDGMENTS

This work is supported by the American Institute for Cancer Research, Washington, DC (grant #9310) and the W. Alton Jones Cell Science Center. We thank Lise Binderup, LEO pharmaceuticals, Ballerup, Denmark, for the generous supply of EB1089 and other vitamin D analogs currently under investigation.

VI. REFERENCES

1. K. Colston, S.K. Chander, A.G. Mackay and R.C. Coombes, Effects of synthetic vitamin D analogs on breast cancer cell proliferation *in vivo* and *in vitro*. *Biochem. Pharm.* 44:693 (1992)
2. C. Chouvet, U. Berger and R.C. Coombes, 1,25 Dihydroxyvitamin D3 inhibitory effect on the growth of two human breast cancer cell lines (MCF-7, BT-20). *Journal of Steroid Biochemistry* 24:373 (1986)
3. J. Eisman, R.L. Sutherland, M.L. McMenemy, J.C. Fragonas, E.A. Musgrove and G. Pang, Effects of 1,25-dihydroxyvitamin D3 on cell cycle kinetics of T47D human breast cancer cells. *J. Cellular Physiology* 138: 611 (1989)
4. G.T. Williams, Programmed Cell Death: Apoptosis and Oncogenesis. *Cell* 65:1097 (1991)
5. F.A. Oberhammer, K. Hochegger, G. Froschl, R. Tiefenbacher and M. Pavelka, Chromatin condensation during apoptosis is accompanied by degradation of lamin A + B, without enhanced activity of cdc2 kinase, *J. Cell Biology* 126:827 (1994)
6. T.E. Miller, L.A. Beausang, N. Meneghini and G. Lidgard Cell death and nuclear matrix proteins. In: Apoptosis II: The molecular basis of apoptosis in disease. Cold Spring Laboratory Press, NY (1994)
7. C.H. Yang, E.J. Lambie and M. Snyder NUMA:An unusually long coiled-coil related protein in mammalian nucleus. *J. Cell Biology* 116:1303 (1991)
8. F. Oberhammer, J.W. Wilson, C. Dive, I.D. Morris, J.A. Hickman, A.E. Wakeling P.R. Walker and M. Sikorska M, Apoptotic death in epithelial cells: cleavage of DNA to 300 and/or 50 kb fragments prior to or in the absence of internucleosomal degradation. *EMBO J.* 12:3679 (1993)
9. G.M. Cohen, X.M. Sun, R.T. Snowden, D. Dinsdale and D.N. Skilleter, Key morphological features of apoptosis may occur in the absence of internucleosomal DNA fragmentation. *Biochem. J.* 286:331 (1992)
10. M. Tenniswood, R.S. Guenette, J. Lakins, M. Mooibroek, P. Wong and J.E. Welsh, Active cell death in hormone dependent tissues. *Cancer and Metastasis Reviews* 11:197 (1992)

11. M. Tenniswood, D. Taillefer, J. Lakins, R.S. Guenette, M. Mooibroek, L. Daehlin and J.E. Welsh, Control of gene expression during apoptosis in hormone-dependent tissues. In: Apoptosis II: The molecular basis of apoptosis in disease. Cold Spring Laboratory Press, NY (1994)
12. R.S. Guenette, H. Corbeil, J. Leger, K. Wong, V. Mezl, M. Mooibroek and M. Tenniswood, Induction of gene expression during involution of the lactating mammary gland of the rat. *J. Molecular Endocrinology* 12:47 (1994)
13. N. Kyprianou, H. English, N. Davidson and J. Isaacs, Programmed cell death during regression of the MCF-7 human breast cancer following estrogen ablation. *Cancer Research* 51:162 (1991)
14. A.M. Warri, R.L. Huovinen, A.M. Laine, P.M. Martikainen and P.L. Harkonen, Apoptosis in toremifene-induced growth inhibition of human breast cancer cells *in vivo* and *in vitro*. *Journal of National Cancer Institute* 85:1412 (1993)
15. S. Bardon, F. Vignon, P. Montcourrier and H. Rochefort, Steroid receptor mediated cytotoxicity of an anti-estrogen and an antiprogestin in breast cancer cells. *Cancer Research* 47:1441 (1987)
16. G. Bellomo, M. Perotti, F. Taddei, F. Mirabelli, G. Finardi, P. Nicotera and S. Orrenius, Tumor Necrosis Factor α induces apoptosis in mammary adenocarcinoma cells by an increase in intranuclear free calcium concentration and DNA fragmentation. *Cancer Research* 52:1342 (1992)
17. Armstrong DK, Isaacs JT, Ottaviano YL and Davidson NE (1992) Programmed cell death in an estrogen independent human breast cancer cell line, MDA-MB-468. *Cancer Research* 52:3418-3424
18. M. Simboli-Campbell and J. Welsh, 1,25-Dihydroxyvitamin D_3: Coordinate regulator of active cell death and proliferation in MCF-7 breast cancer cells. Schering Foundation Workshop, vol. 14, Workshop on Apoptosis in hormone dependent cancers, Springer Verlag, Berlin, In press, (1994)
19. J. Welsh, M. Simboli-Campbell and M. Tenniswood, Induction of apoptotic cell death by $1,25(OH)_2D_3$ in MCF-7 breast cancer cells. Proceedings of the Ninth Workshop on Vitamin D, Walter DeGruyter & Co. Berlin, In press (1994)
20. M. Simboli-Campbell, C.J. Narvaez, M. Tenniswood and J. Welsh, 1,25-Dihydroxyvitamin D_3 induces morphological and biochemical markers of apoptosis in MCF-7 breast cancer cells. Submitted for publication.
21. M. Simboli-Campbell and J. Welsh, Comparative effects of $1,25(OH)_2D_3$ and EB1089 on Cell cycle kinetics in MCF-7 cells. Proceedings of the Ninth Workshop on Vitamin D, Walter DeGruyter & Co. Berlin, In press (1994)
22. P.R. Walker, J. Kwast-Welfeld, H. Gourdeau, J. Leblanc, W. Neugebauer and M. Sikorska, Relationship between apoptosis and the cell cycle in lymphocytes:roles of protein kinase C, tyrosine phosphorylation and AP1. *Experimental Cell Research* 207:142 (1993)
23. Y.S. Cho-Chung YS and P.M. Gullino, Mammary tumor regression V. Role of acid ribonuclease and cathepsin. *J. Biol. Chem.* 248:4743 (1973)
24. R.H. Lanzerotti and P.M. Gullino, Activities and quantities of lysosomal enzymes during mammary tumor regression. *Cancer Research* 32:2679 (1972)
25. M. Tenniswood, R.S. Guenette, D. Taillefer and M. Mooibroek, The role of growth factors and extracellular proteases in active cell death in the prostate. Schering Foundation Workshop, vol. 14, Workshop on Apoptosis in hormone dependent cancers, Springer Verlag, Berlin, In press, (1994)
26. J. C. Reed, Bcl-2 and the regulation of programmed cell death. *J. Cell Biology*. 124:1 (1994)
27. S.J. Korsmeyer, J.R. Shutter, D.J. Veis, D. E. Merry and Z. N. Oltvai. Bcl-2/bax: a rheostat that regulates an anti-oxidant pathway and cell death. *Seminars in Cancer Biology* 4:327 (1993)
28. J. Abe, T. Nakano, Y. Nishii, T. Matsumoto, E. Ogata and K. Ikeda, A novel vitamin D3 analog, 22-oxa-1,25 dihydroxyvitamin D_3, inhibits the growth of human breast cancer *in vitro* and *in vivo* without causing hypercalcemia. *Endocrinology* 129:832-837
29. M.A. Anzano, J.M. Smith, M.R. Uskokovic, C.W. Peer, L.T. Mullen, J. L. Letterio, M.C. Welsh, M. W. Shrader, D. L. Logsdon, C.L. Driver, C.C. Brown, A.B. Roberts and M.B. Sporn, 1α25-Dihydroxy-16-ene-23-yne-26,27-hexafluorocholecalciferol (Ro-24-5531), a new deltanoid (vitamin D analogue) for prevention of breast cancer in the rat. *Cancer Research* 54:1653 (1994)
30. K.W. Colston, A.G. Mackay, S.Y. James, L. Binderup, S. Chander and R.C. Coombes. EB1089: a new vitamin D analogue that inhibits the growth of breast cancer cells *in vivo* and *in vitro*. *Biochemical Pharmacology* 44:2273 (1992)
31. M. Koga and R.L. Sutherland, Retinoic acid acts synergistically with 1,25-dihydroxyvitamin D_3 or antioestrogen to inhibit T-47D human breast cancer cell proliferation. *J. Steroid Biochem. Molec. Biol.* 39:445 (1991)

5

VITAMIN D AND PROSTATE CANCER

David Feldman,* Roman J. Skowronski, and Donna M. Peehl

Departments of Medicine and Urology (D.M.P.)
Stanford University School of Medicine
Stanford California 94305-5103

I. INTRODUCTION†

A. Importance of Prostate Cancer

An analysis of the rising incidence and mortality rate of prostate cancer reveals the extraordinary importance of this disease in the world. Adenocarcinoma of the prostate is the most common cancer diagnosed in American males, reaching an estimated incidence of 32%, with 200,000 cases of newly diagnosed cancer cases expected in 1994.[1] In addition, clinically inapparent prostate cancer is an extremely common finding. Over the age of 80, subclinical prostate cancer is found in approximately 60% of all men.[2] Overall, it is estimated that there are 11 million men in the U.S. with lesions within their prostates that are histologically identifiable as cancer.[3] Mortality from prostate cancer represents a considerable problem. It is expected that prostate cancer will account for 13% (38,000 cases) of male cancer deaths in 1994.[1] This makes prostate cancer the second leading cause of cancer-related death in U.S. men after lung cancer. Mortality rates from prostate cancer appear to be on the rise. From 1970 to 1990 the age-adjusted mortality rate from prostate cancer increased approximately 7 % among U.S. caucasians. Since prostate cancer rates increase with age, as the longevity of the population increases, it is projected that prostate cancer will become the leading cause of cancer and cancer death in men. These observations demonstrate that prostate cancer is one of the major adverse factors in the health of the male population in the United States and for that matter, in the rest of the world as well.

B. Epidemiology and Etiology

The etiology of prostate cancer is not well understood. Several risk factors have been identified including age, race and genetic factors.[2] Age is the number one risk factor for

* Correspondence and reprint requests: David Feldman, Division of Endocrinology, Stanford University School of Medicine, Stanford, CA 94305-5103.

† Abbreviations used in this paper: 1,25-D, 1,25-dihydroxyvitamin D_3; VDR, vitamin D receptor; PSA, prostate specific antigen; BPH, benign prostatic hyperplasia.

Diet and Cancer, Edited under the auspices of the American Institute of Cancer Research
Plenum Press, New York, 1995

prostate cancer. The disease is rare before the age of 40, but both the incidence and mortality rates increase geometrically with age. Over 80% of prostate cancers are diagnosed in men over the age of 65. Race is also an important factor. When compared to caucasian American males, African American males tend to develop prostate cancer at an earlier age, have a higher-stage disease at the time of diagnosis, and have increased mortality. Conversely, Native Americans, Hispanics, and Asians have a decreased risk compared to Caucasians. Geographic factors can alter racial trends. For example, when compared to Nigerian males, African American males have a sixfold increased risk of developing clinically detectable prostate cancer. [2] Finally, genetic factors are super-imposed on other risk factors. The highest relative risk of prostate cancer was found in the young (<65 years old) brothers of young (<65 years old) cases within families from the Utah Mormon genealogy.[4] A threefold increase in risk of prostate cancer to the first-degree relatives of cases was found by Woolf,[5] and relative risk of 4 for developing prostate carcinoma in the brothers of the cases compared to their brothers-in-law was found by Meikle et al.[6]

C. Hormonal Factors

Androgens are crucial for the maintenance of virtually all metabolic processes in the prostate including cell proliferation, enlargement and secretory function.[2,7] Several studies have indicated a high correlation between serum testosterone levels and increased risk of development of prostate cancer.[2] Androgen receptors are detectable in nuclear fraction of approximately 80% of prostatic tumors.[8] Circulating androgens may serve as promoting factors of tumor growth and of neoplastic transformation. On the other hand, withdrawal of androgens causes involution and programmed cell death (apoptosis) of normal and malignant prostatic epithelial cells[2] a phenomenon forming the basis for hormonal ablation treatment of the disease.[9] The role of other steroid hormones is unclear. Receptors for estrogens and progesterone have been demonstrated in prostatic tumors, however, no correlation was found between tumor ER levels and response to hormonal therapy.[10]

D. Epidemiology of Prostate Cancer and the Role of Vitamin D

The possibility that vitamin D deficiency might increase the risk of prostate cancer has recently been suggested by Schwartz and Hulka,[11] as has previously been proposed for colorectal cancer.[12,13] The authors based their hypothesis on the findings that prostate cancer mortality rates in the United States are inversely proportional to exposure to ultraviolet (uv) irradiation,[14] which is essential for the synthesis of vitamin D. These considerations led the authors to speculate on possible direct or indirect effects of vitamin D on the growth and differentiation of prostate cells. The finding of decreased risk of clinical prostate cancer in high sunlight areas is very intriguing in the light of our findings of direct actions of vitamin D to inhibit prostate cell growth.[15-18]

A recent report by Corder et al. analyzed levels of vitamin D and metabolites in stored serum in a population of 181 men who subsequently developed prostate cancer.[19] The blood samples were obtained many years previously and controls were selected from blood samples taken the same day from men who did not go on to develop prostate cancer. Mean serum levels of 1,25-dihydroxyvitamin D (1,25-D), the active form of the hormone, were slightly but significantly lower in the cancer patients compared to controls ($p = 0.002$). Risk of prostate cancer decreased with higher levels of 1,25-D, especially in men with low 25-hydroxyvitamin D, the inactive precursor which serves as the major reservoir for 1,25-D in the body. In men older than 57 years of age, 1,25-D was an important predictor of risk for palpable and anaplastic tumors but not for tumors incidently discovered or well differentiated tumors.[19] These data support the relationship of vitamin D status and the risk of developing prostate cancer and suggest that vitamin D is protective.

E. Mechanism of Vitamin D Action

Although a vitamin, vitamin D is further metabolized into a hormone. The active hormonal form of vitamin D, 1,25-D, exerts its action via a specific intracellular vitamin D receptor (VDR) to modulate gene expression in an analogous fashion to the classical steroid hormones.[20,21] The VDR belongs to the steroid-thyroid-retinoic acid receptor gene superfamily,[22-24] and regulates the transcriptional activity of specific genes in response to hormone binding by interaction with DNA hormone response elements located in the promoter regions of target genes.[21] The VDR acts in the gene regulation pathway as dimerized molecules, usually as heterodimers with RXR, a unique form of the retinoid receptor.[21]

F. Current Research on VDR and 1,25-D Actions in Prostate

We have recently developed data which will be described below, indicating that VDR are present in cells derived from normal, benign prostatic hyperplasia (BPH) and malignant prostate specimens.[15-18] We have demonstrated the presence of VDR in fresh tissue removed at surgery as well as in various cultured cell systems including both primary cultures and in three established human cancer cell lines, LNCaP, PC-3 and DU-145 cells.[15,17] Miller et al have also demonstrated VDR in LNCaP and other prostate cells lines.[25] Furthermore, we have shown that treatment with 1,25-D inhibits prostate cancer cell growth *in vitro* and promotes cellular differentiation.[15,17] These findings have led us to propose that 1,25D treatment will inhibit proliferation and stimulate differentiation of prostate cancer cells *in vivo*. We further hypothesize that nutritional insufficiency of vitamin D may predispose to prostate cancer and that vitamin D supplementation may play a beneficial role in the prevention or treatment of prostate cancer.

II. MATERIALS AND METHODS

Human prostate cancer cell lines (LNCaP, PC-3 and DU-145) and primary cultures derived from normal, benign prostatic hyperplasia (BPH) and prostate cancers were grown under conditions described previously.[15,17] Cell proliferation was assessed by determination of DNA content using the diphenylamine assay of Burton.[26] VDR binding was performed with [^{3}H]1,25-D and bound and free hormone was separated by the hydroxylapatite method.[15] RNA was isolated using 4 M LiCl, 8 M urea precipitation as described by Auffray and Rougeon,[27] and Northern blots were performed as described previously.[15,17] The VDR mRNA was probed with a 2.1 kb fragment of the human VDR cDNA,[28] and 24-hydroxylase was probed with a 1.3 kb fragment of the rat 24-hydroxylase cDNA.[29] Blots were exposed to Kodak XAR film (Eastman Kodak, Rochester, NY) at -80°C and developed. The pixel intensity of each 24-hydroxylase or L7 band was scanned by computing densitometer (Molecular Dynamics, Model 300, Sunnyvale, CA) and the data were integrated by scanner software and indexed to the corresponding levels of L7 mRNA. PSA was measured by an automated ELISA assay (Ciba Corning Diagnostics Corp.).[30] Immunocytochemistry employed the anti-VDR rat monoclonal antibody (9A7) and use of the ABC reagent (Vector Laboratories).[15,18]

Table 1. Binding characteristics of VDR in three established human prostate cancer cell lines and epithelial and stromal strains of primary cultures

Cell lines	Kd (M)	Bmax (fmol/mg)
LNCaP	7.5×10^{-11}	26.5
DU-145	5.4×10^{-11}	30.6
PC-3	6.3×10^{-11}	77.7
E-CA-3	1.3×10^{-10}	41.2
E-PZ-2	1.1×10^{-10}	62.4
F-PZ-2	ND[a]	9[b]
F-CA-2	ND	2[b]

The binding in the fibroblastic stromal cells was too low to perform Scatchard analyses so the K_d was not determined and the B_{max} was estimated by single point analysis at saturating concentration. [a]ND - not determined, [b]estimated - using single point assay, E - epithelial, F - fibroblastic, PZ - peripheral zone, CA - carcinoma

III. RESULTS

A. VDR in Prostate

Ligand Binding in Prostate Cells. Our initial experiments were designed to determine whether the prostate was a target organ for 1,25-D. If so, according to the receptor hypothesis, VDR should be present in prostate tissue. We used ligand binding analysis with [^{3}H]1,25-D to investigate the presence of VDR in three human prostate cancer cell lines, LNCaP, PC-3 and DU-145. All three cell lines exhibited VDR.[15] As shown in Table 1, the affinity and number of binding sites, determined by Scatchard analysis, were in the range found in various classical target tissues. Whereas all cell lines had similar affinity for [^{3}H]1,25-D (5 to 7 x 10^{-11} M), the number of sites was highest in PC-3 cells (77.7 fmol/mg protein) and lower in LNCaP cells (26.5 fmol/mg protein) and DU-145 cells (30.6 fmol/mg protein). Additionally, we surveyed a number of normal, BPH and prostate cancer specimens freshly removed during prostate surgery.[17] In all cases but one, ligand binding assays demonstrated the presence of VDR in the specimens and the content ranged between 2 and 79 fmol/mg). We next examined primary cultures of prostate cells, using cells derived from normal, BPH and prostate cancer. The primary cultures techniques that we employed allowed us to individually culture fibroblastic stromal cells and epithelial cells so that each lineage could be analyzed separately.[17] VDR was demonstrated in both stromal and epithelial cells from all three sources, normal, BPH and prostate cancer. Various epithelial cell strains consistently exhibited higher levels of VDR (13 to 79 fmol/mg) than stromal cells (2 to 27 fmol/mg).

Northern Blot Analysis To determine whether VDR mRNA could be detected in these cells, we used Northern blot analysis employing a 2.1 kb cDNA probe of human VDR.[28] As shown in Figure 1 (Panel A), all three cell lines showed a hybridization band corresponding to the VDR mRNA. Also, various primary cultures of prostate exhibited the mRNA for VDR. As a control for VDR, the blot also contains specimens from J_1 and J_2, siblings from a family with Hereditary Vitamin D Resistant Rickets.[24,31] The blot shows the presence of

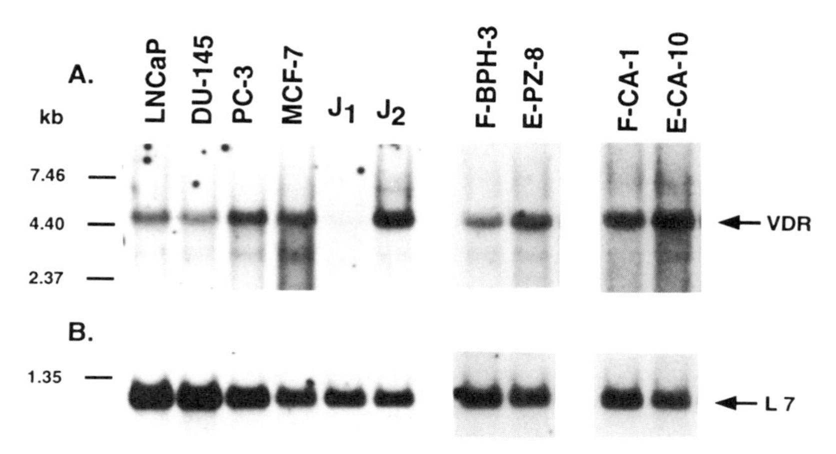

Figure 1. Northern blot analysis of VDR mRNA in prostate specimens. **Panel A**: RNA samples from human cancer cell lines, children from an HVDRR family (J_1, VDR mutant; J_2, unaffected sibling), and samples from primary cultures of prostate cells (F, fibroblast; E, epithelial; BPH, benign prostatic hyperplasia; PZ, peripheral zone of normal prostate; CA, cancer). The hybridization band at 4.6 kb represents the VDR. **Panel B**: L7 is the ribosomal gene used as a control for RNA loading and transfer.

VDR mRNA in the normal child J_2, and its absence in J_1, the child with HVDRR having a mutated VDR. Panel B shows blots of L7. The hybridization band represents a mRNA encoding a ribosomal protein that is expressed consistently throughout the cell cycle and which we use as a control for RNA loading and transfer.[15,17]

Immunocytochemistry. In studies shown elsewhere, immunocytochemistry using the anti-VDR antibody 9A7,[32] demonstrated the presence of VDR in the nuclei of both prostate cancer cell lines [15] and primary cells.[17,18] These findings confirm the ligand binding data and Northern blots and substantiate the presence of VDR in these prostate cells.

B. Functional Responses to 1,25-D Treatment

Growth-Inhibition After demonstrating the presence of VDR, we next turned our attention to the functional consequences of treating prostate cells with 1,25-D. The first response we investigated was the effect of 1,25-D on cell proliferation. As summarized in Table 2, LNCaP and PC-3 cells were growth inhibited while DU-145 cells were not.[15] It is not clear why DU-145 cells were not growth-inhibited since they contain a substantial number of VDR. Although PC-3 cells had the highest content of VDR, LNCaP cells showed greater growth-inhibition. Primary cultures were also growth inhibited, the epithelial cells more than the stromal cells.[17] Since the epithelial cells are grown in serum-free medium, the possible confounding effects of 1,25-D action and serum factors are not operative in this system. Because all cells tested show growth-inhibition except DU-145, we are confident that this effect represents the actual action of 1,25-D on prostate tissue.

24-Hydroxylase mRNA Induction. Next, we examined the prostate cell lines for the induction of 24-hydroxylase mRNA.[29] 24-Hydroxylase is a critical enzyme in the metabolism of 25-hydroxyvitamin D and 1,25-D and its induction has proven to be a useful bioresponse marker induced in response to treatment with 1,25-D in a number of cells that possess the VDR.[24,31] This assay is performed by Northern blot analysis using a 1.3 kb cDNA

Table 2. Summary of functional responses to 1,25-D in prostate cancer cell lines

	LNCaP cells	PC-3 cells	DU-145 cells
VDR content	++	+++	++
growth inhibition	+++	++	+/-
24-hydroxylase induction	—	++	++
PSA stimulation	+++	—*	—*

*PSA not inducible in these cell lines.

probe to rat 24-hydroxylase as previously described.[15] As summarized in Table 2, 24-hydroxylase mRNA was induced in DU-145 cells following 1,25-D treatment. This finding confirms that DU-145 cells are 1,25-D targets as would be predicted by their VDR content. Although they respond in some ways (24-hydroxylase induction), the DU-145 cells do not respond to 1,25-D by exhibiting growth-inhibition (Table 2). On the other hand, LNCaP cells, which are growth-inhibited, do not respond to 1,25-D treatment with the induction of 24-hydroxylase. In the case of PC-3 cells, both responses (growth-inhibition and 24-hydroxylase induction) are present indicating that the patterns of response are variable in different cells which may exhibit one or the other or both responses. The primary cells were not examined for 24-hydroxylase induction.

PSA Induction. The third functional response we examined was the induction of prostate specific antigen (PSA) measured by an automated ELISA assay.[30] PSA is an extremely important clinical biomarker for the presence of prostate tissue and the PSA levels in the blood of patients correlate well with the mass of prostate tissue in the body.[30,33,34] Measurement of PSA is therefore a very useful test in monitoring the course of disease in prostate cancer patients. Of the various cells we have available for study, only LNCaP is capable of producing PSA. As summarized in Table 2, 1,25-D induced PSA in LNCaP cells. The effect was dose-dependent and developed over 48 hours.[15] Neither DU-145 nor PC-3 cells make PSA and 1,25-D treatment did not induce its production.

C. Analogs of 1,25-D

Effect of Vitamin D Analogs on Cultured Prostate Cancer Cells. In recent years, a number of analogs of 1,25-D have been designed that have actions similar to vitamin D to inhibit cell growth and stimulate cellular differentiation but which have a reduced tendency to cause hypercalcemia when administered to intact animals. The mechanism for this dichotomy in actions is not well understood. However, since hypercalcemia is the major side-effect following the therapeutic use of vitamin D preparations in patients, these analogs are attractive as potential drugs that might achieve growth-inhibition of cancer cells without the attendent hypercalcemia.[16,35,36] We have therefore explored the potency of some of these analogs on prostate cancer cells. The findings indicate that some vitamin D analogs exhibit greater binding potency for the VDR than 1,25-D. As shown in Table 3, EB-1089 is one such analog. Other analogs, even with lower binding potency, show increased growth-inhibitory activity compared to 1,25-D.[16] This pattern is demonstrated by several analogs in Table 3. The mechanism for the differences in functional potency are currently being investigated and, in part, may relate to rates of intracellular metabolism of the active molecular species. When the drugs are used in vivo, additional factors such as binding to the vitamin D

Table 3. Relative potencies of 1,25$(OH)_2D_3$ and synthetic analogs for VDR binding, inhibition of proliferation, stimulation of PSA production by LNCaP cells and induction of 25$(OH)D_3$-24-hydroxylase mRNA by PC-3 cells

Analog	VDR (%)	Proliferation (%)	PSA (%)	24-hydroxylase (%)
1,25$(OH)_2D_3$	100	100	100	100
MC-903[a]	40	330	500	330
EB-1089[a]	200	250	500	60
OCT[b]	10	250	330	60
Ro24-2637[c]	7	200	170	140

[a]Leo Pharmaceutical Products Ltd., Ballerup,Denmark, [b]Chugai Pharmaceutical, Tokyo, Japan, [c]Hoffman La-Roche Co., Nutley, N.J.

binding protein and metabolic clearance rate come into play in determining the potency. Since these analogs are less likely to cause hypercalcemia in vivo, they can safely be used in greater dosage than 1,25-D and therefore the combination of greater potency and higher dosage should yield an increased likelihood of substantially greater action on prostate cancer.[16]

IV. DISCUSSION

According to the classical receptor hypothesis, in order for 1,25-D to be capable of acting directly on prostate cells, VDR must be present. The presence of VDR in prostate tissue has been investigated in two earlier studies [37,38] which yielded contradictory results. Our recent data[15,17] and those of Miller et al[25] definitively resolve this issue proving that prostate cells possess VDR. In addition, our data indicate that 1,25-D has important and direct actions on the prostate to inhibit cell proliferation and stimulate 24-hydroxylase expression. Furthermore, 1,25-D stimulates PSA production in LNCaP cells, the only cell line in this group of cells capable of synthesizing PSA. We interpret this action of 1,25-D to be a stimulation of cellular differentiation. The findings indicate that prostate is a 1,25-D target organ and that the actions of 1,25-D within this tissue are inhibition of proliferation and stimulation of differentiation.[15,17]

It is important to emphasize that all cell types were growth-inhibited including both epithelial and stromal elements of the prostate gland.[17] Moreover, the results indicate that, in addition to prostate cancer cells, both normal prostate as well as BPH-derived cells have VDR and are growth-inhibited by the hormone. It will require further study to determine why DU-145 cells, which exhibit VDR and other 1,25-D induced responses, were not growth-inhibited but this may be a very important area of investigation. The receptors and responses are similar in primary cultures of prostate cells as well as in established cancer cells lines.[15,17] These data indicate the important role of vitamin D in prostate as a regulator of growth, both normal and abnormal, as well as an activator of cellular differentiation. In addition, the findings in normal cells and primary cultures support the results obtained in LNCaP and PC-3 cancer cell lines as representing the actual actions of 1,25-D as growth-inhibitory on the prostate.

Based on these findings, we postulate that vitamin D may have protective actions on the development and/or progression of prostate cancer. We believe that these data provide the molecular basis for the epidemiologic findings that indicate that decreased sunlight may predispose to a higher prostate cancer risk, [11,14] and that higher blood levels of 1,25-D may be protective.[19] We further hypothesize that vitamin D supplementation may have beneficial effects on retarding the development and/or progression of prostate cancer.

In addition to the well known effects of 1,25-D on calcium and phosphate homeostasis,[20] the VDR has been demonstrated to be present in a number of other normal and malignant tissues not associated with mineral balance.[39-42] The findings that 1,25-D induced the differentiation of a variety of undifferentiated and malignant cells as well as inhibited rapid cell proliferation, was an important breakthrough in appreciating an expanded scope of 1,25-D action. The human malignant cell lines that were shown to possess VDR and to be differentiated and/or growth-inhibited by 1,25-D treatment included: HL-60 leukemic cells,[43] breast cancer cells,[44,45] malignant melanoma cells,[46] colon carcinoma cells [47-49] as well as our recent data with prostate cancer cells.[15-18] Interestingly, lower levels of VDR were positively associated with malignant transformation of colon tissue.[50] In addition, it has been suggested that VDR could serve as a marker of differentiation of colorectal adenocarcinoma.[49,50]

The antiproliferative effects of 1,25-D also have been demonstrated *in vivo.* 1,25-D inhibited the growth of xenografts of human melanoma and colonic cancers in immune-supressed mice.[45] Moderately elevated plasma concentration of vitamin D were associated with a large reduction in the incidence of colorectal cancer in the recent studies of Garland et al.[12,13,51] Mammary tumors induced by nitrosomethylurea in rats are inhibited by 1,25-D.[44] 1,25-D also suppresses the proliferation and promotes the differentiation of a number of non-malignant tissues *in vitro* [52] including normal osteoblasts,[53] myoblasts,[54] keratinocytes,[55] and intestinal cells.[49,56] In a very recent study, use of a vitamin D analog (Ro 24-5531) lessened tumor incidence and extended tumor latency in a rat breast cancer study.[35] Another vitamin D analog (EB-1089) caused a dose-dependent significant inhibition and even regression of the N-methyl-nitrosourea-induced rat mammary tumor in vivo.[36] A clinical trial using EB-1089 to treat breast cancer is currently under way in the United Kingdom.

V. SUMMARY

Our findings demonstrate the presence of VDR in various human prostate cancer cell lines and in primary cultures derived from normal, BPH and prostate cancer. In addition, 1,25-D induced several bioresponses in these cells including growth inhibition and PSA stimulation. Based on examples in many different malignant cells as well as our data in prostate cells, that vitamin D is anti-proliferative and promotes cellular maturation, it seem clear that vitamin D must be viewed as an important cellular modulator of growth and differentiation in addition to its classical role as regulator of calcium homeostasis. In this respect, vitamin D has the potential to have beneficial actions on various malignancies including prostate cancer. Its ultimate role in prostate cancer remains to be determined, but 1,25-D may prove useful in chemoprevention and/or differentiation therapy. We believe the data currently available provide the basis for an optimistic view on the possible use of vitamin D to treat prostate cancer in patients and that further investigation is clearly warranted to better define its potential therapeutic utility.

ACKNOWLEDGMENTS

This work was supported by grant 92B29 from the American Institute for Cancer Research, grant DK 42482 from the National Institutes of Health and a grant from the Lucas Foundation.

REFERENCES

1. C.C. Boring, T.S. Squires, T. Tong, and S. Montgomery, Cancer statistics, 1994, *CA Cancer J Clin*. 44:7 (1994).
2. J.M. Kozlowski, and J.T. Grayhack, Carcinoma of the prostate, in: Adult and Pediatric Urology, J. Y. Gillenwater, J. T. Grayhack, S. S. Howards and J. W. Duckett, eds. Mosby Year Book, St Louis (1991).
3. H. Carter, and D. Coffey, Prostate cancer: The magnitude of the problem in the United States., in: A Multidisciplinary Analysis of Controversies in the Management of Prostate Cancer., D. Coffey Resnick,M.,Dorr,F., et al., eds. Plenum Press, New York (1988).
4. L. Cannon, D. Bishop, M. Skolnick, S. Hunt, J. Lyon, and C. Smart, Genetic epidemiology of prostate cancer in the Utah Mormon genealogy., *Cancer Survey*. 1:47 (1983).
5. C. Woolf, An investigation of the familial aspects of carcinoma of the prostate, *Cancer*. 13:739 (1960).
6. W. Meikle, J. Smith, and D. West, Familial factors affecting prostatic cancer risk and plasma steroid levels, *Prostate*. 6:121 (1985).
7. J.D. McConnell, Physiologic basis of endocrine therapy for prostatic cancer, *Urol Clin North Amer*. 18:1 (1991).
8. P. Ekman, and J. Brolin, Steroid receptor profile in human prostate cancer metastases as compared with primary prostatic carcinoma, *Prostate*. 18:147 (1991).
9. C. Huggins, R.E. Stevens, and C.V. Hodges, Studies on prostatic cancer.II. The effects of castration on advanced carcinoma of the prostate gland, *Arch Surg*. 43:209 (1941).
10. R.M. Wolf, S.L. Schneider, J.E. Pontes, L. Englander, J.P. Karr, G.P. Murphy, and A.A. Sandberg, Estrogen and progestin receptors in human prostatic carcinoma, *Cancer*. 55:2477 (1985).
11. G.G. Schwartz, and B.S. Hulka, Is Vitamin D deficiency a risk factor for prostate cancer?(Hypothesis), *Anticancer Res*. 10:1307 (1990).
12. C.F. Garland, G.W. Comstock, F.C. Garland, K.J. Helsing, E.K. Shaw, and E.D. Gorham, Serum 25-Hydroxyvitamin D and colon cancer: Eight-year prospective study., *Lancet*. 2:1176 (1989).
13. C.F. Garland, F.C. Garland, and E.D. Gorham, Can colon cancer incidence and death rates be reduced with calcium and vitamin D?, *Am J Clin Nutr*. 54(Suppl.1):193 (1991).
14. C.L. Hanchette, and G.G. Schwartz , Geographic patterns of prostate cancer mortality: Evidence for a protective effect of ultraviolet radiation, *Cancer*. 70:2861 (1992).
15. R.J. Skowronski, D.M. Peehl, and D. Feldman, Vitamin D and prostate cancer: 1,25 Dihydroxyvitamin D_3 receptors and actions in human prostate cancer cell lines, *Endocrinology*. 132:1952 (1993).
16. R.J. Skowronski, D.M. Peehl, and D. Feldman, Actions of vitamin D_3 analogs on human prostate cancer cell lines: Comparison with 1,25-dihydroxyvitamin D_3, *Endocrinology* 136:20 (1995).
17. D.M. Peehl, R.J. Skowronski, G.K. Leung, S.T. Wong, T.A. Stamey, and D. Feldman, Antiproliferative effects of 1,25-Dihydroxyvitamin D_3 on primary cultures of human prostatic cells, *Cancer Res*. 54:805 (1994).
18. D. Peehl, R.J. Skowronski, and D. Feldman, Role of vitamin D receptors in prostate cancer, in: Intl. Symp."Sex Hormones and Antihormones in Endocrine Dependent Pathology: Basic and Clinical Aspects, Elsevier, Amsterdam, in press (1994).
19. E.H. Corder, H.A. Guess, B.S. Hulka, G.D. Friedman, M. Sadler, R.T. Vollmer, B. Lobaugh, M.K. Drezner, J.h. Vogelman, and N. Orentreich, Vitamin D and prostate cancer: A prediagnostic study with stored sera, *Cancer Epidemiology, Biomakers & Prevention*. 2:467 (1993).
20. A.W. Norman, J. Roth, and L. Orci, The vitamin D endocrine system: Steroid metabolism, hormone receptors, and biological response (calcium binding proteins)., *Endocr Rev*. 3:331 (1982).
21. J.W. Pike, Vitamin D_3 receptors: structure and function in transcription, *Ann Rev Nutr*. 11:189 (1991).
22. R.M. Evans, The steroid and thyroid hormone receptor superfamily, *Science*. 240:889 (1988).
23. B. O'Malley, The steroid receptor superfamily: more excitement predicted for the future, *Mol Endocrinol*. 4:363 (1990).
24. D. Feldman, and P.J. Malloy, Hereditary 1,25 dihydroxyvitamin D resistant rickets: Molecular basis and implications for the role of $1,25(OH)_2D_3$ in normal physiology, *Mol Cell Endocrinol*. 72:C57 (1990).
25. G.J. Miller, G.E. Stapleton, J.A. Ferrara, M.S. Lucia, S. Pfister, T.E. Hedlund, and P. Upadhya, The human prostatic carcinoma cell line LNCaP expresses biologically active, specific receptors for 1α,25-dihydroxyvitamin D_3, *Cancer Res*. 52:515 (1992).
26. K. Burton, A study of conditions and mechanisms of the diphenyl amine colorimetric estimation of deoxyribonucleic acid, *Biochem J*. 62:315 (1956).
27. C. Auffray, and F. Rougeon, Purification of mouse immunoglobulin heavy-chain messenger RNAs from total myeloma tumor RNA., *Eur J Biochem*. 107:303 (1980).

28. A.R. Baker, D.P. McDonnell, M. Hughes, T.M. Crisp, D.J. Mangelsdorf, M.R. Haussler, J.W. Pike, J. Shine, and B.W. O'Malley, Cloning and expression of full-length cDNA encoding human vitamin D receptor., *Proc Natl Acad Sci USA*. 85:3294 (1988).
29. Y. Ohyama, M. Noshiro, and K. Okuda, Cloning and expression of cDNA encoding 25-hydroxyvitamin D_3 24-hydroxylase, *FEBS Lett*. 278:195 (1991).
30. T.A. Stamey, H.C.B. Graves, N. Wehner, M. Ferrari, and F. Freiha, Early detection of residual prostate cancer after radical prostatectomy by an ultrasensitive assay for prostate specific antigen, *J Urol*. 149:787 (1993).
31. P.J. Malloy, Z. Hochberg, D. Tiosano, J.W. Pike, M.R. Hughes, and D. Feldman, The molecular basis of hereditary 1,25-dihydroxyvitamin D_3 resistant rickets in seven related families, *J Clin Invest*. 86:2071 (1990).
32. J.W. Pike, Monoclonal antibodies to chick intestinal receptors for 1,25-dihydroxyvitamin D_3, *J Biol Chem*. 259:1167 (1984).
33. T.A. Stamey, and J.N. Kabalin, Prostate specific antigen in the diagnosis and teratment of adenocarcinoma of the prostate.I.Untreated patients, *J Urol*. 141:1070 (1989).
34. H.-P. Schmid, J.E. McNeal, and T.A. Stamey, Observations on the doubling time of prostate cancer, *Cancer*. 71:2031 (1993).
35. M.A. Anzano, J.M. Smith, M.R. Uskokovic, C.W. Peer, L.T. Mullen, J.J. Letterio, M.C. Welsh, M.W. Shrader, D.l. Logsdon, C.L. Driver, C.C. Brown, A.R. Roberts, and M.B. Sporn, 1α,25-Dihydroxy-16-ene-23yne-26,27-hexafluorocholecalciferol (Ro24-5531) a new deltanoid (vitamin D analogue) for prevention of breast cancer in the rat, *Cancer Res*. 54:1653 (1994).
36. K.W. Colston, A.G. Mackay, S. Chander, L. Binderup, and R.C. Coombes, Novel vitamin D analogues suppress tumor growth *in vivo*, in: Vitamin D: Gene Regulation, Structure-Function Analysis and Clinical Application: Proceedings of the Eighth Workshop on Vitamin D, Berlin, de Gruyter, 465 (1991).
37. G. Schleicher, T.H. Privette, and W.E. Stumpf, Distribution of soltriol [1,25$(OH)_2$-Vitamin D_3] binding sites in male sex organs of the mouse: An autoradiographic study., *J Histochem Cytochem*. 37:1083 (1989).
38. M.R. Walters, D.L. Cuneo, and A.P. and Jamison, A possible significance of new target tissues for 1,25-dihydroxyvitamin D_3, *J Steroid Biochem*. 19:913 (1983).
39. K. Colston, J.M. Colston, Fieldsteel, and D. Feldman, 1,25-Dihydroxyvitamin D_3 receptors in human epithelial cancer cell lines, *Cancer Res*. 42:856 (1982).
40. B.C. Osmundsen, H.F.S. Huang, M.B. Anderson, S. Christakos, and M.R. Walters, Multiple sites of action of the vitamin D endocrine system: FSH stimulation of testis 1,25-dihydroxyvitamin D_3 receptors, *J Steroid Biochem*. 34:339 (1989).
41. H. Reichel, H.P. Koeffler, and A.W. Norman, The role of the vitamin D endocrine system in health and disease, *N Engl J Med*. 320:980 (1989).
42. W.E. Stumpf, M. Sar, F.A. Reid, and e. al., Target cells for 1,25-dihydroxyvitamin D_3 in intestinal tract, stomach, kidney, skin, pituitary, and parathyroid, *Science*. 206:1188 (1979).
43. E. Abe, C. Miyayra, H. Sakagami, M. Takeda, K. Konno, T. Yamazaki, S. Yoshiki, and T. Suda, Differentiation of mouse myeloid leukemia cells induced by 1α,25-dihydroxyvitamin D_3., *Proc Natl Acad Sci USA*. 78:4990 (1981).
44. K.W. Colston, U. Berger, and R.C. Coombes, Possible role for vitamin D in controlling breast cancer cell proliferation, *Lancet*. 1:188 (1989).
45. J.A. Eisman, D.H. Barkla, and P.J. Tutton, Suppression of *in vivo* growth of human cancer solid tumor xenografts by 1,25-dihydroxyvitamin D_3, *Cancer Res*. 47:21 (1987).
46. K. Colston, J.M. Colston, and D. Feldman, 1,25-Dihydroxyvitamin D_3 and malignant melanoma: The presence of receptors and inhibition of cell growth in culture, *Endocrinology*. 108:1083 (1981).
47. A. Brehier, and M. Thomassat, Human colon cell line HT-29: Characterisation of 1,25-dihydroxyvitamin D_3 receptor and induction of differentiation by the hormone, *J Steroid Biochem*. 29:265 (1988).
48. K.D. Harper, R.V. Iozzo, and J.G. Haddad, Receptors for and bioresponses to 1,25-dihydroxyvitamin D in a human colon carcinoma cell line (HT-29), *Metabolism*. 38:1062 (1989).
49. X. Zhao, and D. Feldman, Regulation of vitamin D receptor abundance and responsiveness during differentiation of HT-29 human colon cancer cells, *Endocrinology*. 132:1808 (1993).
50. P. Lointier, F. Meggouh, P. Dechelotte, D. Pezet, C. Ferrier, J. Chipponi, and S. Saez, 1,25-Dihydroxyvitamin D_3 receptors and human colon adenocarcinoma, *Br J Surg*. 78:435 (1991).
51. C.F. Garland, and F.C. Garland, Do vitamin D and sunlight reduce the likelihood of colon cancer?, *Int J Epidemiol*. 9:227 (1980).
52. H.A.P. Pols, J.C. Birkenhager, J.A. Foekens, and J.P.T.M. Van Leeuwen, Vitamin D: A modulator of cell proliferation and differentiation, *J Steroid Biochem*. 37:873 (1990).

53. T.A. Owen, M.S. Aronow, L.M. Barone, B. Bettencourt, G.S. Stein, and J.B. Lian, Pleitropic effects of vitamin D on osteoblast gene expression are related to the proliferative and differentiated state of the bone cell phehnotype: Dependency upon basal levels of gene expression, duration of exposure, and bone matrix competency in the normal rat osteoblast cultures, *Endocrinology*. 128:1496 (1991).
54. E.M. Costa, H.M. Blau, and D. Feldman, 1,25-dihydroxyvitamin D_3 receptors and hormonal responses in cloned human skeletal muscle cells, *Endocrinology*. 119:2214 (1986).
55. S. Pillai, D.D. Bikle, and P.M. Elias, 1,25-dihydroxyvitamin D production and receptor binding in human keratinocytes varies with differentiation., *J Biol Chem*. 263:5390 (1988).
56. T. Suda, T. Shinki, and N. Takahashi, The role of vitamin D in bone and intestinal cell differentiation, *Ann Rev Nutr*. 10:195 (1990).

6

CHOLINE AND HEPATOCARCINOGENESIS IN THE RAT

Steven H. Zeisel,* Kerry-Ann da Costa, Craig D. Albright, and Ok-Ho Shin

Department of Nutrition
School of Public Health and School of Medicine
The University of North Carolina at Chapel Hill
Chapel Hill, North Carolina 27599

ABSTRACT

Rats fed a choline deficient diet develop foci of enzyme-altered hepatocytes with subsequent formation of hepatic tumors. This is the only nutritional deficiency that, in itself, causes cancer. We suggested that carcinogenesis is triggered, in part, because of abnormalities in cell signals which regulate cell proliferation and cell death. Because choline deficient rats develop fatty liver (choline is needed for hepatic secretion of certain lipoproteins), we examined whether an important lipid second messenger involved in proliferative signaling, 1,2-*sn*-diacylglycerol, accumulated in liver and resulted in the prolonged activation of protein kinase C. We observed that 1,2-*sn*-diacylglycerol accumulated in the plasma membrane from the non-tumor portion of livers of rats fed a choline deficient diet, and that unsaturated free fatty acids, another activator of protein kinase C, also accumulated in deficient livers. Protein kinase C in the hepatic plasma membrane and nucleus of choline deficient rats was elevated for months; this is the only model system which exhibits such prolonged activation of protein kinase C. Premalignant, abnormal hepatic foci were detected only in the deficient rats, and 15% of deficient rats (none of the controls) had hepatocellular carcinoma at 1 year on the diet. In rats, an early event in choline deficiency is an increase in the rate of cell death. In liver from choline deficient rats, we observed an increase in the numbers of liver cells with fragmented DNA (characteristic of programmed cell death; apoptosis). We used a cell culture model (immortalized rat hepatocytes) to study the effects of choline deficiency on apoptosis. Liver cells grown in a choline deficient medium became depleted of choline, accumulated triacylglycerol and 1,2-*sn*-diacylglycerol, and had increased DNA fragmentation and other morphologic and biochemical changes associated with apoptosis. This model has great potential as a tool for studying the underlying link between choline deficiency and the regulation of the balance between cell proliferation and cell death. We suggest that choline deficiency altered the cell proliferation signals mediated by protein

* Address reprint requests to S. H. Zeisel. Tel: (919) 966-7218; fax: (919) 966-7216

Diet and Cancer, Edited under the auspices of the American Institute of Cancer Research
Plenum Press, New York, 1995

kinase C within liver, and altered cell apoptosis. These changes in cell signaling may be the triggering events which result in hepatic carcinogenesis.

Acknowledgment: This work has been funded by a grant from the American Institute for Cancer Research.

INTRODUCTION

Choline is an important dietary source of methyl-groups and a precursor of phosphatidylcholine, the major phospholipid in mammalian membranes (1). Choline deficient rats accumulate fat within their livers (2). They also develop hepatocarcinomas in the absence of any known carcinogen (3-10). Choline deficiency also enhances the carcinogenicity of chemicals such as aflatoxin, diethylnitrosamine and ethionine (7, 9).

The mechanism for the cancer-promoting effect of choline deficiency is not completely elucidated (11-14). Proposed mechanisms include genetic damage due to lipid peroxide formation, hypomethylation of DNA, and abnormal regulation of cell proliferation. Our research focuses on protein kinase C-mediated signals involved in cell proliferation and on signaling involved in apoptosis, a unique cellular process for programmed cell death. We have found that choline deficiency perturbs both types of signaling within the liver cell, and we suggest that these abnormalities may be the reason that cancer eventually develops.

CARCINOGENESIS

There are several mechanisms suggested for the cancer-promoting effect of a choline deficient diet. There is a progressive increase in cell proliferation, related to regeneration after parenchymal cell death in the choline deficient liver (3, 7, 12). We have observed that the increased rate of cell death associated with choline deficiency is due to apoptosis (15) (see discussion below). Cell proliferation, with associated increased rate of DNA synthesis, could be the cause of greater sensitivity to chemical carcinogens (11). Stimuli for increased DNA synthesis, such as hepatectomy and necrogenic chemicals, increase carcinogenesis. Methylation of DNA is important for the regulation of expression of genetic information. Hypomethylation of DNA, observed during choline deficiency (despite adequate dietary methionine), may be responsible for carcinogenesis (4, 16). Another proposed mechanism derives from the observation that, when rats eat a choline deficient diet, increased lipid peroxidation occurs within liver (17). Lipid peroxides in the nucleus could be a source of free radicals that could modify DNA and cause carcinogenesis. Recently, we have observed that choline deficiency perturbs PKC signal transduction, thereby promoting carcinogenesis.

PROTEIN KINASE C MEDIATED SIGNAL TRANSDUCTION

Choline phospholipid-mediated signal transduction plays an important role in cell proliferation (18). Stimulation of membrane-associated receptors activates neighboring phospholipases, resulting in the formation of breakdown products that are signaling molecules either by themselves (i.e., they stimulate or inhibit the activity of target macromolecules), or after conversion to signaling molecules by specific enzymes such as protein kinase C isoforms (PKC) which modify cellular activities and gene expression (Figure 1). Much of signaling research has focused on minor membrane phospholipid components, particularly phosphatidylinositol derivatives. However, metabolism of choline phospholipids, especially phosphatidylcholine and sphingomyelin, forms biologically active molecules that can am-

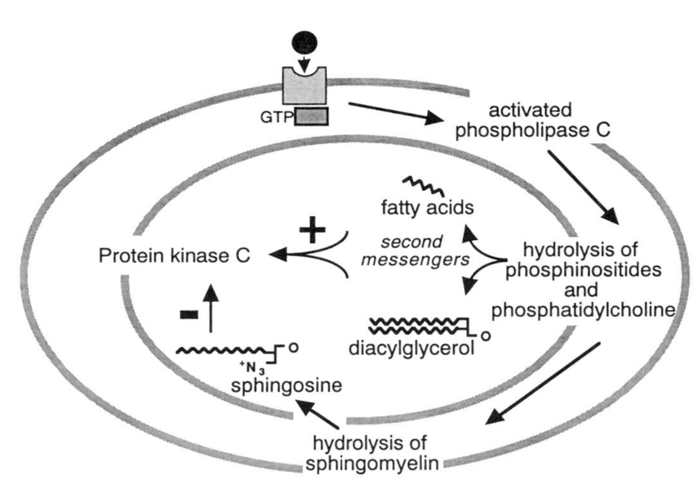

Figure 1. Signal transduction via protein kinase C. Receptor activation results in phospholipase C-mediated hydrolysis of phospholipids. This generates second messengers which activate (diacylglycerol and fatty acids) or inhibit (sphingosine) protein kinase C.

plify external signals or that can terminate the signaling process by generating inhibitory second messengers (18).

Signal transduction via receptor mediated hydrolysis of phospholipids has been extensively reviewed elsewhere (18-21). Phosphatidylcholine hydrolysis generates a series of messengers that sustain the PKC phosphorylation cascade (21). These second messengers include 1,2-*sn*-diacylglycerol and certain unsaturated free fatty acids. Hydrolysis of sphingomyelin generates messengers that terminate the cascade (the metabolite, sphingosine, is a potent inhibitor of PKC that acts by blocking diacylglycerol-mediated activation) (22).

Sustained activation of PKC is essential for triggering cell differentiation and proliferation (23). PKC signals impinge on several known intracellular control circuits (24). The targets for phosphorylation by PKC include receptors for insulin, epidermal growth factor and many proteins involved in control of gene expression (25, 26).

CHOLINE AND HEPATIC DIACYLGLYCEROL

Choline deficiency causes massive fatty liver, and we have observed that the lipids 1,2-*sn*-diacylglycerol (2, 27, 28) and unsaturated free fatty acids (28) accumulate in choline deficient liver. Both families of lipids are activators of PKC (29-33) (see earlier discussion). In plasma membrane from livers of choline deficient rats, 1,2-*sn*-diacylglycerol reaches values higher than those occurring after stimulation of a receptor linked to phospholipase C activation (e.g., vasopressin receptor) (see Figure 2). Oleic acid is increased 8-fold in liver at 6 weeks on a deficient diet (see Figure 3). These elevated second messengers result in a stable activation of PKC and/or an increase in the total PKC pool in the cell (2, 28) with changes in several PKC isotypes in liver (at 6 weeks of choline deficiency, amounts of PKC α and δ increased 2-fold and 10-fold, respectively). Changes in plasma membrane PKC were sustained for 12 months (see Figure 4) (28). In receptor-mediated activation, cytosolic PKC is translocated to membranes, is activated, and then is rapidly degraded (34). We saw no down regulation of PKC activation mediated by choline deficiency. While prolonged activation of PKC by 1,2-*sn*-diacylglycerol without down-regulation of the enzyme has been

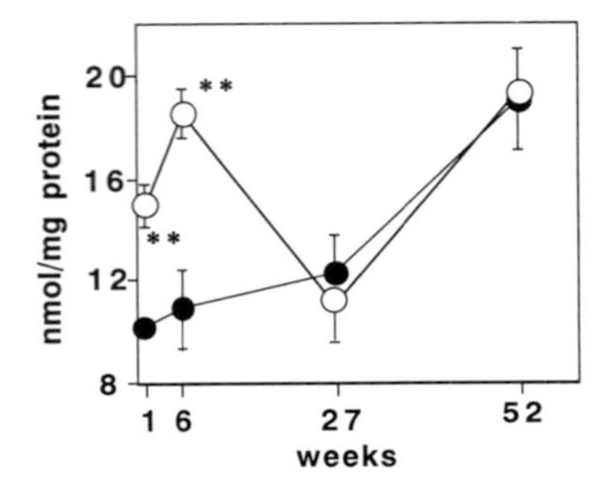

Figure 2. Plasma membrane 1,2-*sn*-diacylglycerol in liver from rats fed a choline deficient diet. Rats were fed defined diets containing choline (control) or devoid of choline (deficient) for 52 weeks. Liver was collected and plasma membrane prepared. 1,2-*sn*-diacylglycerol was extracted and measured using a radioenzymatic assay. Data are expressed as mean ± SEM (n=6-15/group). Statistical differences were determined using unpaired t-Test. **= p< 0.01 different from control at same time point. Data were previously reported (2, 28).

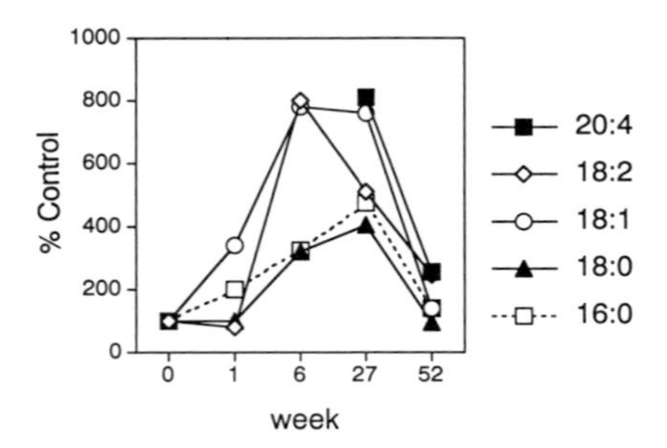

Figure 3. Free fatty acids in liver from rats fed a choline deficient diet for 52 weeks. Rats were fed defined diets containing choline (control) or devoid of choline (deficient) for 52 weeks. Liver was collected and free fatty acids measured using gas chromatography. Data are expressed as means (n=6-15/group). All data points at 6 and 27 weeks are significantly different from control (P< 0.05). Data were previously reported (28).

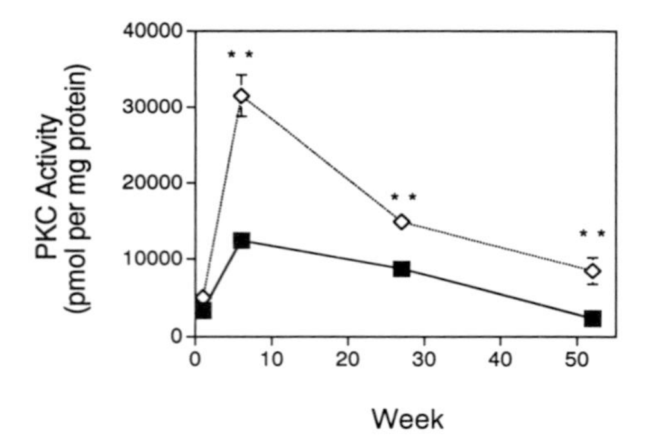

Figure 4. Plasma membrane protein kinase C activity in liver from rats fed a choline deficient diet for 52 weeks. Rats were fed defined diets containing choline (control) or devoid of choline (deficient) for 52 weeks. Liver was collected and plasma membrane prepared in the presence of protease inhibitors. Protein was extracted and its activity measured using a radioenzymatic assay. Data are expressed as mean ± SEM. Statistical differences were determined using Student's unpaired t-Test. Data were previously reported (2, 28).

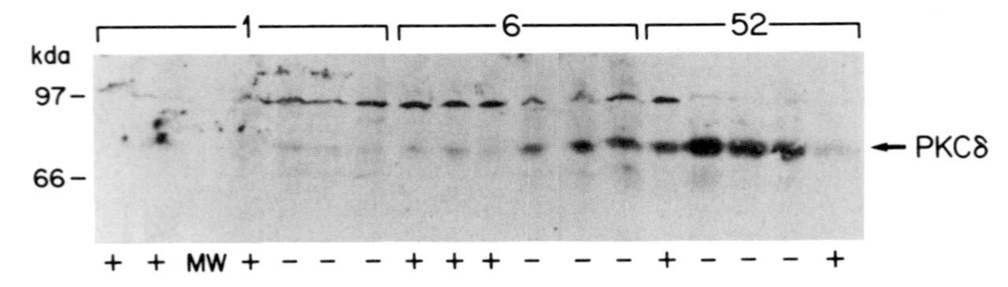

Figure 5. Protein kinase C isoforms in nuclei from choline deficient liver. Rats were treated as described in figure 3 legend. Nuclear protein kinase C protein was separated by Western blot, transferred to nitrocellulose paper, and probed with specific PKC antisera. Data were previously reported (37).

Table 1. Triacylglycerol and 1,2-*sn*-diacylglycerol in liver from Zucker rats

	Lean Zucker Control (nmol/mg protein)	Obese Zucker (nmol/mg protein)	p value
Liver triacylglycerol	157 ± 20	1,737 ± 187	< 0.01 (n=6/group)
Liver 1,2-sn-diacylglycerol	2.9 ± 0.2	12.2 ± 1.3	< 0.01 (n = 5/group)
Liver plasma membrane 1,2-sn-diacylglycerol	9.3 ± 1.0	22 ± 1.8	<0.01 (n=5/group)

Male Zucker obese rats, and control Zucker lean rats (Charles River), 12 weeks old, were fed Prolab Rodent Chow 3000 (Agway Country Food Inc., Syracuse, NY) for 5 weeks. Hepatic lipids were extracted (55), and triacylglycerol was separated using thin layer chromatography (hexane/ethylether/acetic acid, 50:50:1), and quantitated using gas chromatography (2). Subcellular fractions were prepared and diacylglycerol was measured after thin layer chromatography using a radioenzymatic assay (2). Data are expressed as mean ± SEM. Statistical differences were determined by unpaired t-Test.

reported in other studies (35, 36), we are reporting PKC activation that has been sustained for months. It is known that sustained activation of PKC is essential for long-term cellular responses such as cell proliferation and differentiation (23). Nuclear PKC δ is also elevated in liver for months after rats were exposed to a choline deficient diet (see Figure 5) (37).

Our studies were performed on areas of the liver that appeared normal, and not on tumor tissue. Choline deficient liver clearly contained abnormal foci which would have been included in analyses. Therefore, the changes we report may be due to diet, or to diet and to processes involved in carcinogenic transformation. However, abnormal foci never accounted for more than about 1% of liver tissue, and it is unlikely that changes in this small percent of cells were responsible for our observations.

We have started to examine whether the accumulation of 1,2-*sn*-diacylglycerol occurs because rats develop fatty liver, or because of some other specific effect of choline deficiency. Zucker obese rats accumulated massive amounts of fat in their liver, yet were not choline deficient. Hepatic plasma membrane from obese Zucker rats had more than twice the amount of 1,2-*sn*-diacylglycerol associated with them than did equivalent membrane from lean Zucker rats (see Table 1). Thus, it is possible that many forms of fatty liver may be associated with accumulation of 1,2-*sn*-diacylglycerol and sustained activation of PKC.

PKC and Cancer

The accumulation of diacylglycerol and subsequent activation of PKC within liver during choline deficiency may be the critical abnormality that eventually contributes to the development of hepatic cancer in these animals (2). Chronic elevation of 1,2-*sn*-diacylglycerol can induce DNA synthesis, cause cell proliferation, and lead to cell transformation, presumably via abnormal PKC signaling (38, 39). Abnormalities in PKC-mediated signal transduction may trigger carcinogenesis (26). Many mitogens activate PKC (23) and the phorbol esters (diacylglycerol analogs) are potent tumor promoters (25). The expression of oncogenes can perturb PKC-mediated signal transduction in a manner similar to that which we described for choline deficiency. In *erb*B-transformed fibroblasts, 1,2-*sn*-diacylglycerol accumulates because diacylglycerol kinase activity does not remove it (38). Transformation by *ras* or *src* oncogenes translocates PKC to the plasma membrane (36). Diacylglycerol concentrations are elevated *in vivo* in *ras*-transformed liver of neonatal transgenic mice (40). NIH 3T3 cells transformed with Ha-*ras*, Ki-*ras*, v-*src*, or v-*fms* oncogenes have increased diacylglycerol levels as well as tonic activation and partial down regulation of PKC (41).

Degradation of phosphatidylcholine produces these elevated 1,2-*sn*-diacylglycerol levels (35, 36). Transfection of fibroblasts so that they overexpress PKC (α, β_1 or ε) activity causes them to become transformed and tumorigenic (24, 42-45). Thus, many observations suggest that choline deficiency may causes cancer by increasing 1,2-*sn*-diacylglycerol concentration with subsequent sustained activation of PKC.

APOPTOSIS

The commitment to becoming a cancer cell occurs relatively early during choline deficiency; after 6 months of deficiency the return to a normal diet does not prevent cancer from developing (3). For this reason we have studied the early events that occur in liver when rats are made choline deficient. The most obvious of these phenomena is that choline deficiency induces repeated rounds of cell death and compensatory cell proliferation in liver (11, 12, 46, 47). This cell turnover may select for the survival and growth of putative preneoplastic hepatocytes (48) and may promote the induction of liver cancer in the absence of known carcinogens (5, 10).

There are two types of cell death - massive tissue damage resulting in necrosis, and apoptosis, the active destruction of cells so that breakdown products are compartmentalized and removed in an orderly manner. Apoptosis is a physiological form of cell death that has been described in hormone dependent tissues (e.g., mammary after prolactin is removed during weaning). It can be induced by positive modulators (e.g. Mullerian duct inhibiting substance (MIS) or tumor necrosis factor (TNF) in adipocytes). It also can be induced in organs with cells that retain their proliferative potential but must maintain homeostatic control of cell numbers (e.g. hepatocytes) (49). Apoptosis requires specific gene transcription and protein synthesis (50). The morphologic changes classically associated with apoptosis include: (i) chromatin condensation and nuclear pyknosis, (ii) fragmentation of the nucleus, and (iii) progressive degeneration of the apoptotic bodies, which may retain residual nuclei and cytoplasmic organelles (50, 51). The irreversible commitment to apoptosis involves the activation of an endonuclease which cleaves transcriptionally active nuclear DNA (51), and not mitochondrial DNA (52), into internucleosomal fragments. In parallel to nuclear condensation, cytoplasmic condensation involves the disruption of desmosomal contacts between cells and disruption of cytoskeleton. These processes occur in minutes. During the fragmentation process the cytoplasm and organelles of the cells undergoing apoptosis are repackaged without leakage of the cellular components into the extracellular fluid. The removal of apoptotic bodies by normal neighboring cells and their subsequent degradation by lysosomes takes hours to complete (49).

We found that the increased rate of cell death that occurs in early choline deficiency is due to the induction of apoptosis. In slices of liver collected from rats fed a choline deficient diet for 6 weeks, DNA fragmentation and apoptosis were detected using a direct immunoperoxidase method (ApopTag, Oncor, Inc., Gaithersburg, MD) to visualize the incorporation of digoxigenin-labeled nucleotide into free 3'-OH ends of DNA, a biochemical change specifically associated with the morphological appearance of apoptosis (53). More than 10% of cells in choline deficient liver were apoptotic, versus 3% of cells in control liver (15). In order to study the mechanisms associated with choline deficiency-induced apoptosis, we developed a model system using immortalized, non-tumorigenic, CWSV-1 rat hepatocytes in culture. This cell line was derived from Fischer 344 rat hepatocytes and immortalized with simian virus 40 DNA (54) (the cells were a generous gift from Dr. Harriet C. Isom). In these cells choline deficiency recapitulates what is seen *in vivo* in liver - depleted pools of intracellular choline, phosphocholine and phosphatidylcholine, and increased intracellular triacylglycerol and 1,2-*sn*-diacylglycerol (see Table 2). To determine whether apoptosis is

Table 2. Choline pools in CWSV1 hepatocytes grown in media deficient in choline content

Metabolite	Control (nmol per mg DNA)	Deficient (nmol per mg DNA)	p value
Choline	11.5 ± 0.7	1.1 ± 0.1	< 0.01
Phosphocholine	141 ± 3	8 ± 2	< 0.01
Phosphatidylcholine	1298 ± 127	841 ± 37	< 0.05
1,2-*sn* -diacylglycerol	4.1 ± 0.2	5.8 ± 0.6	< 0.05
triacylglycerol	132 ± 17	418 ± 21	< 0.01

Immortalized CWSV-1 rat hepatocytes were grown in medium containing 21.5 μM choline (control) or 2.15 μM choline (deficient) for 4 days. Cells were harvested in methanol, and biochemical parameters measured (2, 56). Data are expressed as mean ± SEM. Statistical differences were determined by unpaired t-Test.

induced by choline deficiency, cells were plated in 70 mM choline sufficient serum-free medium for 4 days and then cultured in choline sufficient or choline deficient (5 μM or 0 μM choline) medium for 3 days. Using the digoxigenin-11-dUTP method to detect apoptosis, we found that fragmentation of genomic DNA occurred in 3%, 43%, and 67% of cells grown for 72 hrs in 70 , 5, and 0 μM choline, respectively (see Figure 6) (15). Cells undergoing DNA fragmentation retained the ability to exclude trypan blue and had viable mitochondria. This is consistent with the induction of a physiological, non-necrotic form of cell death by choline deficiency. The initiation of DNA fragmentation was followed by the progressive, dose-dependent accumulation of end-stage apoptotic cells exhibiting nuclear chromatin condensation, pyknosis, and fragmentation of the nucleus found within intact cytoplasm (see Figure 7). Ultrastructurally these apoptotic cells contained intact cell organelles.

SUMMARY

We have identified two molecular mechanisms that may be perturbed in choline deficient liver - PKC signal transduction and regulation of apoptosis. These changes may contribute to the process of carcinogenesis that occurs in choline deficiency.

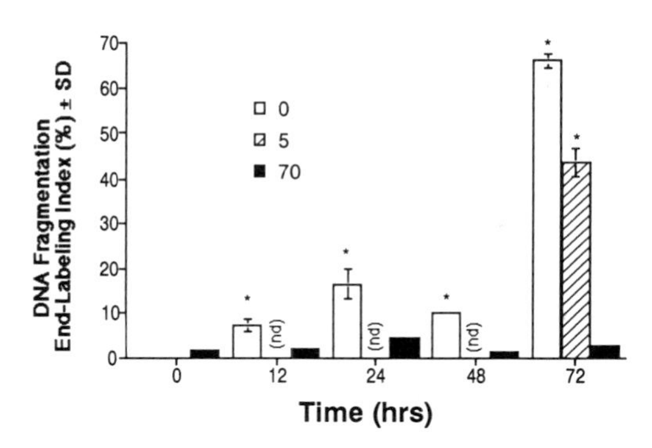

Figure 6. DNA fragmentation in choline deficient CWSV1 cells. Effect of choline deficiency on DNA fragmentation. The height of the bars shows the percentage of cells with nuclear incorporation of digoxigenin-11-dUTP after 72 hrs in culture. Open bars = 0 μM choline, cross-hatched bars = 5 μM choline, black bars = 70 μM choline; nd = not done. Data are expressed as mean ± SD; n = 3/treatment (t-test: * = $p < .05$). Data were previously reported (15).

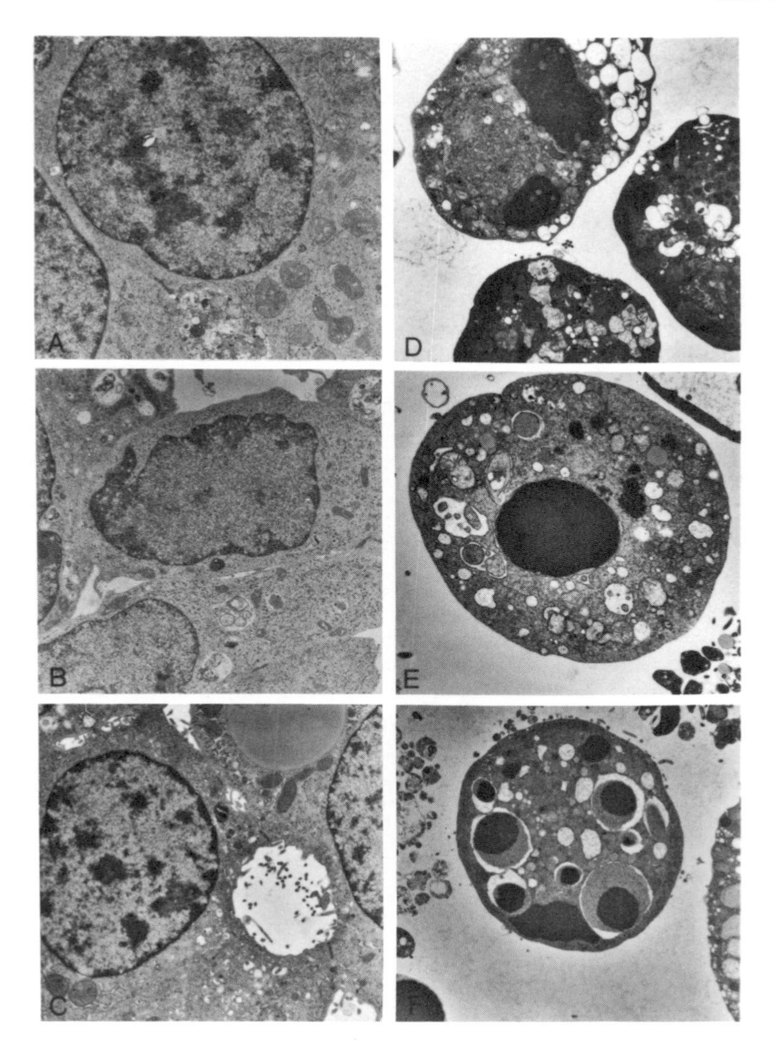

Figure 7. Apoptosis in choline deficient CWSV1 cells. Ultrastructural morphology of attached (A-C) and detached (D-F) CWSV-1 cells following treatment for 72 hrs with 70 μM (A and D), 5 μM (B and E) , and 0 μM choline-supplemented medium (C and F) . Detached, apoptotic cells are found in all treatment groups, but increase in number over time, inversely proportional to the choline concentration. Attached and floating cells contain mitochondria. Magnification: (A-C) X 4,000; (D-F) X 5,000. Data were previously reported (15).

REFERENCES

1. Zeisel, S. H. (1981). Dietary choline: biochemistry, physiology, and pharmacology. *Ann. Rev. Nutr.* **1**, 95-121.
2. daCosta, K., Cochary, E. F., Blusztajn, J. K., Garner, S. C. and Zeisel, S. H. (1993). Accumulation of 1,2-*sn*-diradylglycerol with increased membrane-associated protein kinase C may be the mechanism for spontaneous hepatocarcinogenesis in choline deficient rats. *J. Biol. Chem.* **268**, 2100-2105.
3. Chandar, N. and Lombardi, B. (1988). Liver cell proliferation and incidence of hepatocellular carcinomas in rats fed consecutively a choline-devoid and a choline-supplemented diet. *Carcinogenesis*. **9**, 259-263.
4. Locker, J., Reddy, T. V. and Lombardi, B. (1986). DNA methylation and hepatocarcinogenesis in rats fed a choline devoid diet. *Carcinogenesis*. **7**, 1309-1312.

5. Ghoshal, A. K. and Farber, E. (1984). The induction of liver cancer by dietary deficiency of choline and methionine without added carcinogens. *Carcinogenesis*. **5**, 1367-1370.
6. Mikol, Y. B., Hoover, K. L., Creasia, D. and Poirier, L. A. (1983). Hepatocarcinogenesis in rats fed methyl-deficient, amino acid-defined diets. *Carcinogenesis*. **4**, 1619-1629.
7. Newberne, P. M. and Rogers, A. E. (1986). Labile methyl groups and the promotion of cancer. *Ann. Rev. Nutr.* **6**, 407-432.
8. Lombardi, B., Chandar, N. and Locker, J. (1991). Nutritional model of hepatocarcinogenesis. Rats fed choline-devoid diet. *Dig. Dis. Sci.* **36**, 979-984.
9. Rogers, A. E. and Newberne, P. M. (1980). Lipotrope deficiency in experimental carcinogenesis. *Nutr. Cancer*. **2**, 104-112.
10. Yokoyama, S., Sells, M. A., Reddy, T. V. and Lombardi, B. (1985). Hepatocarcinogenic and promoting action of a choline-devoid diet in the rat. *Cancer Res.* **45**, 2834-2842.
11. Ghoshal, A. K., Ahluwalia, M. and Farber, E. (1983). The rapid induction of liver cell death in rats fed a choline-deficient methionine-low diet. *Am. J. Pathol.* **113**, 309-314.
12. Chandar, N., Amenta, J., Kandala, J. C. and Lombardi, B. (1987). Liver cell turnover in rats fed a choline-devoid diet. *Carcinogenesis*. **8**, 669-673.
13. Banni, S., Salgo, M. G., Evans, R. W., Corongiu, F. P. and Lombardi, B. (1990). Conjugated diene and trans fatty acids in tissue lipids of rats fed an hepatocarcinogenic choline-devoid diet. *Carcinogenesis*. **11**, 2053-2057.
14. Wainfan, E. and Poirier, L. A. (1992). Methyl groups in carcinogenesis: effects on DNA methylation and gene expression. *Cancer Res.* **52**, 2071s-2077s.
15. Albright, C. D., Bethea, T. C., da Costa, K.-A. and Zeisel, S. H. (1995). Choline deficiency induces apoptosis in SV40-immortalized CWSV-1 rat hepatocytes in culture. *submitted*.
16. Dizik, M., Christman, J. K. and Wainfan, E. (1991). Alterations in expression and methylation of specific genes in livers of rats fed a cancer promoting methyl-deficient diet. *Carcinogenesis*. **12**, 1307-1312.
17. Rushmore, T., Lim, Y., Farber, E. and Ghoshal, A. (1984). Rapid lipid peroxidation in the nuclear fraction of rat liver induced by a diet deficient in choline and methionine. *Cancer Lett.* **24**, 251-5.
18. Zeisel, S. H. (1993). Choline phospholipids: signal transduction and carcinogenesis. *FASEB J.* **7**, 551-557.
19. Berridge, M. J. (1989). Inositol trisphosphate, calcium, lithium, and cell signaling. *JAMA*. **262**, 1834-1841.
20. Taylor, C. W. and Marshall, I. (1992). Calcium and inositol 1,4,5-trisphosphate receptors: A complex relationship. *Trends Biochem Sci*. **17**, 403-407.
21. Exton, J. H. (1990). Signaling through phosphatidylcholine breakdown. *J. Biol. Chem.* **265**, 1-4.
22. Merrill, A. H., Jr. and Schroeder, J. J. (1993). Lipid modulation of cell function. *Annu. Rev. Nutr.* **13**, 539-559.
23. Nishizuka, Y. (1992). Intracellular signaling by hydrolysis of phospholipids and activation of protein kinase C. *Science*. **258**, 607-614.
24. Stabel, S. and Parker, P. J. (1991). Protein kinase C. *Pharmac. Ther.* **51**, 71-95.
25. Nishizuka, Y. (1986). Studies and perspectives of protein kinase C. *Science*. **233**, 305-312.
26. Weinstein, I. B. (1990). The role of protein kinase C in growth control and the concept of carcinogenesis as a progressive disorder in signal transduction. *Adv. Second Messenger Phosphoprotein Res.* **24**, 307-316.
27. Blusztajn, J. K. and Zeisel, S. H. (1989). 1,2-*sn*-diacylglycerol accumulates in choline-deficient liver. A possible mechanism of hepatic carcinogenesis via alteration in protein kinase C activity? *FEBS Lett.* **243**, 267-270.
28. da Costa, K.-A., Garner, S. C., Chang, J. and Zeisel, S. H. (1995). Effects of prolonged (1 year) choline deficiency and subsequent refeeding of choline on 1,2,-sn-diradylglycerol, fatty acids and protein kinase C in rat liver. *Carcinogenesis,* **16**, 327-334.
29. Khan, W. A., Blobe, G. C. and Hannun, Y. A. (1992). Activation of protein kinase C by oleic acid. *J. Biol. Chem.* **267**, 3605-3612.
30. Khan, W. A., Blobe, G., Halpern, A., Taylor, W., Wetsel, W. C., Burns, D., Loomis, C. and Hannun, Y. A. (1993). Selective regulation of protein kinase C isoenzymes by oleic acid in human platelets. *J. Biol. Chem.* **268**, 5063-5068.
31. Lester, D. S., Collin, C., Etcheberrigaray, R. and Alkon, D. L. (1991). Arachidonic acid and diacylglycerol act synergistically to activate protein kinase C in vitro and in vivo. *Biochem. Biophys. Res. Comm.* **179**, 1522-1528.
32. Shinomura, T., Asaoka, Y., Oka, M., Yoshida, K. and Nishizuka, Y. (1991). Synergistic action of diacylglycerol and unsaturated fatty acid for protein kinase C activation: Its possible implications. *Proc. Natl. Acad. Sci., USA*. **88**, 5149-5153.

33. Yoshida, K., Asaoka, Y. and Nishizuka, Y. (1992). Platelet activation by simultaneous actions of diacylglycerol and unsaturated fatty acids. *Proc. Natl. Acad. Sci. USA.* **89**, 6443-6446.
34. Nishizuka, Y. (1986). Perspectives on the role of protein kinase C in stimulus-response coupling. *J. Natl. Cancer Inst.* **76**, 363-370.
35. Price, B. D., Morris, J. D., Marshall, C. J. and Hall, A. (1989). Stimulation of phosphatidylcholine hydrolysis, diacylglycerol release, and arachidonic acid production by oncogenic *ras* is a consequence of protein kinase C activation. *J. Biol. Chem.* **264**, 16638-16643.
36. Diaz-Laviada, I., Larrodera, P., Diaz-Meco, M., Cornet, M. E., Guddal, P. H., Johansen, T. and Moscat, J. (1990). Evidence for a role of phosphatidylcholine-hydrolysing phospholipase C in the regulation of protein kinase C by *ras* and *src* oncogenes. *Embo J.* **9**, 3907-3912.
37. da Costa, K.-A., Hannun, Y. A. and Zeisel, S. H. (1995). Nuclear protein kinase C in livers of choline deficient rats. *in preparation.*
38. Kato, M., Kawai, S. and Takenawa, T. (1989). Defect in phorbol acetate-induced translocation of diacylglycerol kinase in *erb*B-transformed fibroblast cells. *FEBS Lett.* **247**, 247-250.
39. Johansen, T., Bjorkoy, G., Overvatn, A., Diaz-Meco, M. T., Traavik, T. and Moscat, J. (1994). NIH 3T3 cells stably transfected with the gene encoding phosphatidylcholine-hydrolyzing phospholipase C from *Bacillus cereus* acquire a transformed phenotype. *Mol. Cell. Biol.* **14**, 646-654.
40. Wilkison, W. O., Sandgren, E. P., Palmiter, R. D., Brinster, R. L. and Bell, R. M. (1989). Elevation of 1,2-diacylglycerol in ras-transformed neonatal liver and pancreas of transgenic mice. *Oncogene.* **4**, 625-628.
41. Wolfman, A., Wingrove, T. G., Blackshear, P. J. and Macara, I. G. (1987). Down-regulation of protein kinase C and of an endogenous 80-kDa substrate in transformed fibroblasts. *J. Biol. Chem.* **262**, 16546-16552.
42. Megidish, T. and Mazurek, N. (1989). A mutant protein kinase C that can transform fibroblasts. *Nature.* **342**, 807-811.
43. Krauss, R., Housey, G., Johnson, M. and Weinstein, I. B. (1989). Disturbances in growth control and gene expression in a C3H/10T1/2 cell line that stably overproduces protein kinase C. *Oncogene.* **4**, 991-998.
44. Persons, D. A., Wilkison, W. O., Bell, R. M. and Finn, O. J. (1988). Altered growth regulation and enhanced tumorigenicity of NIH 3T3 Fibroblasts transfected with protein kinase C-1 c DNA. *Cell.* **52**, 447-458.
45. Cacace, A. M., Guadagno, S. N., Krauss, R. S., Fabbro, D. and Weinstein, I. B. (1993). The epsilon isoform of protein kinase C is an oncogene when overexpressed in rat fibroblasts. *Oncogene.* **8**, 2095-2104.
46. Abanobi, S. E., Lombardi, B. and Shinozuka, H. (1982). Stimulation of DNA synthesis and cell proliferation in the liver of rats fed a choline-devoid diet and their suppression by phenobarbital. *Cancer Res.* **42**, 412-5.
47. Giambarresi, L. I., Katyal, S. L. and Lombardi, B. (1982). Promotion of liver carcinogenesis in the rat by a choline-devoid diet: role of liver cell necrosis and regeneration. *Brit. J. Cancer.* **46**, 825-9.
48. Harris, C. C. and Sun, T. (1984). Multifactoral etiology of human liver cancer. *Carcinogenesis.* **5**, 697-701.
49. Bursch, W., Kleine, L. and Tenniswood, M. (1990). The biochemistry of cell death by apoptosis. *Biochem Cell Biol.* **68**, 1071-1074.
50. Wyllie, A. H. (1987). Cell death. *Int. Rev. Cytol.* **17 (Suppl)**, 755-785.
51. Arends, M. J., Morris, R. G. and Wyllie, A. H. (1990). Apoptosis: the role of endonuclease. *Am. J. Pathol.* **136**, 593-608.
52. Murgia, M., Pizzo, P., Sandona, D., Zanovello, P., Rizzuto, R. and Virgilio, F. D. (1992). Mitochondrial DNA is not fragmented during apoptosis. *J. Biol. Chem.* **267**, 10939-10941.
53. Wijsman, J. H., Jonker, R. R., Keijzer, R., Vande Velde, C. J. H., Cornelisse, C. J. and Dierendonck, J. H. V. (1993). A new method to detect apoptosis in paraffin sections: in situ end-labeling of fragmented DNA. *J. Histochem. Cytochem.* **41**, 7-12.
54. Woodworth, C. D., Secott, T. and Isom, H. C. (1986). Transformation of rat hepatocytes by transfection with simian virus 40 DNA to yield proliferating differentiated cells. *Cancer Res.* **46**, 4018-4026.
55. Bligh, E. G. and Dyer, W. J. (1959). A rapid method of total lipid extraction and purification. *Can. J. Biochem. Physiol.* **37**, 911-917.
56. Pomfret, E. A., daCosta, K., Schurman, L. L. and Zeisel, S. H. (1989). Measurement of choline and choline metabolite concentrations using high-pressure liquid chromatography and gas chromatography-mass spectrometry. *Analyt. Biochem.* **180**, 85-90.

7

DIETARY EFFECTS ON GENE EXPRESSION IN MAMMARY TUMORIGENESIS

Polly R. Etkind

Departments of Oncology and Medicine
Montefiore Medical Center and
The Albert Einstein College of Medicine
Bronx, New York, 10467

I. INTRODUCTION

Although a large number of epidemiological studies in humans[1-6] and dietary experiments in rats[7-14] and in mice[15-19] have been published on the relationship of dietary fat to breast cancer incidence, this important health question has not been resolved and has become instead increasingly controversial. Animal studies have strongly suggested that there is a mammary tumor promoting effect of diets high in total fat (approximately 20% fat by weight or 40% fat in calories, which is similar to the American diet) as compared to diets low in total fat (approximately 5% fat by weight or 10% fat in calories, which is similar to the Japanese diet)[20]. In the animal studies, diets with a greater proportion of unsaturated fats (specifically corn oil) were more effective than diets rich in saturated fats in increasing mammary tumor incidence and/or reducing the latent period of mammary tumor appearance[21-23]. The epidemiological data in humans have been for the most part based on comparison of breast cancer incidence and fat consumption in different countries worldwide[5]. Recently, however, two prospective epidemiological studies[24,25] involving a large cohort of women have suggested a lack of evidence for a relationship between a high fat diet and human breast cancer for middle-aged and post- menopausal women. These studies did not, however, rule out that dietary fat may have an effect during childhood and adolescence. Also, differences in breast cancer risk due to a low fat diet may be manifested only at fat levels far below what is considered low fat in a Western diet.

These conflicting results have contributed to the uncertainties in this complex and important health issue of the relationship of dietary fat to breast cancer. What is needed to resolve these conflicting results is an understanding of how dietary fat affects the incidence and progression of breast cancer *in vivo* at the molecular/genetic level. However, studies to determine the effects of a high fat diet at the molecular/genetic level are difficult to design because in both the human system as well as in most mouse models the mechanisms by which tumor initiation and progression take place are unknown, and therefore the steps at which high levels of unsaturated fatty acids may act to increase the incidence of breast tumors remain a matter of conjecture.

Diet and Cancer, Edited under the auspices of the American Institute of Cancer Research
Plenum Press, New York, 1995

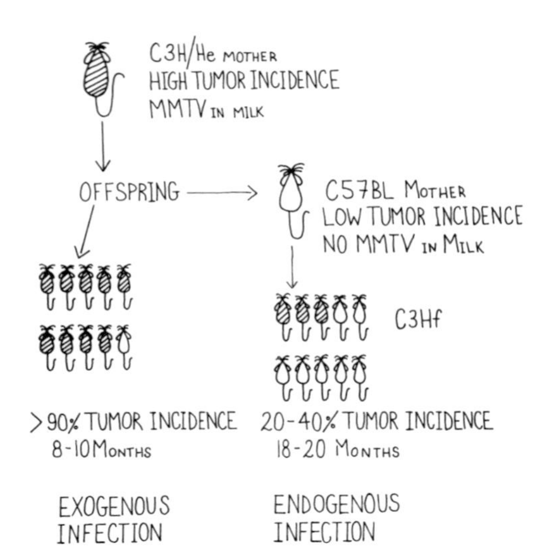

Figure 1. Mouse mammary tumorigenesis can be the result of either exogenous or endogenous MMTV infection. The exogenous virus is passed from mother to offspring via the milk and at eight to ten months of age ninety percent of the female offspring develop breast tumors. Mice foster-nursed on mothers whose milk does not contain MMTV develop breast tumors at an incidence of twenty to forty percent at eighteen to twenty months of age. These late developing mammary tumors are the result of infection by endogenous virus which is produced by activation of expression of MMTV proviral DNA present at the Mtv-1 locus in the C3Hf mouse genome.

However, in the C3H Heston foster-nursed mouse (C3Hf mouse) the initial event of mammary tumorigenesis is known. This initial event is the hormonally controlled expression of a single Mouse Mammary Tumor Virus (MMTV) endogenous proviral DNA present at the Mtv-1 locus[26-28]. The C3Hf mouse is thus a unique and important in vivo model for determining the effect of diet on specific gene expression in the pathway to mammary tumorigenesis. Steps in mammary tumorigenesis subsequent to the initial activation of the Mtv-1 locus are also known for the C3Hf mouse.

Although discovered as an exogenous agent of mammary carcinoma[29], Mouse Mammary Tumor Virus (MMTV) is also transmitted vertically as endogenous proviruses present in the germline of all inbred mouse strains such as the C3H Heston[28,30]. The exogenous virus is usually passed from mother to offspring via the milk and at eight to ten months of age ninety percent of the female offspring develop breast tumors[29,31] (Figure 1). Foster-nursing of C3H/He offspring on low mammary tumor incidence strains of mice (i.e. the C57BL mouse whose milk does not contain MMTV) results in a much lower incidence (20-40 %) and later development (18-20 months) of mammary tumors in the C3Hf mouse[26] (Figure 1).

The late occurring mammary tumors of the C3Hf mouse which develop independent of exogenous MMTV infection are the result of the expression, after a number of pregnancies, of the MMTV endogenous proviral DNA located at the Mtv-1 locus[28,32,33]. Although the C3Hf mouse contains 3 copies of the MMTV genome at different locations in the cell genome, it is only the MMTV proviral sequence located at the Mtv-1 locus which becomes transcribed as a result of pregnancy, and which is responsible for tumor formation[32]. C3Hf mice who do not go through pregnancy do not transcribe the Mtv-1 locus, do not make endogenous MMTV, and do not develop mammary tumors. Human DNA contains full-length proviral-like sequences which are homologous to MMTV endogenous proviral DNA sequences[34-36]. What role these human MMTV-like endogenous retroviral-like sequences may play in human breast cancer is unknown at the present time. Nevertheless, the C3Hf mouse represents an important model of specific gene activation initiating the process of mammary tumorigenesis.

Specific steps in breast tumor development in the C3Hf mouse are outlined in Table 1. The transcription of the specific MMTV proviral DNA present at the Mtv-1 locus results in the presence of MMTV specific RNA transcripts in second parity lactating mammary glands

Table 1. Specific steps in breast tumor development in the C3Hf Mouse

1. Transcription of the MMTV proviral DNA present at the Mtv-1 locus into 9.0 and 3.8 Kb RNAs for endogenous virus production.
2. Production of endogenous MMTV which infects the mouse mammary gland.
3. Reverse transcription in the MMTV infected mammary gland cells of MMTV RNA into proviral DNA for integration into the host cell genome.
4. Expression of the WNT and INT protooncogenes as a result of the integration of MMTV proviral DNA

of the C3Hf mouse[28]. These transcripts are responsible for the production of endogenous MMTV which can be detected in low levels in third parity lactating mammary glands of the C3Hf mouse[28]. Mammary glands of multiparous C3Hf mice as well as mammary tumors of these mice contain newly integrated MMTV proviral DNA at new sites in the mammary gland DNA of the same mouse as a result of infection by the newly synthesized endogenous MMTV[37]. Activation of WNT-1 and/or INT-2 protooncogenes which occurs in these mice as a result of the new MMTV proviral DNA integrations in the domain of these protooncogenes is an early step in mammary tumor development in these mice[38].

The C3Hf mouse, in which the initial and early events of mammary tumorigenesis are understood on a molecular/genetic level, thus offers a unique model system in which one can study the effects of a high fat diet on the activation of specific genes in vivo. Using this model we have studied the effect of a high fat diet (23.5% corn oil) on the expression of the endogenous proviral DNA sequence present at the Mtv-1 locus, the locus whose transcription is responsible for mammary tumor development in the C3Hf mouse. We have also determined the effect of a high fat diet on the number and latency period before development of C3Hf mouse mammary tumors which are the result of endogenous MMTV infection. Current work in my laboratory is determining the effect of a high fat diet on the expression in C3Hf mouse mammary tumors of the WNT-1 and INT-2 protooncogenes whose transcription is an early step in C3Hf mouse mammary tumorigenesis. We have previously shown that the percent of tumors expressing either WNT-1 and/or INT-2 differs between those tumors caused by exogenous MMTV infection and those tumors caused by endogenous infection in C3Hf mice on low fat diets[38].

II. MATERIALS AND METHODS

At 21 days of age C3Hf mice obtained from our inbred mouse colony were weaned and placed on either a 5% or 23.5 % corn oil diet as detailed in Table 2. The 5% corn oil diet is based on the American Institute of Nutrition standard reference diet with modification of the source of carbohydrate[39]. The 23.5% corn oil diet is based on the recommendations of the committee on Laboratory Animal Diets of the National Academy of Sciences, with minor modifications[40]. The diets were formulated by Dr. Leonard Cohen of the American Health Foundation in Valhalla, New York. All animals were weighed bi-weekly. Breast tumors were characterized by histological examination of tissue at the American Health Foundation.

Mice were allowed to breed and were sacrificed by carbon dioxide inhalation at specific parities or when they had developed mammary tumors. After one year of breeding, female mice were retired as breeders and monitored for tumor formation. The lactating mammary glands of first, second, and third parity low fat and high fat diet C3Hf mice and

Table 2. Diet protocol

	Percent Composition of Semipurified Diet	
	5.0% fat	23.5% fat
Ingredients		
Casein, Vitamin-free	20.00	23.50
DL-Methionine	0.30	0.35
Corn Starch	52.00	33.00
Dextrose	13.00	8.30
Alphacel	5.00	5.90
Fat (corn oil)	5.00	23.50
Mineral Mix, AIN	3.50	4.10
Vitamin Mix, AIN	1.50	1.70
Choline Bitartrate	0.20	0.24
Caloric Density (Kcals/gm)	3.87	4.72

breast tumors of high fat and low fat diet C3Hf mice were removed surgically after the mice were sacrificed. All 5 pairs of mammary glands were removed, pooled, weighed, and immediately processed for RNA. RNA was isolated from mammary glands and breast tumors by the quanidine isothiocyanate procedure[38,41].

For qualitative analysis of MMTV specific RNA transcripts in the mammary glands of first, second, and third parity C3Hf mice on either a 5% or 23.5% corn oil diet, RNA was fractionated by electrophoresis in 1.2% agarose gels containing 2.2 M formaldehyde, transferred to nitrocellulose by the Northern blotting technique[42], and hybridized under stringent hybridization conditions to MMTV specific radiolabeled probes[43]. Stringent hybridization conditions included prehybridization for 5 hours in 50% formamide, 3X SSC (20x SSC = 3M NaCl,0.3M NaCitrate), 5x Denhardt's solution[44], 300 μg/ml single-stranded salmon sperm DNA, and 0.1 % SDS. Hybridization to P-32 labeled MMTV specific probes was in the same hybridization buffer for 40 hours at 42^0 C. All RNA blots were also hybridized to the radiolabeled single copy gene probe Glyceraldehyde-3-Phosphate Dehydrogenase (GAPDH) to determine that equal amounts of RNA were present in each gel lane used for comparison studies.

Statistical analysis included the use of Fisher's Exact test to evaluate the differences between the number of high fat and low fat diet C3Hf mice expressing the Mtv-1 locus at first parity. The mean ages and litter number of high and low fat diet C3Hf mice who developed mammary tumors were compared using non-parametric Wilcoxon Rank Sum test. The difference in breast tumor incidence between high and low fat diet groups was adjusted using Chi-Square test with continuity adjustment.

III. RESULTS

Mammary glands of multiparous C3Hf mice and mammary tumors of C3Hf mice produce the MMTV specific RNAs of 9.0 and 3.8 kb which result from transcription of the Mtv-1 locus[32,38]. The 9.0 kb transcript serves as the genome for progeny virus and also as messenger RNA which encodes the viral specific gag and gag-pol polypeptide precursors[31]. The 3.8 kb MMTV species is a spliced transcript that encodes only the virion envelope glycoproteins. A 1.7 kb MMTV specific RNA which is always present in C3Hf mammary glands (from first parity) is presumably a spliced transcript that contains sequences derived

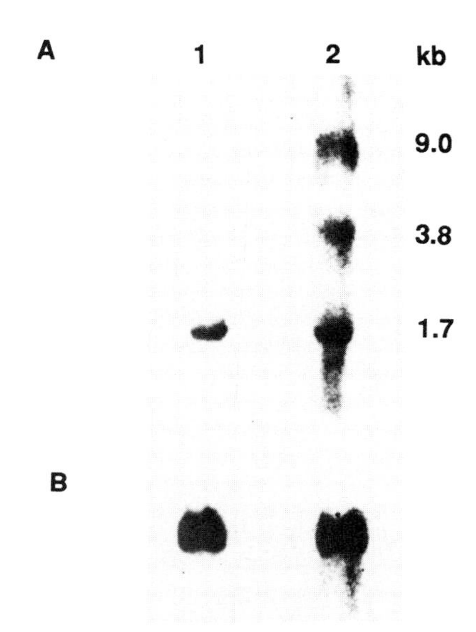

Figure 2. Northern blot of RNA isolated from mammary glands of first parity C3Hf mice on either the low fat diet (lane 1) or high fat diet (lane 2) hybridized to 32-P-labelled MMTV specific probe (panel A) or 32-P-labelled single copy gene probe Glyceraldehyde-3-Phosphate Dehydrogenase (GAPDH), (panel B).

from the long terminal repeat region of an MMTV provirus and is believed to play a role as an endogenous mouse superantigen[45].

Although always present in mammary glands of C3Hf mice at second pregnancy, the 9.0 and the 3.8 kb MMTV specific transcripts are never seen in lactating mammary glands of C3Hf mice before this time. Second parity C3Hf mice contain these MMTV specific transcripts in their lactating mammary glands, while third parity mice contain increasing amounts of these transcripts as well as newly synthesized endogenous MMTV[28,31]. All lactating mammary glands which we have examined of first, second, and third parity C3Hf mice on low fat diets have exhibited this pattern of MMTV specific transcription.

In contrast, however, our examination of the mammary glands of 10 first parity C3Hf mice on the high fat 23.5 percent corn oil diet revealed that 6 of these 10 first parity mice on the high fat diet contained in addition to the 1.7 kb transcript both the 9.0 and 3.8 kb MMTV specific RNA in their lactating mammary glands at first parity. Figure 2, lane A2, represents the typical pattern which we saw in each of the six high fat diet C3Hf mice which showed accelerated transcription of the 9.0 and 3.8 Kb Mtv-1 specific transcripts. We have examined the mammary glands of 15 first parity mice on the low fat diet and all 15 contained only the 1.7 kb MMTV specific RNA as represented in Figure 2, lane A1. Figure 2, lanes B1 and B2, hybridized to the GAPDH single copy gene probe show that equal amounts of RNA are being compared between lanes 1 and 2. Statistical evaluation using the Fisher's Exact test showed the difference in the expression of the Mtv-1 locus at first parity between high and low fat diet mice to be significant ($p < 0.01$).

In addition to the earlier detection of MTV-1 specific transcripts in the lactating mammary glands of C3Hf mice on a high fat diet, mammary tumors also developed earlier and after fewer litters in these mice. Using the non-parametric Wilcoxon Rank Sum test the difference between the mean age at tumor development between high fat and low fat diet mice and the mean number of litters at time of tumor development in the high fat versus low fat diet mice were highly significant ($p < 0.0001$ and < 0.01, respectively). The average age for tumor development in the C3Hf mice on a high fat diet was 11 months, while C3Hf mice on the low fat diet developed mammary tumors at an average age of 17.8 months, which is typical for C3Hf mice. The average number of litters born to each mouse before palpable mammary tumors were detected was 2.1 for the mice on the high fat diet and 4.2 for the mice on the low fat diet. One nulliparous C3Hf mouse on the high fat diet developed a mammary tumor at the age of 6 months. The RNA isolated from the mammary tumor of the nulliparous mouse contained the 9.0, 3.8, and 1.7 kb MMTV specific RNA. C3Hf mice on low fat diets have never been reported to develop mammary tumors before undergoing pregnancy.

Mice on the high fat diet had an incidence of mammary tumor development of 45 percent while the incidence for those on the low fat diet was 35 percent. Using the Chi-Square test with continuity adjustment no statistical significance was found for the incidence of tumors which developed between the two diet groups because of the low numbers of breast tumors which were evaluated. No statistically significant differences in body weight were noted between the diet groups.

IV. DISCUSSION

Our data demonstrate that a high fat diet containing 23.5% corn oil accelerates the expression of the MMTV proviral DNA sequence present at the Mtv-1 locus in C3Hf mice. The expression of this locus, which is normally under hormonal control, is the initial step in C3Hf mouse mammary tumor development. In addition, C3Hf mice on a high fat diet developed mammary tumors earlier and after fewer litters than those mice on the low fat diet.

These results show that a high fat diet can act at the molecular level via effects on gene transcription, and in this case specifically hormonally controlled gene transcription. The question as to whether or not the mammary tumorigenic activities of high fat diets act at the initiation stage or promotional stage (or both) of this tumorigenic process has been addressed by several laboratories[46]. While it is clear from a number of studies that high levels of dietary fat can act at the promotion stage of mammary tumorigenesis [23], it is not resolved whether or not high fat diets can influence the initiation phase of this tumorigenic process. Our data in which 1)the initiating step in mammary tumorigenesis (transcription of the Mtv-1 locus) is accelerated in C3Hf mice on a high fat diet and 2)mamary tumors develop earlier and after fewer litters in mice on the high fat diet demonstrate that high levels of dietetic fat can influence the initiating stage of mouse mammary tumorigenesis.

Transcription of the MMTV proviral DNA is initiated from the promoter region present in the Long Terminal Repeat Regions or LTRs of the proviral DNA[47]. Recent work by DeWille et al has shown that a 25 % corn oil diet can increase v-HA-ras gene mRNA levels in transgenic mice carrying this gene under control of the MMTV LTR[48]. The MMTV LTR contains signals for the initiation and termination of RNA transcription as well as binding sites for the steroid hormones within the Hormone Responsive Region[47]. Transcription of RNA initiated from the promoter region within the LTR of MMTV proviral DNA is subject to both positive (hormone) and negative regulation[49]. With our recent data that high levels of polyunsaturated fats can accelerate hormonally controlled transcription of the Mtv-1 locus in the C3Hf mouse and current knowledge of MMTV RNA transcription, we can begin to study the specific molecular mechanisms by which a high fat diet influences hormonally controlled gene expression.

V. SUMMARY

Studies were undertaken to determine the effect a high fat diet has on the the hormonally controlled transcription of the Mtv-1 locus in C3Hf mice. The expression of this locus is the initiating event in mammary tumor development in the C3Hf mouse. Mice were weaned at 21 days to either a high fat diet containing 23.5 percent corn oil or to a low fat diet containing 5 percent corn oil. Mice were sacrificed at first, second, and third parity, or when they had developed mammary tumors, and their mammary glands and mammary tumors were isolated. RNA was isolated from all mammary glands and breast tumors and analyzed.

The high fat diet accelerated the transcription of the Mtv-1 locus. The transcripts of this locus, which are never seen in C3Hf mouse mammary glands until second parity, were present in first parity mammary glands of 6 out of 10 high fat diet C3Hf mice which were studied. The mammary glands of 15 first parity C3Hf mice which were on the low fat diet were analyzed and none contained the Mtv-1 specific transcripts. In addition, mammary tumor development was detected earlier (11 vs. 17.8 months) and after fewer litters (2.1 vs. 4.2) on the average in high fat diet C3Hf mice. One C3Hf mouse on the high fat diet developed a breast tumor at six months without going through pregnancy.

These results indicate that a high fat diet of 23.5 percent corn oil can accelerate hormonally controlled gene expression specifically linked to mammary tumorigenesis. This work suggests that the mechanism by which mammary tumors develop earlier and after fewer litters in C3Hf mice on a high fat diet is the accelerated transcription of the hormonally controlled MMTV provirus present at the Mtv-1 locus.

ACKNOWLEDGMENTS

This work was supported by Grants 91B11 and 93B22 from the American Institute for Cancer Research, Washington, D.C. The author wishes to thank Lucy Qiu and Kathryn Lumb for excellent technical assistance and Xiaoping Hu for statistical analysis. The author also wishes to thank Dr. Leonard Cohen for his expert help in designing the nutritional experiments and Dr. Leonard Augenlicht for his continual interest and encouragement in these studies.

REFERENCES

1. P. Buell, Changing incidence of breast cancer in Japanese- American women, J. Natl. Cancer Inst. 51:1479 (1973).
2. G.E. Gray, M.C. Pike, and B.E. Henderson, Breast cancer incidence and mortality rates in different countries in relation to known risk factors and dietary practices, British J. Cancer 39:1 (1979).
3. T. Hirayama, Epidemiology of breast cancer with special reference to the role of diet, Prev. med. 7:173 (1978).
4. A. Nomura, B.E. Henderson, and J. Lee, Breast cancer and diet among the Japanese in Hawaii, Am. J. Clin. Nutr. 31:2020 (1978).
5. R.L. Prentice and L. Sheppard, Dietary fat and cancer:consistency of the epidemiologic data and disease prevention that may follow from a practical reduction in fat consumption, Cancer Causes Control 1:81 (1990).
6. E. Wynder and T. Hirayama, Comparative epidemiology of cancers of the United States and Japan, Prev. Med. 6:567 (1977).
7. C. Aylsworth, C. Jone, J. Trosko, J. Meites, and C. Welsch, Promotion of 7,12-dimethylbenz(a)anthracene-induced mammary tumorigenesis by high dietary fat in the rat : possible role of intracellular communication, J. Natl. Cancer Inst. 72:637 (1984).
8. K. Carroll and H. Khor, Effects of dietary fat and dose level of 7,12 dimethylbenz(a)anthracene on mammary tumor incidence in rats, Cancer Res. 30:2260 (1979).
9. K. Carroll and H. Khor, Effects of level and type of dietary fat on incidence of mammary tumors induced in female Spraque-Dawley rats by 7,12-dimethylbenz(a)anthracene, Lipids 6:415 (1971).
10. P. Chan and L. Cohen, Dietary fat and growth promotion of rat mammary tumors, Cancer Res. 35:3384 (1975).

11. P. Chan and T. Dao, Effects of dietary fat on age-dependent sensitivity to mammary carcinogenesis, Cancer Lett. 18:245 (1983).
12. P. Chan, F. Didado, and L. Cohen, High dietary fat, elevation of rat serum prolactin and mammary cancer, Proc. Soc. Exp. Biol. Med. 149:133 (1975).
13. L. A. Cohen, K. Choi, J.H. Weisburger, and D.P. Rose, Effect of varying proportions of dietary fat on the development of N-nitrosomethylurea-induced rat mammary tumors, Anticancer Research 6:215 (1986).
14. L. A. Cohen, D.O. Thompson, K. Choi, R. Karmali, and D.P. Rose, Dietary fat and mammary cancer. II Modulation of serum and tumor lipid composition and tumor prostaglandin by different dietary fats:Association with tumor incidence patterns, J Natl. Cancer Inst. 77:43 (1986).
15. D. Gridley, J. Kettering, J. Slater, and R. Nutter, Modification of spontaneous mammary tumors in mice fed different sources of protein, fat, and carbohydrate, Cancer Lett. 19:133 (1983).
16. P. Pennycuik, A. Fogarty, M. Willcox, M. Ferris, R. Baxter, and A. Johnson, Tumor incidence, growth, reproduction, and longevity in female C3H mice fed polyunsaturated ruminant- derived foodstuffs, Aust J. Biol. Sci. 32:309 (1979).
17. A. Tannenbaum, The genesis and growth of tumors. III. Effect of a high-fat diet, Cancer Res. 2:468 (1942).
18. A. Tannenbaaum and H. Silverstone, Nutrition in relation to cancer, Adv. Cancer Res. 1:451 (1954).
19. S. Waxler, G. Brecher, and S. Beal, The effect of fat- enriched diet on the incidence of spontaneous mammary tumors in obese mice, Proc. Soc. Exp. Biol. Med. 162:365 (1979).
20. J. Silverman, J. Powers, P. Stromberg, J.A. Pultz, and S. Kent, Effects on C3H mouse mammary cancer of changing from a high fat to a low fat diet before, at, or after puberty, Cancer Res. 49:3857 (1989).
21. K. Carroll and G.J. Hopkins, Dietary polyunsaturated fat versus saturated fat in relation to mammary carcinogenesis, Lipids 14:155 (1979).
22. G.J. Hopkins, G.C. Hard, and C.E. West, Carcinogenesis induced by 7,12-dimethylbenz(a)anthracene in C3H-A VyfB mice : influence of different dietary fats, J. Natl. Cancer Inst.60:849 (1978).
23. G.A. Rao and S. Abraham, Enhanced growth rate of transplanted mammary adenocarcinoma induced in C3H mice by dietary linoleate, J. Natl. Cancer Inst. 56:431 (1976).
24. P.A. vandenBrandt, P. vant'Veer, R. Alexandra Goldbohm, E. Dorant, A. Volovics, R.J.J. Hermus, and F. Sturmans, A prospective cohort study on dietary fat and the risks of postmenopausal breast cancer, Cancer Res. 53:75 (1993).
25. W.C. Willett, D.J. Hunter, M.J. Stampfer, G. Colditz, J.E. Manson, D. Spiegelman, B. Rosner, C.H. Hennekens, and F.E. Spiezer, JAMA 268:2037 (1992).
26. H.B. Andervont, The influence of foster nursing upon the incidence of spontaneous mammary cancer in resistant and susceptible mice, J. Natl. Cancer Inst. 1:147 (1940).
27. S. Nandi and C.M. McGrath, Mammary neoplasia in mice, Advances in Cancer Research 17:353 (1973).
28. R. van Nie and A.A. Verstraeten, Studies of genetic transmission of mammary tumor virus by C3Hf mice, Int. J. Cancer 16:922 (1975).
29. J.J. Bittner, Some possible effects of nursing on the mammary gland tumor incidence in mice, Science 84:162 (1936).
30. J.C. Cohen and H.E. Varmus, Endogenous mammary tumor virus DNA varies among wild mice and segregates during inbreeding, Nature (London) 278:418 (1979).
31. D.H. Moore, C.A. Long, A.B. Vaidya, J.B. Sheffield, A.S. Dion, and E.Y. Lasfargues, Mammary tumor viruses, Advances in Cancer Research 29:347 (1979).
32. R. Michalides, R. vanNie, R. Nusse, N.E. Hynes, and B. Groner, Mammary tumor induction loci in GR and DBAf mice contain one provirus of the mouse mammary tumor virus, Cell 23:165 (1981).
33. R. vanNie, A.A. Verstraeten, and J. deMoes, Genetic transmission of mammary tumor virus by GR mice, Int J. Cancer 19:383 (1977).
34. R. Callahan, W. Drohan, S. Tronick, and J. Schlom, Detection and cloning of human DNA sequences related to mouse mammary tumor virus genome, Proc. Natl. Acad. Sci. 79:5503 (1982).
35. F.E.B. May, B. Westley, H. Rochefort, E. Buetti, and H. Diggelman, Mouse mammary tumor virus related sequences are present in human DNA, Nucleic Acids Research 11:4127 (1983).
36. M. Ono, T. Yasunaga, T. Miyata, and H. Ushikuto, Nucleotide sequence of human endogenous retrovirus genome related to the mouse mammary tumor virus genome, J. Virol. 60:589 (1986).
37. J.C. Cohen and H.E. Varmus, Proviruses of mouse mammary tumor virus in normal and neoplastic tissues from GR and C3Hf mouse strains. J. Virol. 35:298 (1980).
38. P.R. Etkind, Expression of the int-1 and int-2 loci in endogenous mouse mammary tumor virus-induced mammary tumorigenesis in the C3Hf mouse, J. Virol. 63:4972 (1989).
39. J.G. Bieri, Second report of the Ad Hoc committee on standards for nutritional studies, J. Nutr. 110:1726 (1980).

40. A.E. Rogers, G.H. Anderson, G.M. Lenhardt, G. Wolfe, and D. Newberne, A semisynthetic diet for long term maintenance of hamsters to study effects of dietary vitamin A, Lab. Anim. Sci. 24:495 (1974).
41. J.M. Chirgwin, A.E. Pryzbyla, R.J. MacDonald, and W.J. Rutter,Isolation of biologically active ribonucleic acid from sources enriched in ribonuclease, Biochemistry 18:5294 (1979).
42. P. Thomas, Hybridization of denatured RNA and small DNA fragments transferred to nitrocellulose, Proc. Natl. Acad. Sci. USA 77:5201 (1980).
43. R.W. Rigby, M. Dieckmann, C. Rhodes, and P. Berg, Labelling deoxyribonucleic acid to high specific activity in vitro by nick translation with DNA polymerase I, J. Mol. Biol. 113:237 (1977).
44. D.T. Denhardt, A membrane-filter technique for the detection of complementary DNA, Biochem. Biophys. Res. Commun. 23:641 (1966).
45. P. Marrack, E. Kushnir, and J. Kappler, A maternally inherited superantigen encoded by a mammary tumor virus, Nature 349:524 (1991).
46. C.W. Welsch, Interrelationship between dietary fat and endocrine processes in mammary gland tumorigenesis, in: Dietary Fat and Cancer, A. Rogers, D. Birt, E. Mettlin, and C. Ip, eds. Alan R. Liss,Inc., New York (1986),p.623.
47. H. Punta, W.H. Gunzburg, B. Salmons, B. Grover, and P. Herrlich, Mouse mammary tumor virus:a proviral gene contributes to the understanding of eucaryotic gene expression and mammary tumorigenesis, J. Gen. Virol. 66:931 (1985).
48. J.W.Dewille, K.Waddell, C. Steinmeyer, and S.J. Farmer, Dietary fat promotes mammary tumorigenesis in MMTV/v-Ha-ras transgenic mice, Cancer Letters 69:59 (1993).
49. S.R. Ross, C.-L.L. Hsu, Y. Choi, E. Mok, and J.P. Dudley, Negative regulation in correct tissue-specific expression of mouse mammary tumor virus in transgenic mice, Mol. and Cell Biol. 10:5822 (1990).

8

EFFECT OF DIETARY FATTY ACIDS ON GENE EXPRESSION IN BREAST CELLS

Z. Ronai,[1] J. Tillotson,[1] and L. Cohen[2]

[1] Molecular Carcinogenesis Program
[2] Section of Nutrition and Endocrinology
American Health Foundation
Valhalla, New York

INTRODUCTION

At present there is considerable controversy surrounding the role of dietary fat in breast cancer (1,2). Epidemiological, anthropological, and animal model studies support the concept that high fat intake increases breast cancer risk (3-6). However, several cohort studies, in consistent with such an hypothesis have recently been reported (1,7). In addition, attention has focused on the role of specific fatty acids (FA) as enhancers, (i.e. linoleic acid (LA)(C18:2, n-6) (9-11)) inhibitors, (i.e. eicosapentaenoic acid (EIA)(C20:5, n-3) (12,13)) or neutral effectors, (i.e. oleic acid (OL)(C18:1, n-9) (8,14)) of breast cancer development and metastatic spread. One weakness of the fat hypothesis is that the mechanism(s) by which the type and/or amount of fat may exert their effects on breast cancer are poorly understood (6,15). In this regard, the effects of specific nutrients, including lipids, on gene expression have recently attracted attention (16-19). In light of the controversy surrounding the fat hypothesis, elucidation of the cellular and molecular mechanisms by which dietary fat exerts its biological effects, may provide the means to resolve this important public health issue.

In a previous study, one of us (L.A. Cohen) reported that diets including 12.5% corn oil and 12.5% menhaden oil, which contains high levels of n-3 FA such as docosahexaenoic acid (DHA 22:6, n-3) or eicosapentaenoic acid (EPA 20:5, n-3) suppress the formation of chemically-induced mammary tumors compared to diets containing pure corn oil (12). These observations prompted us to explore possible molecular mechanisms by which n-3 and n-6 fatty acids exert their inhibitory effects on tumor growth and incidence. Our working hypothesis was that exposure to certain dietary fatty acids would result in altered gene expression which in turn may affect cell cycle control, cell proliferation, and/or DNA repair capacity; the latter could also yield different frequencies of genomic alterations. Accordingly, in the present study, we have explored mechanisms by which specific dietary fatty acids may exert their effects at the genomic level using both *in vivo* and *in vitro* breast cancer model systems.

Diet and Cancer, Edited under the auspices of the American Institute of Cancer Research
Plenum Press, New York, 1995

MATERIALS AND METHODS

In Vivo Feeding Studies

Thirty female inbred F344 rats (Charles River Breeding Laboratories, Wilmington MA)(28 d old) were maintained on the standard NIH-07 diet for 7 d. Two groups of 15 animals each were then fed AIN semi-purified diets containing either high fat (HF); 23% corn oil, CO) or 11.5% menhaden oil and 11.5% CO (see ref 20 for details). At day 50 of age, animals were injected with N-nitrosomethylurea (NMU) (Ash Stevens, Detroit MI) (37.5 ng/kg) via the tail vein.

DNA Preparations

At 3,5,9 and 13 wk 2-6 animals in each group were sacrificed and each mammary fat pad was removed and kept at -80°C. DNA preparation was performed using DNA lysing buffer followed by standard phenol-chloroform extraction and ethanol precipitation.

Amplification of H-*ras* Sequences

Specific oligonucleotides spanning a 102 bp region of rat H-*ras* gene were used for standard amplification procedure. To enrich the fraction of mutant *ras* alleles genomic DNA was digested with Mnl I restriction endonuclease. Analysis after PCR amplification was performed following RE digest and separation of fragment on PAGE (detailed in ref 20).

IN VITRO CELL CULTURE STUDIES

Cell Culture

NMU cells (ATCC CRL 1743) were maintained in RPMI-1640 supplemented with 10% FBS and antibiotics. MCF 7 cells were maintained in MEM supplemented with bovine insulin and 10% FBS.

Cell Cycle Analysis

Cells were collected by trypsinization and fixed in 75% ethanol prior to staining with DAPI. Fluorescence of individual cells was measured with an ICP-22 flow cytometer.

Metabolic Labeling

Cells (10^4/well) were plated in 24-well plates and allowed to attach overnight in medium containing 10% FBS. Experimental medium was added and cells were incubated for specified time period in methiohine or phosphate free medium before ^{35}S methionine or ^{32}P-orthophosphate (1μCi) were added for 3 h. Cell monolayer was than washed, lysed and centrifuged at 100,000 g for 20 min (See ref 21 for details).

Immunoprecipitation

Supernatants obtained from equal number of cells was incubated with 1.5 μg of the specified antibody to p53 (Pab 421 to aa 370-378, and Pab DO1 to aa - 37-45; Oncogene

Science) overnight at 4°C followed by precipitation using protein G beads. Immunoprecipitated proteins were washed, eluted in SDS-sample buffer and loaded on 10% SDS-PAGE. The gels were dried and the radioactivity in the p53 band was quantified with the aid of a computerized radioimaging system.

RESULTS

In Vivo Studies

Effect of FA on NMU Induced ras Gene Mutations. To test the possibility that certain dietary fatty acids could affect frequency of mutations at the genomic level we have tested whether dietary FA alter the frequency of *ras* gene mutations in mammary glands of MNU treated rats. As such, the incidence of the pre-malignant lesions which could ultimately develop into tumors was expected to be reduced.

Dietary FA could exert such an effect via either the change in growth rate, cell cycle distribution, or through their ability to alter the activities of detoxification enzymes or of repair genes. In all cases the number of DNA adducts are expected to be lower, which should be reflected in the number of mutations. Some supporting evidence already existed for such hypothesis. First, n-6 FA increases proliferation in the ductal tree which would lead to enhanced ability to fix mutations after genetic damage by carcinogen. Therefore, a slower rate of proliferation at this stage may be critical for the ability to reduce "fixation" of mutations (22). In a study conducted by Karmali et al, the effect of these FA was tested on DMBA treated F344 rats. When the rats were pre-fed with the n-6 FA, a significantly higher incidence of *ras* mutation was identified when compared with the animals that were pre-fed with the n-3 FA (23,24).

As a model to test the genomic part of our hypothesis we have adopted a well established animal model system; the ability of NMU to induce rat mammary tumors in Fisher 344 rats. Single injection of NMU results in multiple mammary tumors within 20-25 weeks (see Figure 1), of which 70% were found to harbor H-*ras* codon 12 mutation. As n-3 FA in the form of menhaden oil (combined with corn oil in a 50%-50% mixture) was potent in inhibiting tumor number and incidence (9,12,25,26), this model was also tested for possible changes in *ras* oncogene mutations.

To enable proper analysis of potential effects of LA/CO diet on the frequency of *ras* mutation we were interested to determine the mutant allele frequency in fat pads and "normal appearing" mammary gland tissues at various time points after carcinogen administration, and not only in the mammary tumors. The analysis of *ras* mutation in the normal mammary tissues required the use of a highly sensitive method, since the expected frequencies of *ras* oncogene mutations at the early stages of the mammary carcinogenesis process were expected to be very low. This could be achieved via the use of a method which we have developed in our laboratory, namely the *Enriched PCR* (Figure 2).

The *Enriched PCR* enables detection of low frequent mutations which are characteristic to non-neoplastic lesions. The concept behind this approach is that by eliminating as much as possible the normal alleles of the test gene, the mutant fraction which was originally a minor component, is enriched, and can be preferentially amplified. This, in turn, allowed us to achieve a sensitivity of 10^{-4} (thus, detecting one mutant allele in the presence of 10^{4} normal ones). Briefly, this two stage procedure utilizes a specifically designed primers to amplify the genomic DNA fragment of interest. In the case of the H-*ras* gene of rat, the genomic DNA is first cleaved with MnlI which recognizes the normal but not mutant sequence of codon 12. As a result, the mutant will be preferentially available for amplifica-

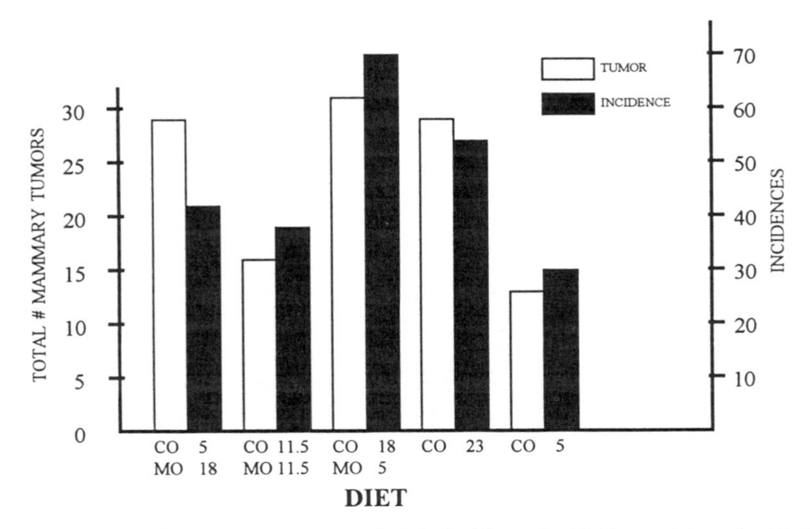

Figure 1. Total number of mammary tumors and their incidence in F344 rats injected with NMU and maintained on 5 different diets. (Figure was adapted from Ref 20).

tion. The amplified fragment consist of a 102 bp region which upon cleavage with MnlI, allows to distinguish between the normal and the mutant alleles via an ethidium bromide staining.

The results of our study are summarized in Table 1, which indicates that there is no difference in the frequency of *ras* mutation between the animals that were fed different diets. Mutations found at similar frequencies at 3 weeks or later regardless of the type of fat used

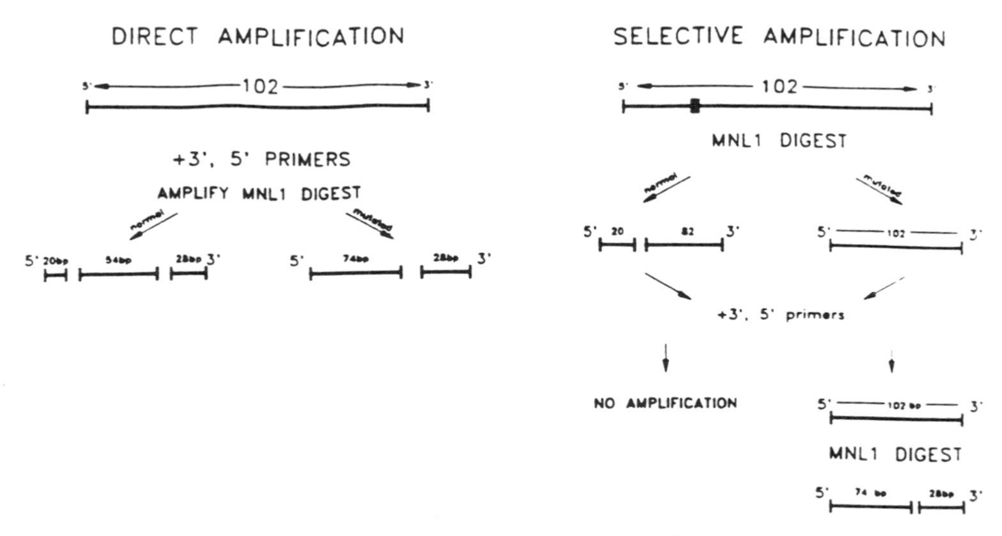

Figure 2. Outline of the Enriched/Selective PCR concept. Genomic DNA is first digested with Mnl which renders only mutant H-*ras* codon 12 alleles available for PCR amplification. The amplified fragment is 102 bp which will generate a characteristic 74 bp band for mutant and 54 bp for normal sequences. (Figure was adapted from Ref 20).

Table 1. Incidence of *ras* mutations in NMU-treated rats fed different dietary fats

Fat type	Amount	Mammary glands with *ras* mutations/total glands tested (%)	Rats with at least one *ras* mutation/ total rats tested (%)
CO	23%	18/44 (41)	11/14 (79)
CO/MO	11.5%/11.5%	18/37 (49)	10/15 (67)

CO, corn oil; MO, menhaden oil.
(Table adapted from ref 20).

in the study (which were sufficient to cause changes of at least 50% in final tumor incidence). This indicates that type of fat does not influence rate of *ras* oncogene mutation and therefore that H-*ras* mutation may not participate in the pathway by which FA alter tumor incidence. Many more mutations were found than the expected number of final tumors, indicating that *ras* mutation is not sufficient for tumor formation. Additional events are required at later times to promote the growth of these *ras*-containing foci. However, in light of recent studies reported by Cha et al (27), *ras* may not have been the right gene to have used for such analysis. Using a highly sensitive method, namely MAMA, (mismatched amplification mutation assay) which enables detection of one mutant allele in 10^5 normal ones, it was demonstrated that H-*ras* codon 12 mutation are pre-existing in normal mammary tissues of F344 rats. Furthermore, the frequency of the mutation increases following NMU administration not due to the direct mutgenetic effect of NMU but, rather, due to the high proliferation of the mammary tissue after NMU administration. The latter may provide an explanation of our ability to detect mutations in greater number of fat pads than the expected number of mammary tumors, as they may have been part of the background mutation. While this newly available information could explain our observation (i.e. no change in the frequency of mutant *ras* allele between different diets) it also points to the possibility that genes which are not pre-existing in a mutant form are those that should be examined in this model. For example, genetic instability reflected by changes in micro satellite repeats (di-tri nucleotide repeats; 28,29).

In Vitro Studies

Effect of Dietary Fatty Acids at the Epigenetic Levels. The second part of our working hypothesis consisted of studies at the epigenetic level. We hypothesized that changes at the epigenetic level could include alteration in expression and activities of key regulatory proteins which would result in altered cell cycle and growth rate. To test this possibility we have used relevant cell culture systems, including rat mammary tumor cell line isolated from NMU-induced tumor in rats (NMU cell line) and human breast cancer derived cell line (MCF-7) as well as a normal human breast cell line (184B5). Our first measurement aimed at identifying possible changes in cell growth when the NMU cells were maintained in each of the different test diets. As shown in Figure 3, a concentration dependent increase in cell growth was noticed with LA supplemented to medium with 1% serum. In contrast, DHA has an opposite effect as it has significantly reduced the number of cells grown within a 6 day culture period (Figure 3). These results were consistent with the dietary effects observed in the animal model (19,22-26).

Changes in growth rate may reflect changes in cell cycle control, which have prompted us to examine possible effects of each of these dietary fatty acids on cell cycle distribution. Shown in Figure 4, when LA (1%) was added to 1% serum it was capable of increasing fraction of cells in the S phase from 23.4 to 31.9% (a 40% increase).

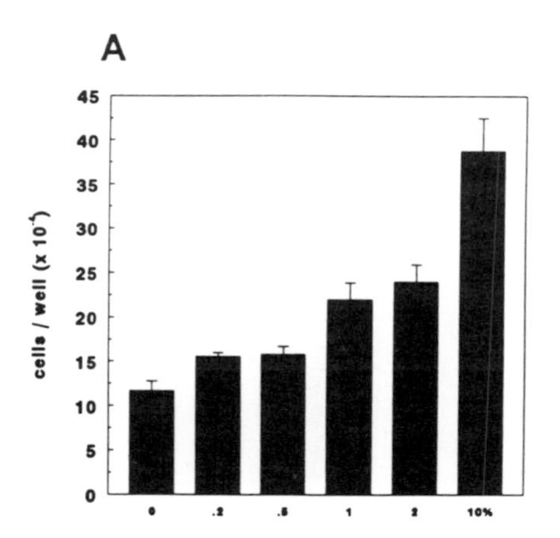

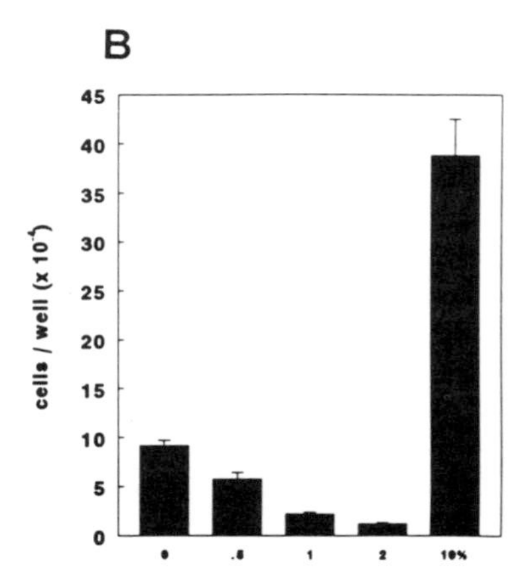

Figure 3. Growth rate of NMU cell lines after 6 days in culture of the indicated concentration of LA (A) or DHA (B) supplemented with 1% serum containing medium. (Figure adapted from Ref 21).

Examination of the effects of FA on incorporation of precursors into DNA, RNA and protein was also undertaken. In agreement with the cell cycle analysis, DNA synthesis was decreased in the low serum medium, but the effect was partially reversed by the addition of FA, especially LA (see ref 21 for details). Taken together, these data demonstrate that linoleic acid is capable of preventing or reversing the G1 cell cycle accumulation that occurs in NMU cells upon withdrawal of serum components. Since it has been shown that G1 block can be induced by increased p53 levels in response to various environmental conditions, we tested whether FA effects are mediated through p53. Analysis of p53 was first performed in the NMU cells. The half life of p53 in the NMU cells was first determined to be 110 min (21). The latter is substantially longer than the half life of wild type p53 proteins, which led us to examine if the p53 contain mutations in its cDNA, yet, none were found (unpublished observations). More recent studies indicate that in mammary cells the half life of the p53 protein is substantially longer than it is in other tissue types (30). Furthermore, cellular proteins such as mdm2 were shown capable of extending the half life of the p53 protein (31).

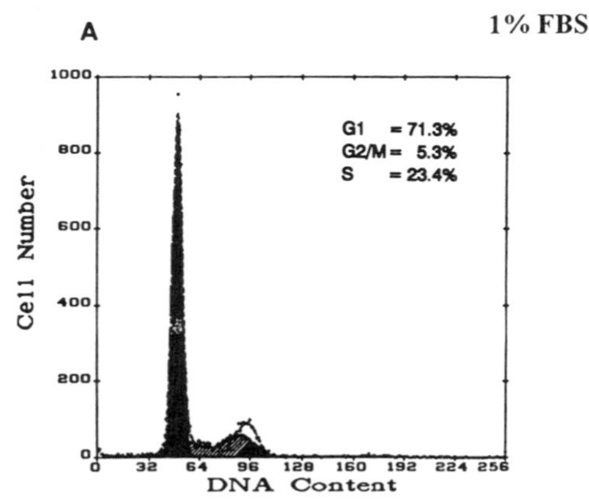

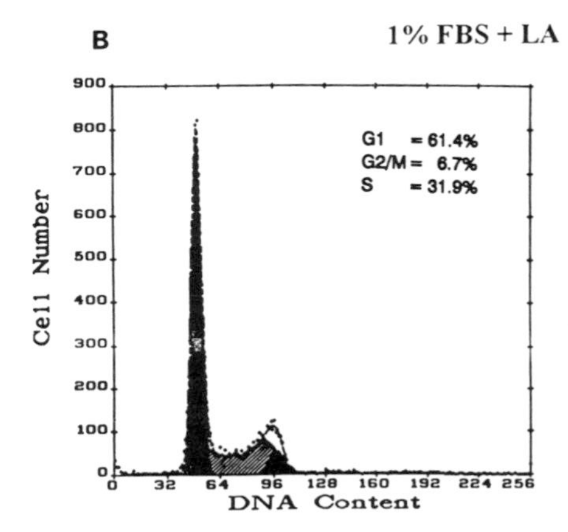

Figure 4. Cell cycle distribution of NMU cell line maintained for 2 days in medium supplemented with 1% FBS or with 1% FBS + LA (1 μg/ml). (Figure adapted from Ref 21).

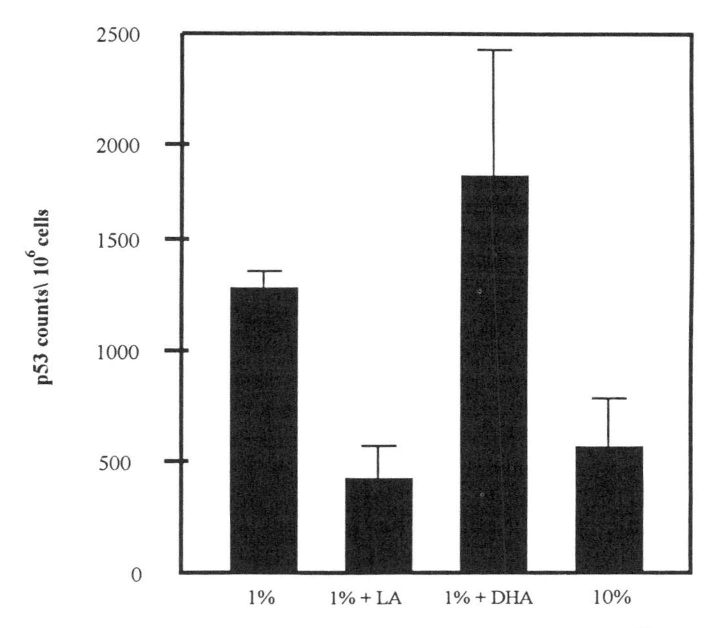

Figure 5. Immunoprecipitation of p53 using Pab 421 from NMU cells that were ^{32}S-methionine labeled following culture for 3 days in the indicated media. (Adapted from Ref 21).

To determine the level of p53 in cells exposed to different FA we have labeled the cellular proteins with ^{35}S and performed immunoprecipitations (IP) using antibodies (Pab 421) directed to aa 370-378 within the p53 gene. The immunoprecipitated material was than separated on SDS-PAGE and the 53 kDa band was quantified with the aid of a radioimaging blot analyzer. Accordingly we cultured the NMU cell line (derived from an NMU-induced tumor) in media with supplements of purified n-3 or n-6 FA.

As demonstrated in Figure 5, NMU cells maintained for 3 days in low serum expressed significantly increased levels of p53 protein as determined via IP with antibody Pab 421. Cells supplemented with LA expressed p53 levels similar to that noticed with high serum. Interestingly, the level of p53 protein in cells maintained in DHA was even higher than that observed with low serum medium. This demonstrates that p53 expression as determined with Ab421 correlates with the growth characteristics noticed in the MNU cells maintained in the different culture media. Additional studies in the human cell MCF 7 (Figure 6) revealed similar pattern when pAb 421 was used.

The availability of antibodies to different forms and regions of p53 (mutant; wt etc) prompted us to perform the above mentioned immunoprecipitated with a different antibody to p53, Pab - D01, which was raised to an N-terminal epitope aa 37-45). To our surprise, the latter antibodies were not able to reproduce the pattern observed with the Pab 421. Repeated experiments that were performed using each of the two antibodies (DO1 and 421) on proteins prepared from MCF 7 and from the normal breast 184B5 cells revealed that Pab while 421 Ab reproduced the pattern shown above in MCF 7 cells, Do1 revealed a different pattern (Figure 7). These differences pointed to the possibility that the discrepancy in the immuno-precipitable pattern generated by the different Ab could reflect certain post translational modifications.

The possible involvement of a post-translational modification in p53 due to the dietary fatty acid treatment was studied using a ^{32}P-orthophosphate labeling of cells main-

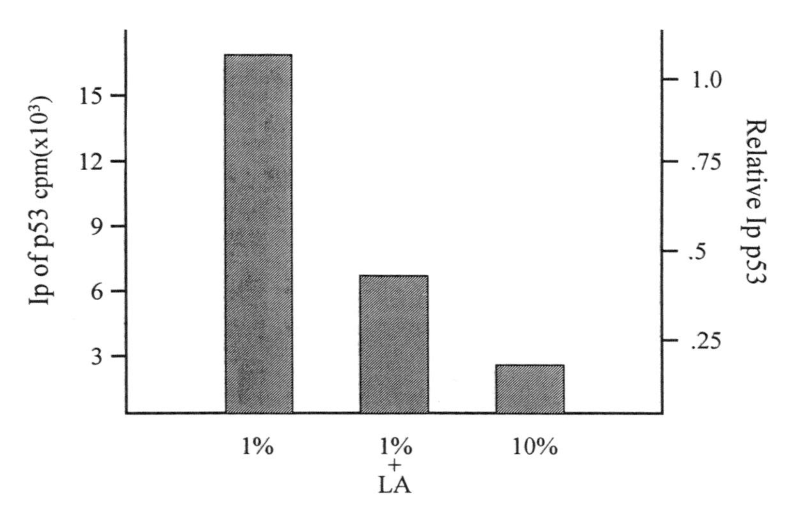

Figure 6. Immunoprecipitation of p53 using Pab 421 from MCF-7 human breast cancer cells. Experiment was performed as indicated in Figure 5.

tained in each of the diets followed by immuno-precipitation and gel analysis. As shown in Figure 8, increased level of ^{32}P-labeled proteins that were immuno-precipitated with antibodies to p53 was noted after LA or OA treatment in both MCF7 and 184B5 cells. The elevated amounts of immunoprecipitated p53 were noticed with both Pab 421 and Pab DO1, which revealed different patterns in our ^{35}S-labeling experiment. This suggest that a small fraction of the overall synthesized protein undergoes phosphorylation following growth in certain dietary fatty acids which are detected only via the ^{32}P-labeling (Figure 8). Nevertheless, even with ^{32}P-labeling we have still observed differences in the amount of immunoprecipitated p53 after DHA treatment, suggesting that DHA may involve a different from of

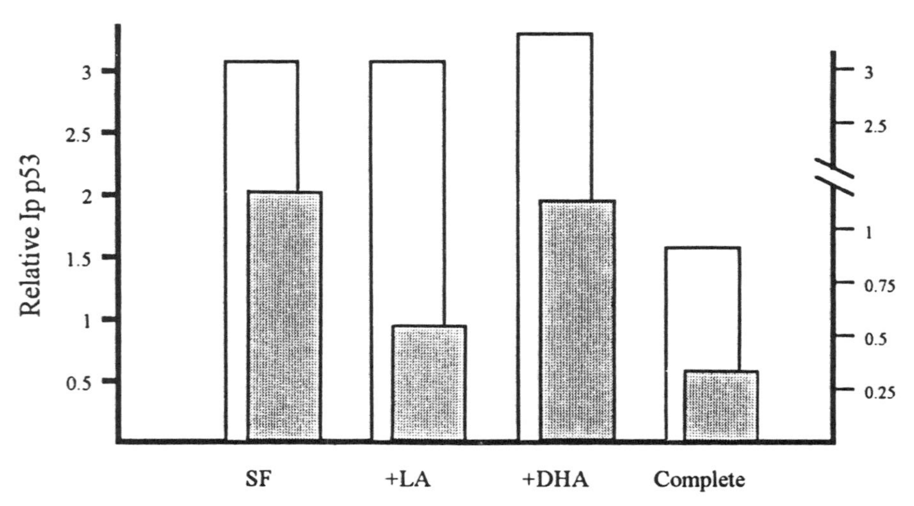

Figure 7. Immunoprecipitation of p53 using Pab 421 (closed bars) or Pab Do1 (empty bars) from MCF-7 cells maintained in the indicated media for 3 days. ^{32}S-labeling was performed for 2 h prior to protein preparations. Values on right axis indicate cpm of immunoprecipitable p53 x 10^3.

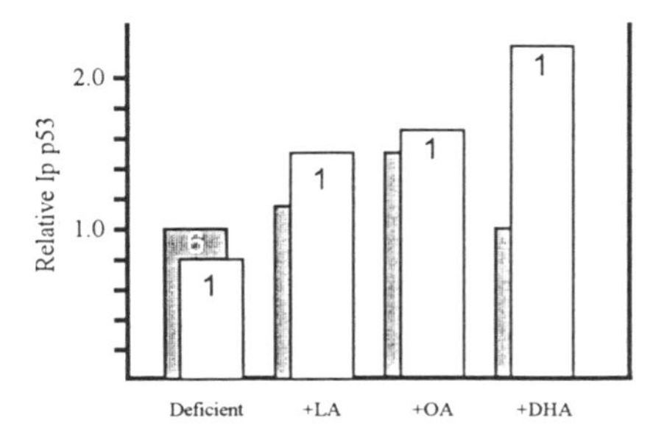

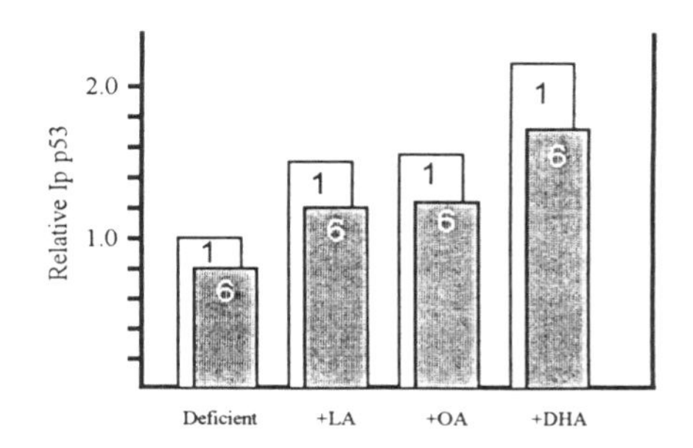

Figure 8. Immunoprecipitation of ^{32}P-labeled proteins from MCF-7 (left) or 184B5 (right) cells maintained in media containing the indicated fatty acids, using Pab 421 (Ab-1; empty bars) or Pab Do1 (Ab-6; closed bars).

post-translational modification. Further investigations should be performed to pinpoint the nature of each of these changes.

Overall, human breast cancer cells are growth stimulated by linoleic acid (n-6), but not by n-3 FA, in a mechanism which could involve p53 control on cell cycle. That changes in the phosphorylation of p53 affects cell cycle control was recently demonstrated (32) and may be implicated in its ability to associate with Bcl 2 (33) which was shown to mediate apoptosis. While the significance of this to *in vivo* tumor growth and progression is yet to be established, these changes could represent part of complex cellular response to dietary feeding which modulates cell cycle and growth rate. The possible effect on other important cellular functions such as DNA repair is yet to be explored.

Finally, it should be noted that a large proportion of human breast cancers appear to have alterations in the p53 gene, including point mutations, loss of heterozygosity, or presence of increased cellular p53 protein levels without obvious mutation of the gene. It will be important to understand the effects of FA administration to cells in each of these categories. For example, in experiments with cultured human cancer cells similar to the ones reported here, Rose and Connolly have demonstrated that many, but not all, human breast and prostate tumor cell lines will proliferate in the presence of LA (34,35). The highest proliferation was seen in cell lines which appear to have point mutations in the p53 gene, and no wild-type p53 protein. Does the mutated p53 protein enhance activity of LA in inducing proliferation? Does the lack of wild-type p53 protein prevent p53-mediated apoptosis which might normally occur during FA-stimulated growth, allowing such growth to be enhanced? Such questions are raised by the current studies, and will be answered following completion of the respective experiments.

EPILOG

The studies toward understanding molecular mechanisms by which dietary fatty acids can mediate their *in vivo* capability to reduce tumor incidence are at the earliest stages. Substantially more efforts needed to allow better understanding of such mechanisms. Our studies pointed to the possible role of p53 phosphorylation, which could mediate cell cycle distribution and altered cell growth, yet, the mechanisms involved in such phosphorylation (i.e. respective protein kinases) in normal vs transformed cells as well as *in vivo* mammary gland obtained cells, are yet to be explored. Futhermore, the multifunctional role of p53 in cell growth is also anticipated to yield additional changes to the cell at the level of transcription and repair. In spite of our inability to identify a different frequency of *ras* mutation in n-6 and n-3 fed animals it is likely that certain genomic alterations, or their rate, would be selectively modified by specific types of FA as tested here. Instability of microsatellite repeats may be one target for such investigation. Finally, future studies will need to address relevance to *in vivo* situations not only in rat but other animal models which best represent the human conditions.

REFERENCES

1. E.L. Wynder, L.A. Cohen, D.P. Rose, and S.D. Stellman, Dietary fat and breast cancer: Where do we stand on the evidence, *J. Clin. Epi.* 47:217-222 (1994).
2. E.L. Wynder, L.A. Cohen, D.P. Rose, and S.D. Stellman, Response to Dr. Walter Willett's Dissent. *J. Clin. Epi.* 47:227-230 (1994).
3. L.A. Cohen, Diet and cancer, *Scientific American* 247:42-49 (1987).
4. S. Sasaki, M. Horacsek, R. Kesteloot, An ecological study of the relationship between dietary fat intake and breast cancer mortality, *Preventive Med.* 22:187-202 (1993).
5. P. Toniolo, E. Riboli, R. Shore, B.S. Pasternack, Consumption of meat, animal products, protein and fat and risk of breast cancer: A prospective cohort study in New York, *Epidemiology* 5:391-397 (1994).
6. L.A. Cohen, Dietary fat and mammary cancer. *In*: Diet, Nutrition and Cancer: A Critical Evaluation. B.S. Reddy and L.A. Cohen (eds.), Vol. 1, Macronutrients and Cancer. CRC Press, Boca Raton, FL, pp 77-100 (1986).
7. P.A. van de Brandt, P. van-Veer, A. Goldbohm, E. Dorant, A. Volovics, R.J.J. Hermus, and F. Sturmans, A prospective cohort study on dietary fat and the risk of postmenopausal breast cancer, *Cancer Res.* 53:77-82 (1993).
8. L.A. Cohen, D.O. Thompson, W. Choi, R. Karmali, and D.P. Rose, Dietary fat and mammary cancer. I. Promoting effects of different dietary fats on N-nitrosomethylurea-induced rat mammary tumorigenesis, *J. Natl. Cancer Inst.* 77:33-42 (1986).
9. E.B. Katz, and E.S. Boylan, Stimulatory effect of a high polyunsaturated fat diet on lung metastasis from the R-13762 mammary adenocarcinoma in female retired breeder rats, *J. Natl. Cancer Inst.* 79:351-358 (1987).
10. N.E. Hubbard, and K.L. Erickson, Enhancement of metastasis from a transplantable mouse mammary tumor by dietary linoleic acid, *Cancer Res.* 47:6171-6175 (1987).
11. J.M. Connolly, and D.P. Rose, Effects of fatty acids on invasion through reconstituted basement membrane ("Matrigel") by a human breast cancer cell line. *Cancer Letts,* 75:137-142 (1993).
12. L.A. Cohen, J.-Y. Chen-Backlund, D. Sepkovic, and S. Sugie, Effect of varying proportions of dietary menhaden and corn oil on experimental rat mammary tumor promotion, *Lipids* 28:449-456 (1993).
13. G. Fernandéz, and J.T. Venkatraman, Role of omega-3 fatty acids in health and disease, *Nutrition Res.* 13 (suppl 1):519-545 (1993).
14. L.A. Cohen, and E.L. Wynder, Do dietary monounsaturated fatty acids play a protective role in carcinogenesis and cardiovascular disease? *Medical Hypoth.* 31:83-89 (1990).
15. L.A. Cohen, D.O. Thompson, K. Choi, R. Karmali, and D.P. Rose, Dietary fat and mammary cancer. II. Modulation of serum and tumor lipid composition and tumor prostaglandins by different dietary fats associated with tumor incidence patterns, *J. Natl. Cancer Inst.* 77:43-51 (1986).
16. C.D. Berdanier, and J.C. Hargrove, Nutrition and Gene Expression, CRC Press Inc., Boca Raton, FL (1994).
17. N.T. Telang, A. Basu, H. Kurihara, M.P. Osborne, and M.I. Modak, Modulation in the expression of murine mammary tumor virus, *ras* proto-oncogene, and of alveolar hyperplasia by fatty acids in mouse mammary explant cultures, *Anticancer Res.* 8:971-976 (1988).
18. J.W. DeWille, K. Waddell, C. Steinmeyer, and S.J. Farmer, Dietary fat promotes mammary tumorigenesis in MMTV/v-Ha-*ras* transgenic mice, *Cancer Letts.* 69:59-66 (1993).
19. E.A. Paisley, J. Kaput, H.J. Mangian, and W.J. Visek, Level of dietary fat and regulation of gene expression in BALB/c mice. Annual Res. Conf., American Inst. Cancer Res., September 1-2, 1994, Washington, DC (Abstr. #24).
20. Z. Ronai, Y. Lau, and L.A. Cohen, Dietary N-3 fatty acids do not affect induction of Ha-ras mutations in mammary glands of NMU-treated rats, *Mol Carcinogen*, 4:120-128 (1991).

21. J.K. Tillotson, Z. Darzynkiewicz, L.A. Cohen, and Z. Ronai, Effects of linoleic acid on mammary tumor cell proliferation are associated with changes in p53 protein expression, *Int J Oncology*, 3:81-87 (1993).
22. C.W. Welsch, Enhancement of mammary tumorigenesis by dietary fat: Review of potential mechanisms, *Am J Clin Nutr*, 45:192-202 (1987).
23. R.A. Karmali, J. Marsh, and C. Fuchs, Effect of omega-3 fatty acids on growth of a rat mammary tumor, *J Natl Cancer Inst*, 73:457-461 (1984).
24. R.A. Karmali, A. Donner, S. Gobel, and T.I. Shimamura, Effect of n-3 and n-6 fatty acids on 7,12-dimethylbenz[a]anthracene-induced mammary tumorigenesis, *Anticancer Res*, 9:1161-1168 (1989).
25. H. Gabor, and S. Abraham, Effect of dietary menhaden oil on tumor cell loss and the accumulation of mass of a transplantable mammary adenocarcinoma in BALB/c mice. *J Natl Cancer Inst*, 76:1223-1229 (1986).
26. R.A. Karmali, Omega-3 fatty acids and cancer: A Review *in:* Lands WEM (ed), Proceedings of the AOCS short course on polyunsaturated fatty acids and eicosanoids. American Oil Chemistry Society, Champaign, IL (1987).
27. R.S. Cha, W.G. Thilly, H. Zarbl et al., N-Nitroso-N-methylurea-induced rat mammary tumors arise from cells with preexisting oncogenic Hras1 gene mutations, *Proc. Natl. Acad. Sci. USA*, 91:3749-3753 (1994).
28. C.J. Yee, N. Roodi, C.S. Verrier, and F.F. Parl, Microsatellite instability and loss of heterozygosity in breast cancer, *Cancer Res*, 54:1641-1644 (1994).
29. R. Wooster, A.M. Cleton-Jansen, N. Collins, J. Mangion, R.S. Cornelis, C.S. Cooper, B.A. Gusterson, B.A. Ponder, A. von Deimling, O.D. Wiestler et al., Instability of short tandem repeats (microsatellites) in human cancers, *Nat Genet*, 6:152-156 (1994).
30. D.M. Barnes, A.M. Hanby, C.E. Gillett, S. Mohammed, S. Hodgson, L.G. Bobrow, I.M. Leigh, T. Purkis, C. MacGeoch, N.K. Spurr et al., Abnormal expression of wild type p53 protein in normal cells of cancer family patient, *Lancet*, 340:259-63 (1992).
31. J.D. Oliner, J.A. Pietonpol, S. Thiagalingam, J. Gyuris, K.W. Kinzler, and B. Vogelstein, Oncoprotein MDM2 conceals the activation domain of tumor suppressor p53, *Nature*, 362:857-860 (1993).
32. M. Fiscella, S.J. Ullrich, N. Zambrano, M.T. Shields, D. Lin, S.P. Lees-Miller, C.W. Anderson, W.E. Mercer, and E. Appella, Mutation of the serine 15 phosphorylation site of human p53 reduces the ability of p53 to inhibit cell cycle progression, *Oncogene*, 8:1519-1528 (1993).
33. S. Haldar, M. Negrini, M. Monne, S. Sabbioni, and C.M. Croce, Down-regulation of bcl-2 by p53 in breast cancer cells, *Cancer Res*, 54:2095-7 (1994).
34. D.P. Rose, and J.M. Connolly, Effects of fatty acids and inhibitors of eicosanoid synthesis on the growth of a human breast cancer cell line in culture, *Cancer Res*, 50:7139-7144 (1990).
35. D.P. Rose, and J. Connolly, Stimulation of growth of human breast cancer cell lines in culture by linoleic acid, *Biochem Biophys Res Comm*, 164:277-283 (1989).

9

LIPOTROPE DEFICIENCY AND PERSISTENT CHANGES IN DNA METHYLATION

Lipotrope Deficiency and DNA Methylation

Judith K. Christman

Department of Biochemistry and Molecular Biology
University of Nebraska Medical Center
Omaha, Nebraska

INTRODUCTION

Dietary deficiency of lipotropes (choline, methionine, folate, vitamin B_{12}) has long been associated with increased incidence of tumors. Copeland and Salmon first reported that long-term feeding of choline deficient diet increased incidence of tumor in the liver and other organs of the rat almost 50 years ago[1]. Subsequent studies demonstrated that the peanut meal they used in the diet was contaminated with aflatoxin[2], and it was soon established that a variety of different diets deficient in choline and/or other lipotropes acted to promote carcinogen induced tumor development in the liver and other organs[3-10]. Later, through feeding of amino-acid defined diets tested for absence of exogenous carcinogens[11], it was possible to prove that lipotrope deficiency alone was sufficient to cause development of tumors in the liver. However, the spectrum of tumors induced by lipotrope deficiency was much narrower than that seen in lipotrope-deficient animals treated with carcinogen and the diet actually seemed to inhibit appearance of some spontaneously arising tumors[11].

The major question as to what underlying mechanism(s) account for the promoting and carcinogenic activity of lipotrope deficient diets remains unanswered. In part, this is due to the complexity involved in identifying which of the many effects of lipotrope deficiency are necessary for induction of tumor development. Lipotropes play a central role in maintaining the levels of both folate-derived one carbon groups and the levels of the primary methyl donor in transmethylation reactions, S-adenosylmethionine, AdoMet.

Because the one-carbon pathways are so interrelated, diets low in methionine and choline can lead to depletion of total folates and severe depletion of folate can lead to decreased levels of AdoMet and ratios of AdoMet/AdoHcy[12-16]. Decreased levels of folates and/or AdoMet have the potential to interfere with DNA replication and repair[17-20], RNA, protein, phospholipid and polyamine synthesis[17,21-24]. They can also affect the rate and extent of methylation of DNA, RNA, proteins, including histones, p21 ras and other membrane associated proteins, and neurohumoral agents[25-27]. Depending on the tissue, these changes could lead either to inhibition or to aberrant stimulation of cell division. In the liver of rats

Diet and Cancer, Edited under the auspices of the American Institute of Cancer Research
Plenum Press, New York, 1995

fed diets deficient in choline and other lipotropes, the most immediate and obvious change observed is accumulation of lipid. Thus, several hypotheses as to the causal factor in hepatocarcinogenesis stress the importance of lipid accumulation. It has been proposed that tumor formation is initiated through damage to DNAs caused by lipid peroxides[28-30]. Promotion of initiated cells could then occur through the mitogenic stimulus provided by recurring cycles of hepatocyte death and replacement[31,32]. Increases in protein kinase C activity due to accumulation of diradyl glycerol and free fatty acids could be important in stimulating cell division and altering gene activity[33]. Finally, failure to maintain levels and patterns of DNA methylation due to reduced levels of AdoMet or decreased ratios of AdoMet/AdoHcy [34-37] could allow enhanced transcription of genes involved in stimulating cell growth or expression of a transformed phenotype. In our studies[37-41], we have attempted to identify which of these events are necessary for oncogenic transformation on the basis of three criteria:

1. They occur early during feeding of a lipotrope deficient diet, i.e. before formation of preneoplastic foci;
2. They persist after restoration of adequate levels of lipotropes to the diet and
3. They are characteristically present in preneoplastic foci and liver tumors[11]. Obviously, this is a reductionist approach, since it is clear that multiple genetic and epigenetic changes are necessary for the progression of normal cells to malignancy and that there may be alternate pathways to the same end. Thus, although it is highly likely that changes that meet all three criteria are of critical importance in tumor formation, it is also possible that some changes necessary in the early stages of progression need not persist and that some changes may only important for the later stages of tumor development.

STABLE CHANGES IN HEPATOCYTES INDUCED BY LIPOTROPE DEFICIENCY

There are a number of reports suggesting that lipotrope deficiency can cause changes in rat liver that are critical for tumor development that are not reversed when normal levels of lipotropes are restored to the diet. Although the minimum period of lipotrope deficiency required for induction of liver tumors in rats remains to be determined, 15 weeks feeding of an amino acid defined (AAD) diet supplemented with homocystine but lacking choline, methionine and vitamin B_{12} (MDD) was sufficient to lead to development of neoplastic nodules in the liver which persisted and grew in size during the next 37 weeks, despite feeding of an adequate diet[42]. Similar reports of persistence and progression of hyperplastic nodules in livers of rats fed different choline-deficient or choline-devoid diets also suggested that at least some tumorigenic changes in the liver have become fixed within as short a time as 3 months[43,44].

The metabolic and biochemical changes that occur within this time frame in livers of rats fed MDD or MDD supplemented with 5 mg/kg folic acid and 100 μg/kg vitamin B_{12} can be summarized as follows:

- Increased rate of DNA synthesis (≤ 1 week)[37,40]
- Increased numbers of mitotic cells (≤ 1 week)[40,42]
- Increased lipid accumulation (≤ 4 days)[40,42]
- Increased activity of DNA MTase (1-2 weeks)[39,41]
- Increased activity of tRNA MTase (1-2 weeks)[39,41,45,46]
- Increased levels of specific mRNAs (≤ 1 week)[38]

- Decreased levels of AdoMet or decreased ratio of AdoMet/AdoHcy ($\leq$ 1 week)[34]
- Decreased levels of specific mRNAs ($\leq$ 1 week)[38]
- Decreased methylation of DNA (1 week)[37,39]
- Decreased methylation of tRNA (5 days)[37,45,47]

Many of these effects are common to lipotrope deficiencies induced by feeding of other differently formulated choline or choline and methionine deficient diets[48-50]. In general, they persist and increase in magnitude during the first month of lipotrope deficiency. However, after approximately three months, elevation in triglyceride content and the increase in rate of liver cell proliferation are reversed[33,43]. AdoMet/AdoHcy ratios also return to normal although AdoMet levels remain low[35]. The 5mC content of liver DNA also remains low and the same loss of methylation at specific CG sites in *c-myc* and *c-Ha-ras* genes in liver DNA that occur within the first month of lipotrope deprivation is still present in the hepatocellular carcinomas that arise many months later[35,38,51]. These findings suggested that alterations in DNA methylation are likely to play an important role in hepatocarcinogenesis and have led us to investigate whether alterations in DNA methylation also persist when adequate lipotropes are restored to the diet of lipotrope deficient rats.

REVERSAL OF SHORT-TERM LIPOTROPE DEFICIENCY

To determine whether any of the effects we have observed after short-term feeding of MDD persist after restoration of adequate levels of lipotropes to the diet, rats were fed MDD for 4 weeks and then fed a diet adequate in lipotropes [CSD, the same basal AAD diet as SD and MDD but supplemented with 2 g/kg choline chloride, 5.2 g/kg D,L methionine, 5 mg/kg folic acid, and 100 μg/kg vitamin B_{12}]. Livers were then examined histologically and biochemically at set intervals. Feeding of CSD caused reversal of many of the effects of lipotrope deficiency within a few weeks, suggesting that simple replenishment of C1 pools is sufficient to restore most normal hepatocyte functions. Levels of mRNA for all of the genes we have studied to date [*c-myc*, *c-Ha-ras*, *c-fos* and EGFR] returned to control levels within 1-3 weeks as did levels of active DNA and t-RNA methyltransferases[39-41]. The rate of DNA synthesis as monitored by BrdUrd accumulation in nuclei of hepatic cells dropped precipitously within 24-48 h and by 1 week was indistinguishable from that found in livers of age matched rats continually fed CSD. The overall extent of methylation of tRNA in the liver, as measured by ability to accept methyl groups in vitro, was also restored to normal within 1-2 weeks. Similar results were obtained when methyl acceptance of DNAs was measured[40,41]. It should be noted, however, that mammalian DNA MTase was used to catalyze methyl transfer in this highly sensitive assay for quantitating unmethylated sites in DNA. Since mammalian DNA methyltransferase is at least 40-fold more active in methylating hemi-methylated sites than completely unmethylated sites[52], what is actually measured is the capacity of the liver cells to maintain methylation of DNA at hemi-methylated sites created during DNA replication, not the overall of 5mC content of DNA. Thus, the finding that within one week of restoring lipotropes to the diet (CSD) of rats previously fed MDD the methyl acceptance of liver DNA is indistinguishable from that of liver DNA from rats continually fed CSD indicates that AdoMet and/or the ratio of AdoMet/AdoHcy rapidly rebound to levels that allow efficient maintenance of methylation of newly synthesized DNA.

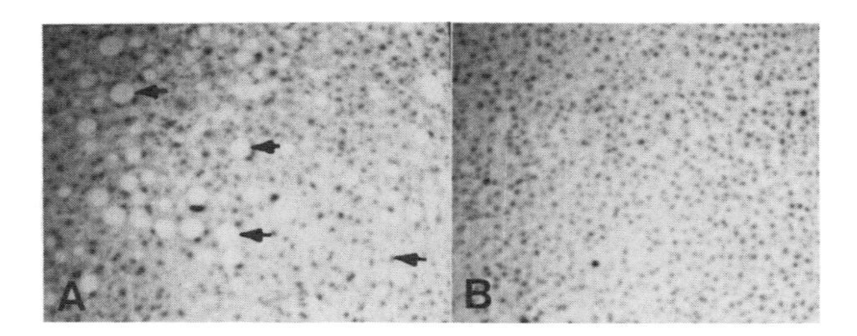

Figure 1. Residual lipid-filled cells (LFC) in liver four weeks after feeding an adequate diet to rats deprived of lipotropes for four weeks. Panel A shows a typical area in the liver of a recovering rat. Arrows point to a few of the LFC. Panel B shows a typical area from the liver of an age matched animal. Sections were fixed and processed as described in Ref. 40.

EFFECTS OF LIPOTROPE DEFICIENCY THAT LAST LONGER THAN ONE MONTH

Only two of the effects of lipotrope deficiency that we have studied persist for more than a week or two after restoration of adequate levels of lipotropes to the diet (CSD), lipid storage and changes in patterns of methylation of specific genes[53].

Lipid Storage

For at least 6 weeks after restoring lipotropes to the diet of rats fed MDD or SD, the livers of these rats contain significantly more lipid-filled cells (LFC) than livers from age matched animals fed CSD continuously . Within a few days of restoring lipotropes to the diet, areas of normal appearing hepatocytes are present in the periductular regions of the liver. After 2-4 weeks, these areas of well organized parenchyma have enlarged, but LFC still remain throughout zone 2 and around the central veins (Fig. 1).

Unlike the LFC present in all regions of the liver during MDD feeding, these cells fail to incorporate BrdUrd into DNA after a 2 h pulse (Data not shown). This suggests that at least some hepatocytes become resistant to the mitogenic effects of lipid accumulation and have lost the capacity to secrete VLDL even when normal lipotrope levels have been restored. (Chen and Christman, unpublished observation). Further investigation will be needed to determine whether the distribution of (LFC) reflects the pattern of replacement of hepatocytes progressing from the periductular to the central vein areas or a gradient of intra-cellular C1 pools. In either case, after 8-12 weeks of CSD feeding, the percentage of LFC is low and no greater than in age-matched animals fed CDS continuously.

Methylation of Specific CCGG Sites

Specific loss of methylation at CCGG sites in liver DNA resulting from feeding of MDD or SD is even more persistent than lipid accumulation. In the majority of rats tested, altered patterns of methylation established during a 4 week period of MDD or SD feeding could be detected for at least a year after subsequent feeding of CSD. Southern blot analysis of HpaII sites in the *c-myc* gene in DNA from livers harvested 4 weeks after restoring adequate lipotropes (CSD) to rats fed either MDD or SD for the previous 4 weeks is shown as an example (Fig. 2).

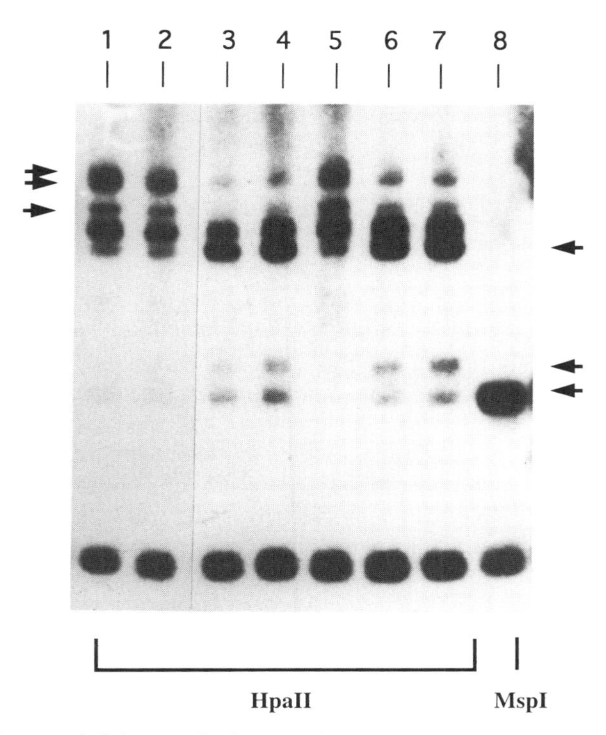

Figure 2. Effect of lipotrope deficiency and subsequent feeding of an adequate diet on methylation of CCGG sites in the *c-myc* gene. HpaII cut liver DNAs from rats fed: Lane 1, MDD (Basal amino acid defined diet supplemented with 9 g/kg D,L-homocystine), 4 d; Lane 2, SD (MDD supplemented with 5 mg/kg folic acid and 100 μg/kg vitamin B_{12}), 4 d; Lane 3, MDD, 4 w; Lane 4, SD, 4 w; Lane 5, CSD (SD without homocystine plus 5 g/kg D,L-methionine and 2 g/kg choline chloride) 4 w; Lane 6, MDD 4 w, CSD 4 w; Lane 7, SD 4 w, CSD, 4 w. Lane 8, DNA from CSD-fed rat cut with MspI. Since MspI cuts at CCGG sites regardless of methylation at the C residue next to G, the fragments in Lane 8 indicate the pattern expected with HpaII if no CCGG sites are methylated. HpaII fragments from rats continually fed an adequate diet are primarily high MW indicating methylation at most CCGG sites. Right pointing arrows indicate HpaII fragments lost due to cleavage of unmethylated CCGG sites induced by dietary lipotrope deficiency. Small HpaII fragments resulting from cleavage at sites which have lost methylation are indicated by left facing arrows. All other details and methods are given in Refs. 39,40.

We have noted similarly persistent loss of methylation at CCGG sites in *c-Ha-ras*, *c-fos* and *p53* genes[39,40]. We have previously detailed an argument supporting the concept that patterns of methylation in adult somatic tissues are most likely to be modified when DNA synthesis occurs under conditions where DNA methylation is inhibited, either through inactivation or inhibition of DNA MTase or, as is most likely in the lipotrope deprivation model, through lack of sufficient concentrations methyl donor[39]. If this mechanism is to account for the changes in methylation that are observed at specific sites in genes, the tissue undergoing alteration in patterns of methylation must contain dividing cells, as two rounds of DNA replication without methylation are required to achieve a completely unmethylated site in half the DNA molecules. Since we have observed complete loss of methylation at some sites in liver DNA, it follows that the majority of cells in the liver have undergone multiple rounds of cell division. The finding that these changes, detected as sensitivity of given site in DNA to cutting with the methylation sensitive restriction endonuclease HpaII, persist for many months, suggests that de novo methylation at these sites is highly inefficient. It also implies that the stem cell population that provides for gradual replacement of the parenchymal cells in the liver has also suffered an essentially permanent loss of methylation at the same CCGG sites.

BIOLOGICAL IMPLICATIONS OF PERSISTENT LOSS OF METHYLATION IN DNA

An overall decrease in the level of DNA methylation in tumor cells as compared to normal tissue is a common finding[54] . In extensive studies of genomic changes during development of human colon cancer, it was noted that one of the earliest detectable molecular changes, occurring in small adenomas, was loss of methyl groups from DNA[55,56]. However, an increase in methylation of specific gene regions[57] and an increase in the level of DNA methyltransferase, the enzyme that catalyzes transfer of methyl groups from AdoMet to carbon 5 of cytosine (C) residues in CpG dinucleotides[41,58] can also occur.

Three mechanisms have been proposed to link these alterations in DNA methylation with events occurring in tumor progression. First, 5mC residues in DNA could serve as hot spots for mutation since 5mC is more readily deaminated than C at physiological temperatures[59]. It has also been reported that a bacterial DNA MTase, HpaII, can act as a deaminase of C residues that it would normally methylate when AdoMet and AdoHcy are absent from the reaction mixture[60]. Although it remains to be demonstrated that sufficient reduction of AdoMet and AdoHcy levels can achieved in mammalian cells to lead to deamination of C residues by DNA MTase, deamination of C or 5mC in CpG sites could account for the high level C:G to A:T transition mutations observed at CpG sites in some genes[61]. Second, since inhibition of methylation of C residues in DNA leads to impaired condensation of chromatin, it has been postulated that decreased levels of methylation in DNA may favor genomic instability due to chromosomal non-disjunction[62]. Finally, alterations in DNA methylation may lead to heritable alterations in gene expression. For a number of tissue specific genes, extensive methylation of C residues in regulatory regions is correlated with decreased efficiency of transcription[63,64] although there are a few genes where extensive methylation is associated with increased gene activity[65,66]. Cytosine methylation is thought to affect transcription by altering the interaction of transcription factors with DNA in regulatory regions or by altering the conformation of chromatin in such a way as to interfere with transcription. Methylation of regulatory sequences in DNA can either enhance or prevent binding of transcription factors or other proteins that recognize these sequences[67-72] and, depending on whether the proteins have a positive or negative effect on transcription, can either enhance or inhibit RNA production. Obviously, if proto-oncogenes or genes encoding inhibitors of the function of suppressor genes are transcribed more efficiently when they are less extensively methylated or if transcription of suppressor genes is blocked by methylation, stable alterations in methylation could lead to a tumorigenic phenotype in cells that have not suffered mutagenic alterations in the primary structure of DNA.

It has been that noted that feeding of a choline deficient diet to diethylnitrosamine initiated rats led to formation of altered hepatic foci containing a greater proportion of aneuploid and tetraploid cells than were found in rats fed a choline-sufficient diet[73]. Mutations in the *p53*[74] gene and amplification of *c-myc* genes[75] have also been detected in hepatocellular carcinomas induced by feeding of choline-deficient diets. Our own studies indicate that lipotrope deficiency rapidly increases levels of mRNAs for a number of growth related genes that also become hypomethylated[38]. Finally, there are a number of indications that these alterations in methylation and in gene expression are common features of hepatocarcinogenesis regardless of the initiating carcinogen or promoting regime[38,51,76-79]. However, at least for the *c-myc*, *c-Ha-ras* and *c-fos* genes, loss of methylation is not sufficient to maintain increased expression once lipotrope sufficiency is restored[39]. This suggests that transcription factors needed for efficient expression of these genes may be present only in mitogenically stimulated or proliferating liver cells.

While these results support the possibility that alterations in DNA methylation induced by lipotrope deficiency could act to promote and/or initiate tumorigenesis by any or even all of these mechanism, there is as yet no direct evidence that loss of methylation in DNA has a direct influence on any of these processes. In terms of evaluating the contribution of altered methylation to critical changes gene expression, the next step will be to determine whether any basic differences exist in gene expression or gene activation in response to specific mitogens and/or carcinogens in livers from rats with stably altered and normal patterns of methylation.

RELEVANCE OF LIPOTROPE DEFICIENCY TO HUMAN CANCER

It is improbable that any human diet could cause a lipotrope deficiency comparable to that induced by experimental diets we and others have used in animal studies. However, brief experimental induction of choline deficiency in humans can lead to signs of incipient liver dysfunction[80]. A significantly increased risk for development of leukemia and other cancers in has been observed in patients recovering from pernicious anemia[81] and deficiency in folate has been epidemiologically liked to development of colorectal, breast and other dysplasias and cancers[82-86]. If hypomethylated sites occur in DNA of stem cell populations and persist in all the progeny of those stem cells as our studies suggest, it is certainly possible that unmethylated sites in specific genes produced by exposure of dividing cell populations to prolonged moderate lipotrope deficiency could contribute to cancer in humans. If persistent changes in DNA methylation are ultimately proven to play a role in development of cancer, our primary challenges will be to prevent dietary deficiencies in lipotropes through recommendations for consumption of lipotrope-rich foods or supplements and to develop methods for enhancing re-establishment of normal patterns of DNA methylation in affected tissues.

ACKNOWLEDGMENTS

I thank Dr. Mei-Ling Chen and Dr. Elsie Wainfan for allowing discussion of their unpublished data. Their contributions, along with those of Dr. Gholamreza Sheikhnejad, Dr. Mark Dizik and Ms. Susana Abileah, have made this work possible. Our reasearch was supported in part by grants from the American Institute for Cancer Research-92A35 (JKC) and 864 (EW) and the Lloyd and Marilyn Smith Fund.

REFERENCES

1. D.H. Copeland, and W.D. Salmon, The occurrence of neoplasms in the liver, lungs, and other tissues of rats as a result of prolonged choline deficiency, *Am. J. Path.* 22: 1059 (1946).
2. P.M. Newberne, Carcinogenicity of aflatoxin-contaminated peanut meal, *in* "Mycotoxins in Foodstuffs," Massachusetts Institute of Technology, Cambridge, MA. (1965).
3. P.M. Newberne, and A.E. Rogers, Labile methyl groups and the promotion of cancer, *Annu. Rev. Nutr.* 6:407 (1980).

4. A.E. Rogers, Variable effects of a lipotrope-deficient high-fat diet on chemical carcinogenesis in rats, *Cancer Res.* 35:2469 (1975).
5. A.E. Rogers, Chemical carcinogenesis in methyl deficient rats, *J. Nutr. Biochem.* 4:666 (1993).
6. A.E. Rogers, and P.M. Newberne, Lipotrope deficiency in experimental carcinogenesis, *Nutr. Cancer* 2:104 (1980).
7. A.E. Rogers, O. Sanchez, F.M. Feinsod, and P.M. Newberne, Dietary enhancement of nitrosamine carcinogenesis, *Cancer Res.* 34:96 (1974).
8. H. Shinozuka, S.L. Kaytal, and B. Lombardi, Azaserine carcinogenesis: Organ susceptibility change in rats fed a diet devoid of choline, *Int. J. Cancer* 22:36 (1978).
9. H. Shinozuka, B. Lombardi, and S. Sell, Enhancement of ethionine liver carcinogenesis in rats fed a choline-devoid diet, *J. Natl. Cancer Inst.* 61:813 (1978).
10. B. Lombardi, and H. Shinozuka, Enhancement of 2-acetylaminofluorene liver carcinogenesis in rats fed a choline-devoid diet, *Int. J. Cancer*, 23:565 (1979).
11. Y.B. Mikol, K.L. Hoover, D. Creasia, and L.A. Poirier, Hepatocarcinogenesis in rats fed methyl-deficient, amino acid-defined diets, *Carcinogenesis* 4:1619 (1983).
12. I. Eto, and C.L. Krumdieck, Role of vitamin B_{12} and folate deficiencies in carcinogenesis, *Adv. Exptl. Biol.* 206:313 (1986).
13. D.E. Horne, R.J. Cook, and C. Wagner, Effect of dietary methyl group-eficiency on folate metabolism in rats, *J. Nutr.* 119:618 (1989).
14. C.L. Krumdieck, Role of folate deficiency in carcinogenesis., *in*: "Nutritional Factors in the Induction and Maintenance of Malignancy," Academic Press, New York (1983).
15. C. Kutzbach, E. Galloway, and E.I. Stokested, Influence of vitamin B_{12} and methionine on levels of folic acid compounds and folate enzymes in rat liver, *Proc. Soc. Exp. Biol. Med.* 124: 801 (1969).
16. S.H. Zeisel, T. Zola, K.A. daCosta, E.A. Pomfret, Effect of choline deficiency on S-adenosylmethionine and methionine concentrations in rat liver. *J. Biochem.* 259:725 (1989).
17. W.C. McMurray, "Essentials of Human Metabolism," Harper and Row, Philadelphia, PA (1983).
18. L.A. Loeb, and T.A. Kunkel, Fidelity of DNA synthesis, *Annu. Rev. Biochem.* 2:429 (1982).
19. B.A. Kuntz, Genetic effects of deoxyribonuceotide pool imbalances, *Environ. Mutagen.* 4:695 (1982).
20. G. Sutherland, The role of nucleotides in human fragile site expression, *Mut. Res.* 200:207 (1988).
21. A.E. Pegg, and H. Hibasami, The role of S-adenosylmethionine in mammalian polyamine synthesis, *in*: "Transmethylation," Elsevier-North Holland, New York (1979).
22. A.J.L. Cooper, Biochemistry of sulfur-containing amino acids, *Annu. Rev. Biochem.* 52:187 (1983).
23. E.P. Kennedy, and S.B. Weiss, The function of cytidine coenzymes in the biosynthesis of phospholipids, *J. Biol. Chem.* 222:193 (1956).
24. J. Bremer, P.H. Figard, and D.M. Greenberg, The biosynthesis of choline and its relation to phospholipid metabolism, *Biochim. Biophys. Acta.* 43:477 (1960).
25. E. Usdin, R.T. Borchardt, and C.R. Creveling, Biochemistry of S-Adenosylmethionine and Related Compounds, Macmillan Press Ltd., London (1982).
26. S. Clarke, Protein isoprenylation and methylation at carboxy-terminal cytsteine residues, *Annu. Rev. Biochem.* 61:131 (1992).
27. T.A. Trautner, Methylation of DNA, *Current Topics in Microbiology and Immunology* 108, Springer-Verlag, Berlin (1984).
28. L.I. Hinrichsen, R.A. Floyd, and O. Sudilovsky, Is 8-hydroxydeoxyguanosine a mediator of carcinogenesis by a choline-devoid diet in the rat liver? *Carcinogenesis* 11: 1879 (1990).
29. M.I. Perera, J.M. Betschart, M.A. Virji, S.L. Kaytal, and H. Shinozuka, Free radical injury and liver tumor promotion. *Toxicol. Pathol.* 15:51 (1987).
30. T.H. Rushmore, D.M. Ghazarian, V. Subrahmanyan, E. Farber, and A.K. Goshal, Probable free radical effects on rat liver nuclei during early hepatocarcinogenesis with a choline-devoid low-methionine diet. *Cancer Res.* 47: 6731 (1987).
31. L.K. Giambarresi, S.L. Katyal, and B. Lombardi, Promotion of liver carcinogenesis in the rat by a choline-devoid diet: role of liver cell necrosis and regeneration. *Br.J. Cancer* 46:825 (1982).
32. A.K. Goshal, M. Ahluwalia, and E. Farber, The rapid induction of liver cell death in rats fed a choline-deficient methionine-low diet. *Am. J. Pathol.* 113: 309 (1983).
33. K. da Costa, E.F. Cochary, J.K. Busztajn, S.C. Garner, and S.H. Zeisel, Accumulation of 1,2-*sn*-diradylglycerol with increased membrane-associated protein kinase C may be the mechanism for spontaneous hepatocarcinogenesis in choline deficient rats. *J. Biol. Chem.* 268:2100 (1993).
34. N. Shivapurkar, and L.A. Poirier, Tissue levels of S-adenosylmethionine and S-adenosylhomocysteine in rats fed methyl-deficient diets for one to five weeks. *Carcinogenesis* 4:1052 (1983).

35. M.J. Wilson, N. Shivapurkar, and L.A. Poirier, Hypomethylation of hepatic nuclear DNA in rats fed with a carcinogenic methyl-deficient diet. *Biochem. J.* 218:987 (1984).
36. J. Locker, T.V. Reddy and B. Lombardi, DNA methylation and hepatocarcinogenesis in rats fed a choline-devoid diet. *Carcinogenesis* 7:1309 (1986).
37. E. Wainfan, M. Dizik, M. Stender, and J.K. Christman, Rapid appearance of hypomethylated DNA in livers of rats fed cancer-promoting, methyl-deficient diets. *Cancer Res.* 49:4094 (1989)
38. M. Dizik, J.K. Christman, and E. Wainfan, Alterations in expression and methylation of specific genes in livers of rats fed a cancer promoting methyl-deficient diet. *Carcinogenesis* 12:1307 (1991).
39. J.K. Christman, G. Sheikhnejad, M. Dizik, S. Abileah and E. Wainfan, Reversibility of changes in nucleic acid methylation and gene expression induced in rat liver by severe dietary methyl deficiency. *Carcinogenesis* 14:551 (1993).
40. J.K. Christman, M-L, Chen, G. Sheikhnejad, M. Dizik, S. Abileah, and E. Wainfan, Methyl deficiency, DNA methylation, and cancer: Studies the reversibility of the effects of lipotrope-deficient diet. *J. Nutr. Biochem.*, 4:672 (1993).
41. J.K. Christman, L. Chen, G. Sheikhnejad, M. Dizik, and E. Wainfan, Regulation of gene expression by DNA methylation: a link between dietary methyl deficiency and hepatocarcinogenesis, *in*: "The Role of Nutrients in Cancer Treatment," A. Roche ed. Report of the 9th Ross Conference on Medical Res. (1991).
42. K.K. Hoover, P.H. Lynch, L.A. Poirier, Profound postinitiation enhancement by short-term severe methionine, choline, vitamin B_{12} and folate deficiency of hepatocarcinogenesis in F344 rats given a single low-dose diethylnitrosamine injection. *JNCI* 73:1327 (1984).
43. N. Chandar, and B. Lombardi, Liver cell proliferation and incidence of hepatocellular carcinomas in rats fed consecutively a choline-devoid and a choline-supplemented diet. *Carcinogenesis* 2:259 (1988).
44. D. Nakae, H. Yoshiji, Y. Mizumoto, K. Horiguchi, K. Shiraiwa, K. Tamura, A. Denda, and Y. Konishi High incidence of hepatocellular carcinomas induced by a choline deficient L-amino acid defined diet in rats. *Cancer Res.* 52:5042 (1992).
45. E. Wainfan, M. Kilkenny, and M. Dizik, Comparison of methyltransferase activities of pair-fed rats given adequate or methyl-deficient diets. *Carcinogenesis*, 9:861 (1988).
46. E. Wainfan, M. Dizik, and M.E. Balis, Increased activity of rat liver N2-guanine tRNA methyltransferase II in response to liver change. *Biochim. Biophys. Acta* 799: 288 (1984).
47. E. Wainfan, M. Dizik, M. Hluboky and M.E. Balis, Altered tRNA methylation in rats and mice fed lipotrope-deficient diets. *Carcinogenesis* 7:473 (1986).
48. A.K. Goshal, M. Ahluwalia, and E. Farber, The induction of liver cancer by a dietary deficiency of choline and methionine without added carcinogen, *Carcinogenesis*, 5:1367 (1984).
49. S. Yokoyama, M.A. Sells, T.V. Reddy, and B. Lombardi, Hepato-carcinogenic and promoting action of a choline-devoid diet in the rat. *Cancer Res.*, 45:2384 (1985).
50. A.E. Rogers, and R.A. MacDonald, Hepatic vasculature and cell proliferation in experimental cirrhosis. *Lab. Invest.* 14:1710 (1965).
51. M.R. Bhave, M.F. Wilson, and L.A. Poirier, c-H-*ras* and c-K-*ras* gene hypomethylation in the livers and hepatomas of rats fed methyl-deficient, amino acid-defined diets. *Carcinogenesis* 9:343 (1988).
52. T.H. Bestor, and V.M. Ingram, Two DNA methyltransferases from murine erythroleukemia cells: purification and sequence specificity. *Proc. Natl. Acad. Sci.* USA 80:5559 (1983).
53. M-L Chen, S. Abileah, E. Wainfan, J.K. Christman, Influence of folate and vitamin B_{12} on the effects of dietary lipotrope deficiency. *Proc. Amer. Assoc. for Cancer Res.* 34: 131 (1993).
54. M.A. Gama-Sosa, V.A. Slagle, R.W. Trewyn, R. Oxenhandler, K.C. Kuo, C.W. Gehrke, and M. Ehrlich, The 5-methylcytosine content of DNA from human tumors. *Nucleic Acids Res.* 11:6883 (1983).
55. S.E. Goelz, B. Vogelstein, S.R. Hamilton, and A.P. Feinberg, Hypomethylation of DNA from benign and malignant human colon neoplasms. *Science* 228:187 (1985).
56. E.R. Fearon, and P.A. Vogelstein, Genetic model for colorectal tumorigenesis. *Cell*, 61:759 (1990).
57. M. Makos, B.D. Nelkin, M.I. Lerman, F. Latif, B. Zbar,and S. Baylin, Distinct hypermethylation patterns occur at altered chromosome loci in human lung and colon cancer. *Proc. Natl. Acad. Sci.* U.S.A., 89:1929 (1992).
58. W.S. el-Deiry, B.D., Nelkin, P. Celano, R.W. Yen, J.P. Falco, S. R. Hamilton, S.B. Baylin, High expression of the DNA methyltransferase gene characterizes human neoplastic cells and progression stages of colon cancer. *Proc. Natl. Acad. Sci.* U.S.A. 88: 3470 (1991).
59. M. Erhlich, X.Y. Zhang, and N.M. Inamdar, Spontaneous deamination of cytosine and 5-methylcytosine residues in DNA and replacement of 5-methylcytosine residues and cytosine residues. *Mutat. Res.* 238:277 (1990).
60. J.C. Shen, W.M. Rideout, and P.A. Jones, High frequency mutagenesis by a DNA methyltransferase. *Cell* 71, 7:1073 (1994).

61. P.A. Jones, J.D. Buckely, B.E. Henderson, R.K. Ross, and M.C. Pike, From gene to carcinogen: a rapidly evolving field in molecular epidemiology. *Can. Res.* 51, 13:3617 (1991).
62. M. Schmid, T. Haaf, and D. Grunert, 5-Azacytidine-induced undercondensations in human chromosomes. *Human Genet.*, 67:257 (1990).
63. H. Cedar, DNA methylation and gene activity. *Cell*, 34:5503 (1988).
64. W. A. Doerfler, DNA methylation and gene activity. *Rev. Biochem.* 52:93 (1983).
65. B. Gellerson, and R. Kempf, Human prolactin gene expression: positive correlation between site-specific methylation and gene activity in a set of human lymphoid cell lines. *Mol. Endocrinol.* 4: 1874 (1990).
66. K. Tanaka, E. Appella, and G. Jay, Developmental Activation of the H-2K gene is correlated with an increase in DNA methylation. *Cell* 35: 457 (1983).
67. J.K. Christman, L. Chen, R. Nicholson, and M. Xu, DNA Methylation: A cellular strategy for regulating expression of integrated hepatitis B virus genes? *in*: Virus Strategies: Molecular Biology and Pathogenesis. Doerfler, W. and Bohm, P. (eds.) VCH Verlagsgesellschaft, Weinheim, New York (1993).
68. M. Comb, and H.W. Goodman, CpG methylation inhibits proenkephalin gene expression and binding of the transcription factor AP-2. *Nucl. Acids Res.* 18:3975 (1990).
69. S.M. Iguchi-Ariga, and W. Schaffner, CpG methylation of the cAMP-responsive enhancer/promoter sequence TGACGTCA abolishes selective factor binding as well as transcriptional activation. *Genes Dev.* 3:612 (1989).
70. R.R. Meehan, J.D. Lewis, S. McKay, E.L. Kleiner, and A.P. Bird, Identification of a mammalian protein that binds specifically to DNA containing methylated CpGs. *Cell* 58: 499 (1989).
71. F. Watt, and P.L. Molloy, Cytosine, methylation prevents binding to DNA of a Hela cell transcription factor required for optimal expression of the adenovirus major late promoter. *Genes Dev.* 2:1136 (1988).
72. R.Y. Wang, X.Y. Zhang, and M. Ehrlich, A human DNA binding protein is methylation-specific and sequence-specific. *Nucl. Acids Res.* 14:1599 (1986).
73. J.H. Wang, L.I. Hinrichsen, C.M. Whitacre, R.L. Cechner, O. Sudilovsky, Nuclear DNA content of altered hepatic foci in a rat liver carcinogenesis model, *Cancer Res.* 50:7571 (1990).
74. M.L. Smith, L. Yeleswarapu, P. Scalamogna, J. Locker, B. Lombardi, p53 mutations in hepatocellular carcinomas induced by a choline-devoid diet in male Fischer 344 rats. *Carcinogenesis*, 14:503-10 (1993).
75. N. Chandar, B. Lombardi, J. Locker, c-myc gene amplification during hepatocarcinogenesis by a choline-devoid diet. *Proc. Natl. Acad. Sci.*, USA 86:2703 (1989).
76. M.S.C. Cheah, C.D. Wallace, R.M. Hoffman, Hypomethylation of DNA in human cancer cells: A site-specific change in the c-*myc* oncogene. *JNCI* 73:1057 (1984).
77. A. Feinberg, B. Vogelstein, Hypomethylation of *ras* oncogenes in primary human cancers. *Biochem. Biophys. Res. Commun.* 111:47 (1983).
78. S. Nambu, K. Inque, and H. Sasaki, Site-specific hypomethylation of the c-*myc* oncogene in human hepatocellular carcinoma *Jpn. J. Cancer* (Gann) 78:695 (1987).
79. P.M. Rao, A. Antony, S. Rajalakshmi, and D.S.R. Sarma, Studies on hypomethylation of liver DNA during early stages of chemical carcinogenesis in rat liver. *Carcinogenesis* 10, 5:933 (1989).
80. S.H. Zeisel, K.A. Da-Costa, P.D. Franklin, E.A. Alexander, J.T. Lamont, N.F. Sheard, A. Beiser, Choline, an essential nutrient for humans, *FASEB J.* 5:2093 (1991).
81. E. Giovannucci, M.J. Stampfer, G.A. Golditz, E.B. Rimm, D. Trichopoulos, B.A. Rosner, F.E. Speizer, and W.C. Willett, Folate, methionine, and alcohol intake and risk of colorectal adenoma. *J. Natl. Can. Inst.* 85, 11:875 (1993).
82. D.C. Heimburger, B. Alexander, R. Birch, C.E. Butterworth, W.C. Bailey, and C.L. Krumdieck, Improvements in bronchial squamous metaplasia in smokers treated with folate and vitamin B12. *J. Am. Med. Assoc.*, 259:1525 (1988).
83. B. A. Lashner, P.A. Heidenreich, G.L. Su, S.V. Kane, and S.B. Hanauer, Effect of folate supplementation on the incidence of dysplasia and cancer in chronic ulcerative colitis. A case-control study. *Gastroenterology*, 97:255 (1989).
84. R.F. Branda, J.P. O'Neill, L.M. Sullivan, and R.J. Albertini, Factors influencing mutation at the *hprt* locus in T-Lymphocytes: Women treated for breast cancer. *Cancer Res.*, 51:6603 (1991).
85. L.A. Brinton, G. Gridley, Z. Hrubec, R. Hoover, and J.F. Fraumeni, Cancer risk following pernicious anemia. *Br. J. Cancer* 59:810 (1989).
86. C.E. Butterworth, K.D. Hatch, H. Gore, H. Meubler, and C.L. Krumdieck, Improvement in cervical displasia with folic acid therapy users of oral contraceptives. *Am. J. Clin. Nutr.*, 35:73 (1982).

10

METHIONINE DEPRIVATION REGULATES THE TRANSLATION OF FUNCTIONALLY-DISTINCT C-MYC PROTEINS

Stephen R. Hann

Department of Cell Biology
Vanderbilt University
School of Medicine
Nashville, Tennessee 37232-2175

ABSTRACT

Numerous studies have demonstrated a critical role for the c-*myc* gene in the control of cellular growth. Alterations of the c-*myc* gene have been found associated with many different types of tumors in several species, including humans. The increased synthesis of one of the major forms of c-Myc protein, c-Myc 1, upon methionine deprivation provides a link between the regulation of oncogenes and the nutritional status of the cell. While deregulation or overexpression of the other major form, c-Myc 2, has been shown to cause tumorigenesis, the synthesis of c-Myc 1 protein is lost in many tumors. This suggests that the c-Myc 1 protein is necessary to keep the c-Myc 2 protein "in check" and prevent certain cells from becoming tumorigenic. Indeed, we have shown that overproduction of c-Myc 1 can inhibit cell growth. We have also shown that c-Myc 1 and 2 proteins have a differential molecular function in the regulation of transcription through a new binding site for Myc/Max heterodimers. We have also recently identified new translational forms of the c-Myc protein which we term Δ-c-Myc. These proteins arise from translational initiation at downstream start sites which yield N-terminally-truncated c-Myc proteins. Since these proteins lack a significant portion of the transactivation domain of c-Myc, they behave as dominant-negative inhibitors of the full-length c-Myc 1 and 2 proteins. The synthesis of Δ-c-Myc proteins is also regulated during cell growth and is repressed by methionine deprivation. Therefore, the synthesis of c-Myc 1 and Δ-c-Myc proteins are reciprocally regulated by methionine availability. We have also found some tumor cell lines which synthesize high levels of the Δ-c-Myc proteins. Taken together, our data suggest that c-Myc function is dependent on the levels of these different translational forms of c-Myc protein which are regulated by the nutritional status of the cell during growth.

Numerous reports have demonstrated a fundamental and diverse role for the *myc* gene in cellular events, including proliferation, differentiation and apoptosis (Cole 1986;

Diet and Cancer, Edited under the auspices of the American Institute of Cancer Research
Plenum Press, New York, 1995

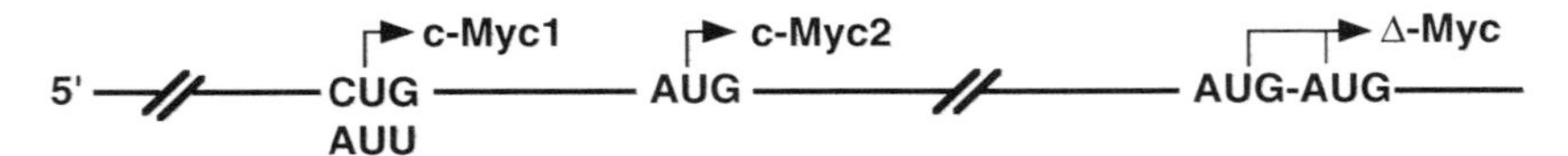

Figure 1. Initiation of c-Myc proteins.

Spencer and Groudine 1991; Askew et al. 1991; Evan et al. 1992). This is dramatically illustrated by the frequent occurrence of a variety of tumors in many species having alterations of *myc* genes and the transduction of c-*myc* sequences by retroviruses (Spencer and Groudine 1991). The diverse biological activity of *myc* is demonstrated by its ability to contribute to cellular proliferation (Spencer and Groudine 1991), inhibit terminal differentiation (Cole 1986), and promote apoptosis (Evan et al. 1992). Despite intensive study, however, the mechanism by which Myc proteins perform such diverse cellular roles is unknown (Luscher and Eisenman 1990).

A distinctive feature of the *myc* gene is that it encodes multiple N-terminally-distinct proteins. Alternative translational forms of the Myc protein exist for all species of c-Myc examined thus far (Hann and Eisenman 1984; Hann et al, 1988), as well as for N-Myc (Ramsay et al. 1986) and L-Myc proteins (Dosaka-Akita et al. 1991). The c-Myc 1 and 2 proteins have been found in all vertebrate species examined (Hann et al. 1988). In mammalian and avian cells, c-Myc 1 protein arises from an upstream non-AUG translational start site and thus contains an N-terminal extension of 14 amino acids compared to c-Myc 2 protein (Hann et al. 1988). Recently we have found that human, murine and avian cells also express smaller-sized c-Myc proteins to the full-length c-Myc 1 and 2 proteins (Spotts and Hann unpublished). These smaller-sized proteins, which we term Δ-c-Myc proteins, arise from translational initiation at a doublet of AUG codons downstream of the initiation sites for c-Myc 1 and 2 yielding proteins lacking the first 100 amino acids of c-Myc 2. Figure 1 diagrams the initiation of the different c-Myc proteins.

REGULATION OF THE ALTERNATIVELY-INITIATED c-Myc PROTEINS

We have previously demonstrated that c-Myc protein expression can be extensively regulated at the level of alternative translational initiation. The AUG-initiated c-Myc 2 protein is predominantly synthesized during cell growth while expression of the upstream, non-AUG-initiated c-Myc 1 protein increases dramatically to levels equal to or greater than that of c-Myc 2 as cells approach high-density growth arrest (Hann et al. 1992). Therefore, the non-AUG initiation of c-Myc can become much more efficient than the AUG initiation. The synthesis of the AUG-initiated c-Myc 2, on the other hand, was minimally affected or decreased. Treatment with depleted medium from high density cells was able to reproduce the activation of c-Myc 1 protein synthesis in low density cells within 5 hrs (Hann et al. 1992). Further analysis of the c-Myc 1-inducing ability of conditioned/depleted medias revealed that methionine deprivation was responsible for this unusual regulation (Hann et al. 1992). This increased efficiency of non-AUG initiation appears to be specific for methionine deprivation rather than a consequence of growth inhibition or stress, since growth inhibition due to deprivation of other essential amino acids, serum deprivation, TGFβ, interferon or terminal differentiation had no effect on the non-AUG initiation of c-Myc (Hann et al. 1992). In contrast to the full-length c-Myc 1 and 2 proteins, the Δ-c-Myc proteins are transiently synthesized during cell growth and are repressed rapidly by methionine deprivation (Spotts and Hann unpublished). These results describe a specific and dramatic regulation of alternative translational initiation in vertebrate cells.

These results raise the question as to how methionine deprivation modulates the translational machinery to allow preferential initiation from the non-AUG codon and repress synthesis of the Δ-c-Myc proteins. It is unclear how non-AUG initiation codons are recognized by the scanning preinitiation complex. Studies which have examined non-AUG initiation in mammalian cells have generally demonstrated that the efficiency of non-AUG initiation is negligible in comparison to AUG initiation (Kozak 1989; Peabody, 1989). The upstream non-AUG of c-Myc is normally only 10-15% as efficient as the AUG in growing cells (Hann et al. 1988). The scanning model predicts that a 43S preinitiation complex consisting of the small 40S ribosomal subunit, Met-tRNA$_i^{met}$, GTP, and the initiation factors eIF-2 and eIF-3 first binds to the 5' end of a capped mRNA (Kozak 1991; Hershey 1991). The preinitiation scanning complex then migrates 3' until it encounters an initiation codon, usually the first AUG, whereupon recognition of the start codon is mediated by eIF2-modulated base pairing with the anticodon in Met-tRNA$_i^{met}$ (Kozak 1991; Hershey 1991). Met-tRNA$_i^{met}$ has been shown to initiate translation at both an AUG and a non-AUG codon (Peabody 1989; Donahue et al. 1988).

One possible explanation for the increased non-AUG initiation of c-Myc is that there is an alteration of initiation factors in the scanning preinitiation complex due to methionine deprivation which directly affects start codon selection. A key initiation factor that is required for formation of the complex and its ability to recognize the start codon is eIF-2 (Hershey 1991). The idea that eIF-2 may be involved in start codon selection and the translational activation of c-Myc 1 is supported by the findings that mutations in the α or β subunit of eIF-2 allow initiation at a non-AUG (UUG) codon of a mutant HIS4 gene in yeast (Donahue et al. 1988). Also, mutations in either the α or β subunits of eIF-2 independently lead to constitutive activation of GCN4 (Williams et al. 1989), a yeast transcriptional activator involved in the biosynthesis of amino acids which is normally translationally activated in response to amino acid starvation (Abastado et al. 1991). In yeast and in higher eukaryotes, amino acid deprivation leads to an increased phosphorylation of eIF2α (Hershey 1991). An aminoacyl-tRNA synthetase may exist in higher eukaryotes with an associated eIF-2 kinase which monitors the levels of uncharged tRNAs, similar to GCN2, a yeast kinase with an associated tRNA synthetase which phosphorylates eIF-2 in response to amino acid deprivation and enhances GCN4 translation (Wek et al. 1989; Ramirez et al. 1991). A partial inactivation of eIF2, which causes only a modest reduction in global protein synthesis, can have a dramatic effect on GCN4 translation (Tzamarias et al. 1989). Since phosphorylation of eIF2α upon amino acid deprivation in higher eukaryotes leads to a global shut-down of protein synthesis (Hershey 1991), the increased efficiency of c-Myc non-AUG initiation would most likely be an early consequence of eIF2 modification before global changes occur. Indeed, c-Myc 1 is induced within hours of methionine deprivation, before any changes are observed in total protein synthesis (Hann et al. 1992). These results demonstrate that a normally inefficient upstream initiation event can become more efficient.

The Δ-c-Myc proteins arise from translational initiation at AUG-initiator codons downstream of the initiation sites for c-Myc 1 and 2 proteins due to a "leaky scanning" mechanism. This is possible since the non-AUG initiator codon for c-Myc 1 protein is normally inefficient in growing cells and the context of the AUG initiator for c-Myc 2 protein is not optimal for initiation (Kozak, 1986). Therefore, the scanning ribosomal preinitiation complex reads through the upstream initiator codons for c-Myc 1 and 2 and initiates translation at the downstream AUG codons which reside in optimal context. When the start codon for c-Myc 2 is changed to an optimal AUG codon, significantly less Δ-c-Myc proteins were synthesized (Spotts and Hann unpublished). Changes in secondary structure stability downstream of the c-Myc 2 AUG start site or in specific initiation factors during growth and methionine deprivation may be critical for regulating the "leaky scanning" past the c-Myc 2 AUG. This regulated "leaky scanning" mechanism allows for specific variations in the ratio of upstream to downstream initiation events during cell growth. Based on the regulation of the alternatively-initiated c-Myc proteins during growth, we propose that two distinct mechanisms may occur to control the synthesis of these proteins. Under high methionine availability leaky scanning past the non-AUG codon (c-Myc 1) appears to occur, leading to higher levels of c-Myc 2 and Δ-c-Myc proteins and lower levels of c-Myc 1 synthesis. Upon reduction of methionine availability, a shift in initiation site preference to the upstream non-AUG codon occurs. However, the transient synthesis of Δ-c-Myc during rapid cell growth when there are no changes in the synthesis of c-Myc 1 suggest that another mechanism also controls the synthesis of the downstream initiated Δ-c-Myc proteins, most likely by enhancing the leaky scanning past the first AUG independent of methionine availability.

DIFFERENTIAL MOLECULAR FUNCTIONS FOR THE ALTERNATIVELY-INITIATED c-Myc PROTEINS

Increasing evidence suggests that members of the *myc* family of proto-oncogenes function as regulators of gene transcription. Myc proteins have many features in common with other transcriptional regulators. The major c-Myc proteins are phosphorylated, localized to the nucleus, and have relatively short half-lives (Hann and Eisenman 1984). Specific molecular functions have been assigned to both the C-terminal and N-terminal regions of the proteins. The C-terminal domain of the Myc proteins is structurally similar to a superfamily of transcription factors having a cluster of basic amino acids necessary for sequence-specific DNA binding adjacent to a α-helix-loop-helix/leucine zipper motif (b-HLH-LZ) (Luscher and Eisenman 1990). Dimerization of Myc with Max, another b-HLH-LZ protein, through the HLH-LZ region of the two proteins (Blackwood and Eisenman 1991), allows sequence-specific binding to the EMS ("*E* box" *M*yc *S*ite) DNA sequence (Blackwell et al. 1990; Prendergast and Ziff 1991). It has been demonstrated that c-Myc protein can stimulate transcription through EMS sequences while excess Max protein antagonizes this transactivation in mammalian and yeast cells (Kretzner et al. 1992; Amati et al. 1993; Amin et al. 1993).

Transcriptional activation by c-Myc protein is also dependent on an intact N-terminal domain as well as the C-terminal domain. The N-terminal region of c-Myc protein functions as a transactivation domain when fused to the DNA binding portion of Gal4 protein (Kato et al. 1990) and deletions of the highly conserved areas within the N-terminal domain (*myc* boxes) reduce its transactivation ability (Kato et al. 1990; Kretzner et al. 1992). In addition, transactivation may be modulated by proteins interacting with the N-terminal domain. Expression of B-myc protein, which resembles the N-terminal third of Myc protein but lacks the C-terminal heterodimerization and DNA binding domains, represses transactivation and blocks transformation of cells by c-Myc protein (Resar et al. 1993). This suggests B-Myc protein blocks transcription and transformation by sequestering factors which interact with the N-terminal domain of c-Myc.

Considering the differential regulation of the alternative translational forms of c-*myc* and their structural differences it is important to determine if they have distinct molecular and biological functions. Are the upstream non-AUG initiation sites merely an insurance or backup mechanism to guarantee translation of important regulatory proteins? Or do the different translational forms have distinct functions, all of which are necessary for normal function? We recently described a functional difference for the alternative forms of the c-Myc protein in the control of transcription. Through an investigation of factors which specifically bind and influence the activity of enhancer elements within the Rous sarcoma virus (RSV) long terminal repeat (LTR), we examined the activity of Myc proteins on the transcriptional activation by a specific enhancer element, EFII. The EFII cis element (Sealy and Chalkley 1987; Sears and Sealy 1992) consists of two nearly direct repeat sequence elements which do not contain the canonical EMS binding sites, but rather contain a consensus binding site ($T^{T}/_{G}NNG^{C}/_{T}AA^{T}/_{G}$) for the C/EBP family of transcription factors. The C/EBP family consists of several related transcription factors which appear to be important in the regulation of growth and differentiation (Cao et al. 1991; Umek et al. 1991). We found that the c-Myc 1 protein is a potent and specific transactivator of the EFII enhancer element through the C/EBP binding site, whereas the c-Myc 2 protein either failed to transactivate the EFII enhancer element or repressed EFII-driven transcription (Hann et al. 1994). In contrast, both c-Myc 1 and 2 proteins are competent to transactivate through the canonical EMS sequence. We have also demonstrated that bacterially-expressed Myc/Max heterodimers, but not Max homodimers, can bind to the C/EBP site in vitro, whereas both Myc/Max heterodimers and Max homodimers can bind to the EMS sequence in vitro (Hann et al. 1994).

Since the Myc proteins have the same C-terminal domain, the region sufficient for specific DNA binding and heterodimerization with Max (Blackwood and Eisenman, 1991), the opposing effects of these two proteins on EFII-driven transcription are most likely a result of differing N-termini. A likely possibility is that the N-terminal extension causes an overall conformational change in the N-terminal region of the Myc protein containing the transactivation domain. Evidence supporting such conformational differences comes from the differential recognition of the two c-Myc proteins with an N-terminal domain antibody. Persson et al. (1984) found that one of their N-terminal domain-specific antibodies only recognized the c-Myc 2 protein whereas their other antibodies recognized both proteins. This conformational difference may allow differential contact with the transcriptional machinery depending on the DNA binding site. The importance of N-terminal interactions is illustrated by several recent findings. B-Myc protein, which only has an N-terminal domain, inhibits the function of c-Myc proteins, suggesting that N-terminal protein-protein interactions are important for c-Myc function (Resar et al. 1993). The TATA-binding protein, TBP (Hateboer et al. 1993; Maheswaran et al. 1994), and the Rb-like protein, p107 (Gu et al. 1994), have recently been shown to interact with the N-terminal domain of c-Myc.

Unlike non-AUG initiation, initiation at downstream AUG codons has been shown to yield N-terminally truncated proteins with predictable activities. For example, the transcriptional activator LAP (C/EBPβ) can be alternatively initiated at a downstream, in-frame initiator codon yielding the inactive repressor protein, LIP (Descombes and Schibler, 1991). There also exists alternative downstream-initiated forms of the thyroid hormone receptor which alter the transcriptional activity of the thyroid hormone receptor (Bigler et al. 1992). The Δ-c-Myc proteins have a truncated transactivation domain but retain the C-terminal dimerization and DNA binding domain. As with full-length c-Myc proteins, the Δ-c-Myc proteins are phosphorylated, appear to be localized to the nucleus and are unstable (Spotts and Hann unpublished). Unexpectably, the Δ-c-Myc proteins preferentially heterodimerize with Max protein in competition with the full-length c-Myc proteins, suggesting that Δ-c-Myc can successfully compete with full-length c-Myc for free Max proteins. Predictably, Δ-c-Myc inhibited transactivation by full-length c-Myc through EFII and EMS-enhancer elements in transient transfection assays in COS cells (Spotts and Hann unpublished). Taken together, these data suggest that the expression of Δ-c-Myc represents an important mechanism by which the transcriptional activities of the full-length c-Myc proteins may be modulated. In addition, these data suggest that alterations of the highly-structured N-terminal

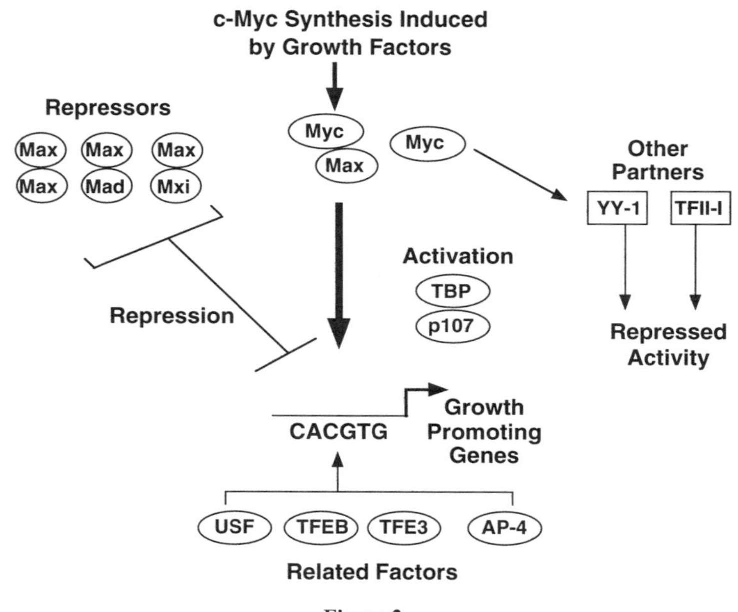

Figure 2.

domain can have significant effects on C-terminal interactions, such as heterodimerization with Max.

IMPLICATIONS FOR THE BIOLOGICAL FUNCTION OF THE Myc PROTEINS

The current model on the function of c-Myc suggests that growth stimulatory genes are transactivated through EMS elements by c-Myc/Max heterodimers. If the amount of c-Myc protein is low, then unphosphorylated Max homodimers (Kretzner et al. 1992; Gu et al. 1993), Mad/Max heterodimers (Ayer et al. 1993) or Mxi/Max heterodimers (Zervos et al. 1993) inhibit the transactivation. This simple model is complicated by the presence of other factors which bind well to the EMS sequence, such as USF (Gregor et al. 1990), TFE3 (Beckmann et al. 1990), TFEB (Carr and Sharp 1990) and AP-4 (Hu et al. 1990), other non-canonical Myc/Max binding sites (Blackwell et al. 1993), multiple forms of the Max protein (Makela et al. 1992) and the interactions with p107 (Gu et al. 1994), TBP (Hateboer et al. 1993; Maheswaran et al. 1994), TFII-I (Roy et al. 1993) and YY-1 (Shrivastava et al. 1993). Figure 2 diagrams the current model of c-Myc protein function. The functional significance of the various interactions has not yet been elucidated.

The regulation of transcription by the Myc proteins through a C/EBP sequence suggests that the overall transcription of genes containing this sequence may be regulated by the relative ratios of the two c-Myc proteins and C/EBP family members in cells. In support of this idea, Freytag and Geddes (1992) have shown that there is opposing regulation of adipogenesis by the C/EBPα and c-Myc 2 proteins. Since the dramatic and specific translational activation of c-Myc 1 protein appears to be controlled by the availability of methionine in the growth media, modulating the levels of c-Myc 1 protein may be one of the cellular responses to control growth in response to limiting nutrients. Figure 3 diagrams

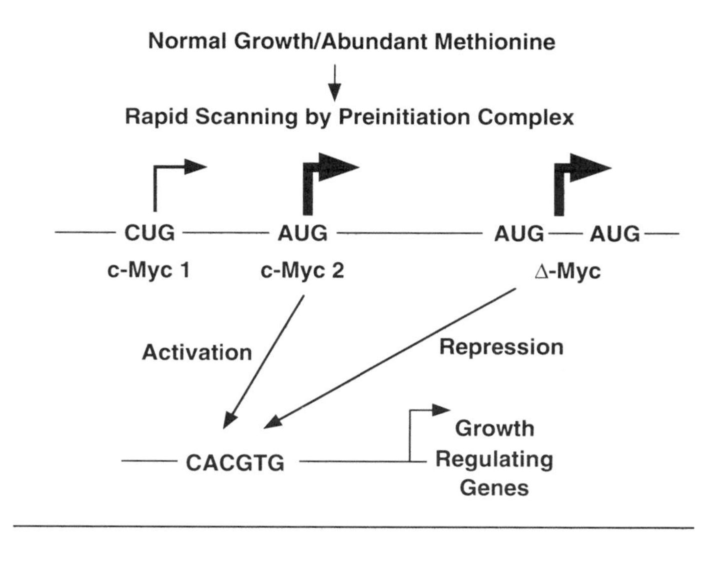

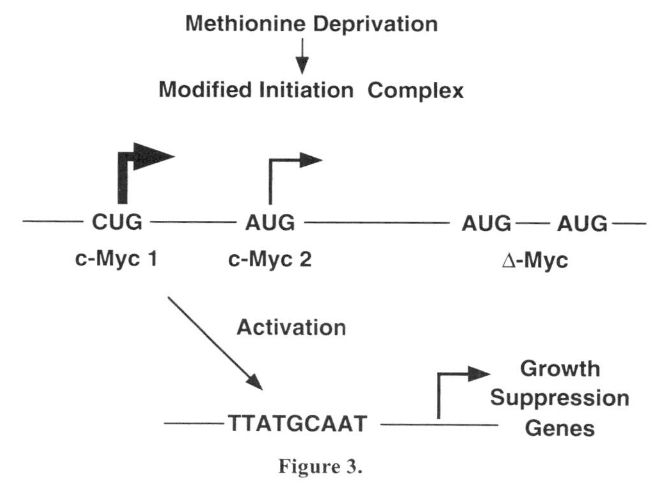

Figure 3.

how methionine deprivation may be involved in the regulation of the alternative and functionally-distinct c-Myc proteins to control the growth of certain cells.

Considering the differential, and at times opposing, regulation and transcriptional activities of the c-Myc proteins, it is likely that they have different biological roles. Our finding that c-Myc 1 and 2 proteins differentially transactivate through another sequence, the C/EBP binding site, yet both transactivate through the EMS sequence, suggests both distinct and overlapping functions for the two proteins. Previous studies have shown a clear role for c-Myc 2 and v-Myc proteins in growth stimulation and oncogenesis (Cole 1986; Spencer and Groudine 1991). In contrast to the ability of c-Myc 2 and v-Myc proteins to stimulate growth, the loss of c-Myc 1 in most Burkitt's lymphoma cell lines (Hann and Eisenman 1984; Hann et al. 1988) and avian bursal lymphoma cell lines (Spotts and Hann unpublished) and the induction of high levels of c-Myc 1 in growth-inhibited cells (Hann et al. 1992) suggest that the c-Myc 1 protein has a growth inhibitory function. Preliminary growth studies with different COS cell subclones overexpressing the c-Myc proteins demonstrate that c-Myc 1 does have a growth inhibitory effect (Hann et al. 1994). Interestingly, overexpression of C/EBPα has also been shown to be growth inhibitory (Umek et al. 1991). However, considering that both the c-Myc 1 and 2 proteins transactivate through EMS sequences, there are some biological functions which may be common to both proteins in some cells. Indeed, both proteins are able to co-transform Rat 1 cells with bcr-abl (Blackwood et al. 1994) and both are able to block differentiation of murine erythroleukemia cells (Spotts and Hann unpublished). Perhaps when there is a disruption of c-Myc 1 protein synthesis as a result of a genetic mutation or rearrangement, as in human Burkitt's lymphomas and avian bursal lymphomas, specific cells lose a growth-inhibiting response to limiting nutrients which contributes to tumorigenicity.

Interestingly, several hematopoietic tumor cell lines having alterations of the c-*myc* locus no longer express c-Myc 1 and some express constitutively high levels of Δ-c-Myc proteins. These lines even express Δ-c-Myc at levels equal to or greater than that of the full-length c-Myc 2 protein. While Δ-c-Myc proteins would not be expected to function alone in cotransformation assays with *ras* (Stone et al. 1987), the occurrence of tumors which express high levels of Δ-c-Myc suggests that these truncated proteins do not interfere with the growth-promoting functions of full-length c-Myc. Moreover, the Δ-c-Myc proteins may actually have a role in promotion of cell growth such as through the inhibition of negative growth regulatory genes. The inhibition of gene expression may be an important component of the mechanism by which many transcription factors and oncogenes including c-*myc* promote cell growth and transformation (reviewed in Luscher and Eisenman, 1990; Lewin, 1991; Foulkes and Sassone-Corsi, 1992).

The ability to express at least three amino-terminally distinct forms of c-Myc protein may be important for the normal function of c-*myc* in cell growth control. These three amino-terminally distinct c-Myc proteins appear to have distinct and apparently opposing biological functions. Furthermore, the properly balanced ratio of their synthesis, which can fluctuate depending on the nutritional environment, may provide a sensitive mechanism for the modulation of c-*myc* function during different stages of cell growth. Therefore, an imbalanced expression of the different c-Myc proteins, as was found in a number of tumor cell lines having alterations of the c-*myc* locus, may directly contribute to the loss of cell growth control associated with tumor development.

ACKNOWLEDGMENTS

This work was supported by U.S. Public Health Service grants CA-47399 and CA-48799 from the National Cancer Instituite and by a grant from the American Institute for Cancer Research.

REFERENCES

J. Abastado, P.F. Miller, and A.G. Hinnebusch, A quantitative model for translational control of the GCN4 gene of Saccharomyces cerevisiae. *The New Biologist* 3:511-524 (1991).

B., Amati, M.W. Brooks, N. Levy, T. Littlewood, G. Evan, and H. Land, Oncogenic activity of the c-Myc protein requires dimerization with Max. *Cell* 72:233-245 (1993).

C. Amin, A. Wagner, and N. Hay, Sequence-specific transactivation by myc and repression by max. *Mol.Biol.Cell.* 13:383-390 (1993).

D. Askew, R. Ashmun, B. Simmons, and J. Cleveland, Constitutive c-myc expression in an IL-3-dependent myeloid cell line suppresses cell cycle arrest and apoptosis. *Oncogene* 6:1915-1922 (1991).

D.E. Ayer, L. Kretzner, and R.N. Eisenman, Mad: A heterodimeric partner for Max that antagonizes myc transcriptional activity. *Cell* 72:211-222 (1993).

H.L. Beckmann, L.K. Su, and T. Kadesch, TFE3: A helix-loop-helix protein that activates transcription through the immunoglobulin enhancer μE3 motif. *Genes and Dev.* 4:167-179(1990).

J. Bigler, W. Hokanson, and R.N. Eisenman, Thyroid hormone receptor transcriptional activity is potentially autoregulated by truncated forms of the receptor. Mol. Cell. Biol. 12:2406-417 (1992).

T.K. Blackwell, L. Kretzner, E.M. Blackwood, R.N. Eisenman, and H. Weintraub, Sequence-specific DNA binding by the c-myc protein. *Science* 250:1149-1151 (1990).

T.K. Blackwell, J. Huang, A. Ma, L. Kretzner, F. Alt, R.N. Eisenman, and H. Weintraub, Binding of Myc proteins to canonical and non-canonical DNA sequences. *Mol. Cell. Biol.* 13:5216-5224 (1993).

E.M. Blackwood, and R.N. Eisenman, Max: A helix-loop-helix zipper protein that forms a sequence-specific DNA-binding complex with myc. *Science* 251:1211-1217 (1991).

E. Blackwood, T. Lugo, L. Kretzner, M. King, A. Street, O. Witte, and R.N. Eisenman, Functional analysis of the AUG- and CUG-initiated forms of the c-Myc protein. *Mol. Biol. Cell*, 5:597-609 (1994).

Z. Cao, R. Umek, and S. McKnight, Regulated expression of three C/EBP isoforms during adipose conversion of 3T3-L1 cells. *Genes Dev.* 5:1538-1552 (1991).

C.S. Carr, and P.A. Sharp, A helix-loop-helix protein related to immunoglobulin E box-binding proteins. *Mol. Cell. Biol.* 10:4384-4388 (1990).

M.D. Cole, The myc oncogene:its role in transformation and differentiation. *Ann. Rev. Genet.* 20:361-384 (1986).

P. Descombes, and U. Schibler, A liver-enriched transcriptional activator protein, LAP, and a transcriptional inhibitory protein, LIP, are translated from the same mRNA. *Cell* 67:569-579 (1991).

T.F. Donahue, A.M. Cigan, E.K. Pabich, and B.C. Valavicius, Mutations at a Zn(II) finger motif in the yeast eIF-2β gene alter ribosomal start-site selection during the scanning process. *Cell* 54:621-632 (1988).

H. Dosaka-Akita, R.K. Rosenberg, J.D. Minna, and M.J. Birrer, A complex pattern of translational initiation and phosphorylation in L-Myc proteins. *Oncogene* 6:371-378 (1991).

G. Evan, A.H. Wyllie, C. Gilbert, T. Littlewood, H. Land, M. Brooks, C. Waters, L. Penn, and D. Hancock, Induction of apoptosis in fibroblasts by c-myc protein. *Cell* 69:119-128 (1992).

N.S. Foulkes, and P. Sassone-Corsi, More is better: activators and repressors from the same gene. *Cell* 68:411-414 (1992).

S.O. Freytag, and T.J. Geddes, Reciprocal regulation of adipogenesis by myc and C/EBPα. *Science* 256:379-382 (1992).

P.D. Gregor, M. Sawadogo, and R.G. Roeder, The adenovirus major late transcription factor USF is a member of the helix-loop-helix group of regulatory and binds to DNA as a dimer. *Genes and Dev.* 4:1730-1740 (1990).

W. Gu, K. Cechova, V. Tassi, and R. Dalla-Favera, Opposite regulation of gene transcription and cell proliferation by c-Myc and Max. *Proc. Natl. Acad. Sci.* 90:2935-2939 (1993).

W. Gu, K. Bhatia, I. Magrath, C. Dang, and R. Dalla-Favera, Binding and suppression of the Myc transcriptional activation domain by p107. *Science* 264:251-254 (1994).

S.R. Hann, M. Dixit, R. Sears, and L. Sealy, The alternatively-initiated c-Myc proteins differentially regulate transcription through a noncanonical DNA binding site. *Genes and Dev.* 8:2441-2452 (1994).

S.R. Hann, and R.N. Eisenman, Proteins encoded by the human c-myc oncogene: differential expression in neoplastic cells. *Mol. Cell. Biol.* 4:2486-2497 (1984).

S.R. Hann, M.W. King, D.L. Bentley, C.W. Anderson, and R.N. Eisenman, A non-AUG translational initiation in c-myc exon 1 generates an N-terminally distinct protein whose synthesis is disrupted in Burkitt's lymphomas. *Cell* 34:185-195 (1988).

S.R. Hann, K. Sloan-Brown, and G. Spotts, Translational activation of the non-AUG-initiated c-*myc* 1 protein at high cell densities due to methionine deprivation. *Genes and Dev.* 6:1229-1240 (1992).

G. Hateboer, H. Timmers, A. Rustgi, M. Billaud, L. Van 'T Veer, and R. Bernards, TATA-binding protein and the retinoblastoma gene product bind to overlapping epitopes on c-Myc and adenovirus E1a protein. *Proc. Natl. Acad. Sci.* 90:8489-8493 (1993).

J.W.B Hershey, Translational control in mammalian cells. *Ann. Rev. Biochem.* 60:717-755 (1991).

Y.F. Hu, B. Luscher, A. Admon, N. Mermod, and R. Tjian, Transcription factor AP-4 contains multiple dimerization domains that regulate dimer specificity. *Genes and Dev.* 4:1741-1752 (1990).

G.J. Kato, J. Barrett, M. Villa-Garcia, and C.V. Dang, An amino-terminal c-myc domain required for neoplastic transformation activates transcription. *Mol. Cell. Biol.* 10:5914-5920 (1990).

M. Kozak, Point mutations define a sequence flanking the AUG initiator codon that modulates translation by eukaryotic ribosomes. *Cell* 44:283-292 (1986).

M. Kozak, Context effects and inefficient initiation at non-AUG codons in eukaryotic cell-free translation systems. *Mol. Cell. Biol.* 9:5073-5080 (1989).

M. Kozak, Regulation of translation in eukaryotic systems. *Ann. Rev. Cell Biol.* 8:197-225 (1991).

L. Kretzner, E.M. Blackwood, and R.N. Eisenman, Myc and Max proteins possess distinct transcriptional activities. *Nature* 359:426-429 (1992).

B. Lewin, Oncogenic conversion by regulatory changes in transcription factors. *Cell* 64:303-312 (1991).

B. Luscher, and R.N. Eisenman, New light on myc and myb. Part I. myc. *Genes and Dev.* 4:2025-2035 (1990).

S. Maheswaran, H. Lee, and G. Sonenshein, Intracellular association of the protein product of the c-myc oncogene with the TATA-binding protein. *Mol. Cell. Biol.* 14:1147-1152 (1994).

T.P. Makela, P.J. Koskinen, I. Vastrik, and K. Alitalo, Alternative forms of Max as enhancers or suppressors of Myc-Ras cotransformation. *Science* 256:373-376 (1992).

D.S. Peabody, Translation initiation at non-AUG triplets in mammalian cells. *J. Biol. Chem.* 264:5031-5035 (1989).

H. Persson, L. Hennighausen, R. Taub, W. DeGrado, and P. Leder, Antibodies to human c-myc gene product: evidence of an evolutionarily conserved protein induced during cell proliferation. *Science* 225:687-693 (1984).

G.C. Prendergast, and E.B. Ziff, Methylation-sensitive sequence-specific DNA binding by the c-myc basic region. *Science* 251:186-189 (1991).

M. Ramirez, R.C. Wek, and A.G. Hinnenbusch, Ribosome association of GCN2 protein kinase, a translational activator of the GCN4 gene of Saccharomyces cerevisiae. *Mol. Cell. Biol.* 11:3027-3036 (1991).

G. Ramsay, L. Stanton, M. Schwab, and J.M. Bishop, The human proto-oncogene N-myc encodes nuclear proteins that bind DNA. *Mol. Cell. Biol.* 6:4450-4457 (1986).

L.M. Resar, C. Dolde, J.F. Barrett, C.V. and Dang, B-Myc inhibits neoplastic transformation and transcriptional activation by c-Myc. *Mol. Cell. Biol.* 13:1130-1136 (1993).

A. Roy, C. Carruthers, T. Gutjahr, and R.G. Roeder, Direct role for Myc in transcription initiation mediated by interactions with TFII-I. *Nature* 365:359-361 (1993).

L. Sealy, and R. Chalkley, At least two nuclear proteins bind specifically to the Rous sarcoma virus long terminal repeat enhancer. *Mol. Cell. Biol.* 7:787-798 (1987).

R.C. Sears, and L. Sealy, Characterization of nuclear proteins that bind the EFII enhancer sequence in the Rous sarcoma virus LTR. *J. Virol.* 66:6338-6352 (1992).

A. Shrivastava, S. Saleque, G.V. Kalpana, S. Artandi, S.P. Goff, and K. Calame, Inhibition of transcriptional regulator Yin-Yang-1 association with c-Myc. *Science* 262:24796-24804 (1993).

C.A. Spencer, and M. Groudine, Control of c-myc regulation in normal and neoplastic cells. *Adv. Cancer Res.* 56:1-48 (1991).

J. Stone, T. DeLange, G. Ramsay, E. Jakobovits, J.M. Bishop, H. Varmus, and W. Lee, Definition of regions of human c-myc that are involved in transformation and nuclear localization. *Mol. Cell. Biol.* 7:1967-1709 (1987).

D. Tzamarias, I. Roussou, and G. Thireos, Coupling of GCN4 mRNA translational activation with decreased rates of polypeptide chain initiation. *Cell* 57:947-954 (1989).

R.M. Umek, A.D. Friedman, and S.L. McKnight, CCAAT/enhancer binding protein: a component of a differentiation switch. *Science* 251:288-292 (1991).

R.C. Wek, B.M. Jackson, and A.G. Hinnebusch, Juxtaposition of domains homologous to protein kinases and histidyl-tRNA synthetases in GCN2 protein suggests a mechanism for coupling GCN4 expression to amino acid availability. *Proc. Natl. Acad. Sci. USA* 86:4579-4583 (1989).

N.P. Williams, A.G. Hinnebusch, and T.F. Donahue, Mutations in the structural genes for eukaryotic initiation factors 2α and 2β of Saccharomyces cerevisiae disrupt translational control of GCN4 mRNA. *Proc. Natl. Acad. Sci. USA* 86:7515-751 (1989).

A.S. Zervos, J. Gyuris, and R. Brent, Mxi1, a protein that specifically interacts with Max to bind Myc/Max recognition sites. *Cell* 72:223-232 (1993).

11

PROGRESSIVE LOSS OF SENSITIVITY TO GROWTH CONTROL BY RETINOIC ACID AND TRANSFORMING GROWTH FACTOR-BETA AT LATE STAGES OF HUMAN PAPILLOMAVIRUS TYPE 16-INITIATED TRANSFORMATION OF HUMAN KERATINOCYTES

Kim E. Creek,[1] Gemma Geslani,[2] Ayse Batova,[2] and Lucia Pirisi[3]

[1] Children's Cancer Research Laboratory
Department of Pediatrics, and Department of Pathology
University of South Carolina School of Medicine
Columbia, South Carolina 29208
[2] Department of Chemistry and Biochemistry
University of South Carolina
Columbia, South Carolina 29208
[3] Department of Pathology
University of South Carolina School of Medicine
Columbia, South Carolina 29208

I. INTRODUCTION

Human papillomavirus (HPV), especially HPV types 16, 18, and 33, play an important role in the development of cervical cancer.[1-3] We,[4,5] as well as others,[6-9] have shown that DNA from these HPV types immortalize cultured human foreskin keratinocytes (HKc) and human cervical cells. HPV16-immortalized HKc (HKc/HPV16) represent an attractive *in vitro* model system to study the process of multistep carcinogenesis in human cells, and for identifying agents that can modify or prevent HPV-induced transformation, since they undergo malignant conversion through four phenotypically defined "steps" or stages. Although immortal, HKc/HPV16 are not tumorigenic and retain many features characteristic of normal HKc, including a requirement for epidermal growth factor (EGF) and bovine pituitary extract (BPE) for proliferation.[4,5] In the next step of this progression model, growth factor independent sublines, HKc/GFI, that no longer require EGF and BPE for proliferation, are selected from HKc/HPV16.[5,10] Next, sublines that no longer respond to calcium and serum-induced differentiation, HKc/DR, are selected from HKc/GFI.[5] In the final step,

Diet and Cancer, Edited under the auspices of the American Institute of Cancer Research
Plenum Press, New York, 1995

malignant conversion of HKc/DR is achieved by transfection with viral Harvey *ras* (v-Ha-*ras*) or herpes simplex virus type 2 DNA.[11,12] The steps from immortalized HKc/HPV16 to malignant conversion are highly reproducible and at each step the cells progressively acquire phenotypic and biochemical characteristics similar to those found in cell lines established from human tumors.[10]

We have previously shown that HKc/HPV16 are more sensitive than normal HKc to growth and differentiation control by the active metabolite of vitamin A, all-*trans*-retinoic acid (RA).[13-15] In addition, RA treatment of HKc/HPV16 reduced steady state levels of the HPV16 oncogenes E6 and E7 at both the mRNA and protein levels.[13-15] Also, RA, at physiologic levels (1 nM), inhibited HPV16-mediated immortalization of normal HKc by about 95%.[14,15] Overall, these results provided a biochemical basis for several epidemiological studies that have associated a diet poor in vitamin A and/or its precursor, ß-carotene, to an increased cervical cancer risk.[16-19] Namely, our studies suggest that endogenous RA may serve as a natural inhibitor of proliferation of cells harboring HPV, possibly through an inhibition of the expression of the HPV16 oncogenes E6 and E7.

The precursor to vitamin A, ß-carotene, and vitamin A derivatives, the retinoids, have shown potential both in the chemoprevention of cancer in humans,[20-25] and in the treatment of premalignant lesions such as cervical intraepithelial neoplasia (CIN). In phase I and phase II clinical trials RA reversed cervical dysplasia in some patients, demonstrating the feasibility of retinoids in the chemoprevention and treatment of cervical malignancies.[26-31] The most recent study found that RA was effective in treating stage II CIN but was without effect in patients with more advanced lesions.[32,33] However, 13-*cis*-RA was found to be effective in advanced cervical lesions, squamous cell carcinoma, when used in combination with interferon alpha.[34,35]

Recent studies suggest that RA inhibition of growth of cells harboring HPV and inhibition of HPV16 early gene expression may be mediated through the cytokine transforming growth factor-beta (TGF-ß), a potent inhibitor of HKc proliferation.[36-40] Glick *et al.* [41] were the first to show that RA treatment of mouse keratinocytes induces TGF-ß production and we have recently reported that RA treatment of normal HKc and HKc/HPV16 induces the secretion of TGF-ß types 1 and 2.[42] Woodworth *et al.*[43] found that TGF-ß inhibits E6 and E7 mRNA in several HPV16-immortalized human genital epithelial cell lines. Therefore, the inhibition of HPV E6 and E7 expression by RA that we[13-15] and others[44] have reported may be mediated through TGF-ß.

In the present studies we have used the HPV16-initiated *in vitro* model of multistep carcinogenesis of HKc to compare the sensitivity of early to late passage HKc/HPV16, HKc/GFI, and HKc/DR to growth control by RA and TGF-ß. These studies found a loss of RA and TGF-ß sensitivity as the cells progressed in culture. Overall, loss of growth inhibition by RA paralleled loss of TGF-ß sensitivity. The loss of RA and TGF-ß sensitivity observed during progression of HPV16-immortalized HKc suggests that retinoids may be most effective in the prevention of cervical cancer, and in the treatment of HPV-induced lesions at early preneoplastic stages.

II. MATERIALS AND METHODS

A. Materials

RA was from Eastman Kodak Co. or Sigma Chemical Co. [^{3}H]Thymidine (specific activity, 65 Ci/mmol) was purchased from ICN and Na^{125}I (13.0 mCi/µg) was from Amersham Corp. YM 10 ultrafiltration membranes were from Amicon. A rabbit anti-porcine platelet TGF-ß1 antibody (purified IgG) was from R&D Systems. This antibody cross-reacts and neutralizes both human TGF-ß1 and TGF-ß2. Porcine TGF-ß1 and TGF-ß2 were also from R&D Systems. Recombinant human TGF-ß1 was a gift from Genentech, Inc. Cesium trifluoroacetate was from Pharmacia LKB Biotechnology, Inc.

B. Cell Culture and Cell Lines

Normal HKc were isolated from newborn foreskins as described previously[4,5] except the epidermis was separated from the dermis by incubation overnight at 4° C in 0.25% trypsin (Gibco/BRL) instead of collagenase. Isolation and characterization of the immortalized HKc/HPV16 lines has been described in detail in previous publications.[4,5] These cell lines were obtained by transfecting normal HKc strains, each derived from a different individual, with the plasmid pMHPVd; a head-to-tail dimer of the full length HPV16 DNA cloned into the *Bam*H1 site of the vector pdMMTneo, which carries a gene for resistance to the antibiotic G418.[4] The different immortalized lines were selected with G418 and were designated HKc/HPV16d-1 to -5.

Growth factor independent HKc lines, HKc/GFI, were established by maintenance of the various HKc/HPV16 lines in complete medium (see below for media composition) lacking BPE and EGF. The establishment and growth characteristics of the HKc/GFI lines have been described previously.[5,10] Differentiation resistant HKc/HPV16, HKc/DR, were selected from HKc/HPV16d-1 by maintenance in medium containing 5% fetal bovine serum and 1 mM calcium chloride.[5]

Normal HKc and HKc/HPV16 were cultured in serum-free MCDB153-LB medium, supplemented with hydrocortisone (0.2 μM), insulin (5 μg/ml), transferrin (10 μg/ml), triiodothyronine (10 nM), $CaCl_2$ (0.1 mM), EGF (5 ng/ml), and BPE (35-50 μg protein/ml) with medium changes every 48 h.[4,5] This medium will be referred to as complete MCDB153-LB medium (CM). HKc/GFI were maintained in CM without EGF and BPE, which will be referred to as growth factor depleted medium (GFDM). HKc/DR were cultured in CM containing 5% FBS and 1 mM calcium chloride. Media were changed every 48 h. Cells were routinely split 1:10 so that each passage represents 3 to 4 population doublings.

C. Probes

A 1.04-kb human TGF-ß1 cDNA, subcloned in pSP64, was provided by Dr. Arjun Singh of Genentech, Inc. The cDNA was separated from vector sequences by *Eco*RI digestion followed by agarose gel electrophoresis, and purified with a Gene Clean Kit (BIO 101). A 2.3-kb human TGF-ß2 cDNA (pPC21) was provided by Dr. Anthony Purchio of Oncogen.[45] The cDNA was separated from vector sequences following *Eco*RI digestion, and a 1.08-kb TGF-ß2 fragment was then isolated, following digestion of the purified insert with *Ava*II. A human ß2-microglobulin cDNA probe, labeled by random priming, was used as a control for RNA loading.

D. Clonal Growth Assay

Normal HKc and different passage HKc/HPV16d-1 were plated at low density (1,000 cells/60-mm culture dish) in CM. Cells were fed 1 and 6 days after plating with 8 ml/dish of CM containing various concentrations of RA. Retinoic acid was added to the medium in dimethyl sulfoxide (DMSO) and the controls contained DMSO only. The final DMSO concentration was 0.1%. Colonies were fixed in methanol and stained with Giemsa 10 to 12 days after plating. The total area of the colonies, relative to the area of the dish, was determined by computerized image analysis.[10,13]

E. Mass Culture Growth Assay

HKc/HPV16d-1 at different passages and HKc/GFI were plated at a density of 20,000 cells/35-mm culture dish in their respective media and re-fed 24 h after plating in media containing DMSO only (0.1%), or the indicated concentration of RA. Cell number was determined in triplicate dishes for each RA concentration 6 days after plating, and every other day up to about 2 weeks in culture, by trypsinizing and counting cells in a hemocytometer.

F. Conditioned Media Collections

Conditioned media were collected from normal HKc, HKc/HPV16, and HKc/GFI cultured in the absence or the presence of the indicated concentrations of RA, with media changes every 24 h. Conditioned media collections were made in CM lacking BPE, due to the presence of TGF-ß in BPE. Initiation of RA treatment was begun when the cells were one-third to one-half confluent. The conditioned media were clarified by centrifugation and concentrated (30 to 40-fold) by ultrafiltration using an Amicon YM10 membrane filter. Both the ultrafiltration membrane and the filter unit were rinsed with Dulbecco's phosphate buffered saline containing 0.1% bovine serum albumin (BSA) prior to use.

G. Acid-Activation of Conditioned Media

Most cells synthesize and secrete TGF-ß predominantly as a latent biologically inactive precursor, which can be activated *in vitro* to a biologically active form by transient acidification.[46-48] Therefore, the following protocol was used to activate latent TGF-ß present in concentrated conditioned media. The media were brought to pH 1.5 with 5 N HCl (5-7 μl/500μl conditioned media) for 1 h at room temperature. The media were then neutralized with a volume of 5 N sodium hydroxide containing 1 M HEPES equal to that of added 5 N HCl. Controls received a premixed solution of acid and base.

H. Radioiodination of TGF-ß1

Recombinant human TGF-ß1 was labeled with ^{125}I using a modified chloramine-T method described by Frolik and DeLarco.[49] Specific activity ranged from 19-67 μCi/μg TGF-ß1.

I. TGF-ß Radioreceptor Competition Assay

Total TGF-ß present in conditioned media was quantified using the TGF-ß radioreceptor competition assay described by Wakefield.[50] Briefly, HKc/HPV16d-1 (100,000 cells/well) were plated in 24-well clusters and cultured in CM for 48 h. Cells were re-fed 24 h after plating. Cells were washed once (1 ml) with binding buffer (Dulbecco's Modified Eagles Medium containing 0.1% BSA and 25 mM HEPES buffer, pH 7.4) followed by two 1 h incubations with binding buffer (1 ml) at 37° C, prior to a final wash with binding buffer (1 ml). Binding buffer (200 μl) containing ^{125}I-TGF-ß1 (30-40 pM) and concentrated conditioned media or unlabeled porcine TGF-ß1 were added to wells and incubated at room temperature for 2 h. Cell were then washed 4 times with 1 ml of ice-cold Hank's buffered saline solution containing 0.1% BSA and lysed in 0.75 ml of solubilization buffer (1% Triton X-100, 10% glycerol, 20 mM HEPES, pH 7.4). Radioactivity was determined in an aliquot of the lysate in a Beckman gamma counter. Nonspecific binding, measured in the presence of excess (10 nM) unlabeled recombinant human TGF-ß1, averaged 9 ± 5% of total binding.

TGF-ß concentrations in the conditioned media samples were calculated from a standard inhibition of ^{125}I-TGF-ß1 binding curve generated by adding increasing amounts (5-100 pM) of unlabeled porcine TGF-ß1 to the binding assay.

J. Northern Blot Analysis for TGF-ß1 and TGF-ß2

Total RNA was extracted from cells according to the guanidine thiocyanate-cesium trifluoroacetate technique, modified as described by Yasumoto *et al.*[51] and polyadenylated RNA selected by oligo(dT)-cellulose (Collaborative Research) affinity chromatography. Formaldehyde-agarose electrophoresis of polyadenylated RNA was performed in 1.2% agarose/2.2 M formaldehyde gels and RNA was transferred to Gene-Screen (Du Pont-New England Nuclear) filters by electroblotting. Polyadenylated RNA was UV-crosslinked to the filter by exposing the blot to a germicidal shortwave UV lamp for 10 min. The blot was then baked for 2 h at 80° C. Hybridizations with random-primed ^{32}P-labeled human TGF-ß1 and TGF-ß2 specific DNA probes were conducted under stringent conditions: 5X SSPE (1X SSPE is 0.15 M NaCl - 10 mM NaH_2PO_4, pH 7.4 - 1 mM EDTA)- 50% formamide - 10% dextran sulfate - 5X Denhardt solution - 1% SDS - 100 µg/ml yeast tRNA, at 42° C for 16 h. The filters were washed twice in 2X SSC (1X SSC is 0.15 M NaCl, 0.015 M sodium citrate, pH 7.0)- 0.5% SDS for 20 min at room temperature, once in 0.25X SSC, 0.1% SDS at 42° C for 20 min and then once in 0.1X SSC - 0.1% SDS at 42° C for 20 min. Filters were exposed to Kodak XAR-2 film for 7 h for TGF-ß1 and 3-4 days for TGF-ß2, at -80° C with an enhancing screen.

K. [^{3}H]Thymidine Uptake Assay

Normal HKc, different passage HKc/HPV16d-1, or HKc/DR were plated (50,000 cells/well) into 24-well clusters (Costar) in their respective media. At 24 h after plating, the cells were re-fed with CM (1 ml) containing the indicated concentrations of TGF-ß1 or TGF-ß2, and 24 h later the cells were pulsed for 16 h with [^{3}H]thymidine (0.5 µCi/well). The cells were then washed 3 times with 1 ml of CM, followed by 2 washes with 1 ml of ice-cold 10% trichloroacetic acid. The cells were solubilized in 0.5 ml of 1 N sodium hydroxide containing 0.1% SDS. [^{3}H]Thymidine present in 0.4 ml of each sample was determined in 5 ml of an aqueous compatible scintillation cocktail containing 0.4 ml of 1 N HCl.

III. RESULTS

A. HPV16-Immortalized Human Keratinocytes as a Model for Multistep Human Cell Carcinogenesis *In Vitro*

We have developed a four-stage *in vitro* model of HPV-mediated transformation of normal HKc that reflects many key individual steps that are thought to occur during the multistep process of human cell carcinogenesis *in vivo*. In this model, normal HKc, immortalized by transfection with HPV16 DNA (HKc/HPV16), undergo malignant conversion through the four phenotypically defined "steps" or stages illustrated in Figure 1. In the first step (immortalization), HKc/HPV16 lines are established by transfection of normal HKc with HPV16 DNA.[4] Like normal HKc, HKc/HPV16 retain a requirement for EGF and BPE for proliferation in serum-free MCDB153-LB medium. In addition, HKc/HPV16 respond to serum- and calcium-induced terminal differentiation, in a manner similar to normal HKc. In

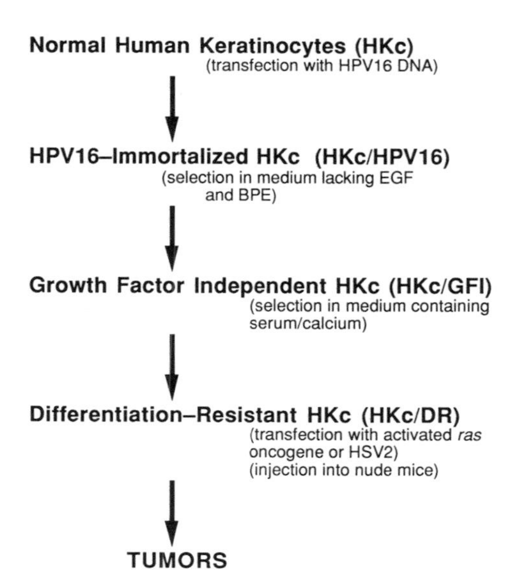

Figure 1. Schematic diagram of *in vitro* model of multistep carcinogenesis of human papillomavirus type 16-immortalized human keratinocytes.

the second step (growth factor independence), growth factor independent lines (HKc/GFI), able to proliferate in the absence of EGF and BPE, are established from HKc/HPV16 by selection in MCDB153-LB medium lacking EGF and BPE.[5,10] In the third step, (differentiation-resistance, DR), differentiation resistant lines (HKc/DR), which no longer respond to serum- and calcium-induced terminal differentiation, are selected from HKc/GFI by growth in MCDB153-LB medium containing 5% fetal bovine serum and 1.0 mM calcium chloride.[5,10] In the final step, malignant conversion was achieved by others by transfection of HKc/DR with v-Ha-*ras* or herpes simplex virus type 2 DNA.[11,12] This model reflects recognized steps that occur during the process of transformation *in vivo*. For example, only "high-risk" HPV types, those associated with malignancies *in vivo*, immortalize HKc or cervical cells *in vitro*.[6] Also, the loss of growth factor requirements (for EGF and BPE) in HKc/GFI parallels the reduced growth factor/serum requirements often found in human tumor-derived cell lines established from a wide variety of neoplasms, which also often show a lack of differentiation, like HKc/DR.[52-62] Interestingly, it has been found that cervical cancers, like HKc/GFI,[10] express increased levels of the EGF receptor.[63-65]

B. Loss of Sensitivity to Growth Control by Retinoic Acid during Progression of HPV16-Immortalized Cells

We previously reported that early passage HKc/HPV16 are about 100-fold more sensitive than normal HKc to growth inhibition by RA in both clonal and mass culture growth assays.[13,14] Northern blot analysis showed that RA treatment of HKc/HPV16 inhibited the expression of mRNA for the HPV16 oncogenes E6 and E7 in a dose- and time-dependent manner.[13,14] RA also inhibited the expression of other early open reading frames of HPV16, including E2 and E5.[14,15] Inhibition of E6 and E7 mRNA expression by RA was paralleled by reduced protein levels of E6 and E7.[14,15]

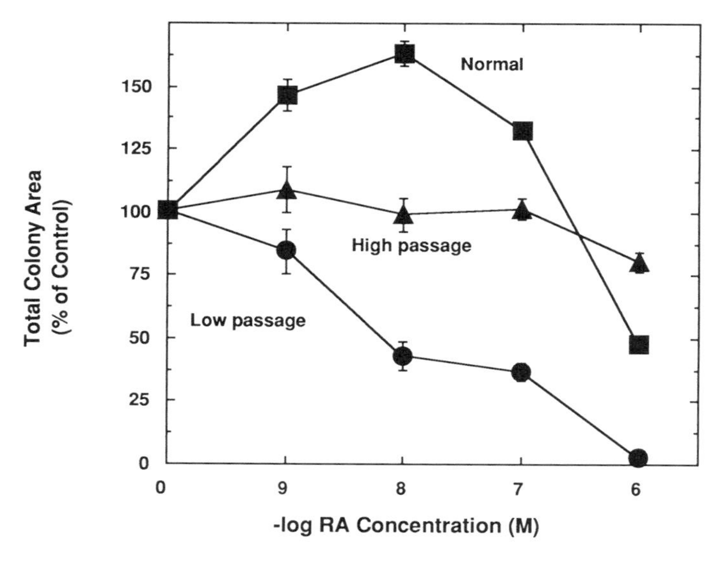

Figure 2. Effect of retinoic acid on clonal growth of normal HKc and low and high passage HKc/HPV16. Normal HKc (■), low passage HKc/HPV16d-1 (less than passage 35) (●), and high passage HKc/HPV16d-1 (greater than passage 90) (▲) were plated at 1,000 cells/60-mm culture dish and re-fed the following day with complete MCDB153-LB medium (8 ml) containing the indicated concentrations of RA. The cells were re-fed as above 6 days after plating and stained with Giemsa 5 days later. The total area of the colonies, relative to the surface area of the culture dish, was determined by computerized image analysis. The area of the dish occupied by colonies in cultures not receiving RA was normalized to 100%. The data are the means ± SD of at least three dishes per experimental condition.

We used a clonal growth assay to compare the sensitivity of low and high passage HKc/HPV16 to growth control by RA. In this procedure, cells are plated at low density and allowed to attach before the initiation of RA treatment, thus, plating efficiency is not affected by RA and the area of the colonies is a direct measure of the extent of growth. The clonal growth assay was quantified using an image analysis system which determines the area of the culture dish occupied by colonies, relative to the total area of the culture dish. As shown in Figure 2, normal HKc were growth stimulated by low RA concentrations and inhibited about 50% by 1 μM RA. Low passage (less than 35) HKc/HPV16 were very sensitive to growth inhibition by RA and exhibited almost 60% growth inhibition by 10 nM RA. In contrast, growth of high passage (greater than 90) HKc/HPV16 was not inhibited by 10 nM RA and was inhibited only about 20% by 1 μM RA (Figure 2). Interestingly, although 1 μM RA inhibited the growth of both high passage HKc/HPV16 and normal HKc, high passage HKc/HPV16, unlike normal HKc, showed no growth stimulation in response to low RA concentrations (Figure 2).

We also compared the influence of RA on the growth of low and high passage HKc/HPV16 and HKc/GFI in mass culture. In the mass culture growth assays, the cells are plated at high density and cell numbers determined at different times (between one and two weeks) in cultures treated with various concentrations of RA. In agreement with the clonal growth assays, high passage HKc/HPV16 were much less sensitive to growth inhibition by RA than low passage HKc/HPV16 (Figure 3). HKc/GFI exhibited about the same sensitivity to RA as high passage HKc/HPV16 (Figure 3).

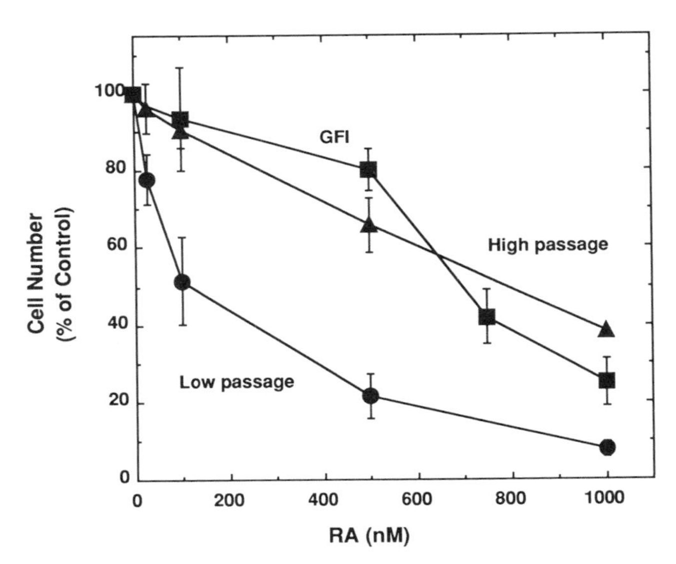

Figure 3. Effect of retinoic acid on mass culture growth of low and high passage HKc/HPV16 and HKc/GFI. Low passage HKc/HPV16d-1 (less than passage 25), (●) high passage HKc/HPV16d-1 (greater than passage 145) (▲), and HKc/GFI (■) were plated at a density of 20,000 cells/35-mm culture dish in their respective media. Cells were re-fed 24 h after plating and every 48 h thereafter in media without RA or containing the indicated RA concentration. Triplicate dishes for each experimental group were trypsinized and cell numbers determined by counting in a hemocytometer 6 days after plating and every other day up to about 2 weeks in culture. Each data point represents the average ± SD of the percent of cells found in dishes treated with RA compared to those grown in the absence of RA.

C. Retinoic Acid Induction of Transforming Growth Factor-Beta Secretion in Normal and HPV16-Immortalized Human Keratinocytes

Since similar cellular responses are elicited by RA and the cytokine TGF-ß, we reasoned that TGF-ß may act as a local mediator of RA action. In other words, could RA be inhibiting the proliferation of normal HKc and HPV16-immortalized HKc through an induction in the synthesis and secretion of the negative growth factor TGF-ß? It has previously been shown that normal HKc proliferation is strongly inhibited by picomolar concentrations of both TGF-ß1 and TGF-ß2.[36-40] To determine whether RA modulated TGF-ß secretion by normal HKc and HKc/HPV16 we used a specific quantitative TGF-ß radioreceptor competition assay[50] to measure total TGF-ß in conditioned media collected from normal HKc and HKc/HPV16 treated with increasing concentrations of RA. RA induced TGF-ß secretion in a dose-dependent manner in both normal HKc and HKc/HPV16 (Figure 4). Significant induction of TGF-ß by RA was observed at 1 to 10 nM. Induction of TGF-ß secretion by 100 nM RA was about 2-fold for normal HKc and about 5-fold for HKc/HPV16.

The time course of TGF-ß induction by RA was also determined. HKc/HPV16 were cultured without or with RA (100 nM) for 4 days, with media changes every 24 h. TGF-ß levels in conditioned media collected at the end of each 24 h period were then determined by the radioreceptor competition assay. Significant induction (1.6-fold) of TGF-ß secretion by RA was found in media collected during the first 24 h of RA treatment (Figure 5). Maximal induction (2.7-fold) of secreted TGF-ß levels were observed after 3 days of RA treatment

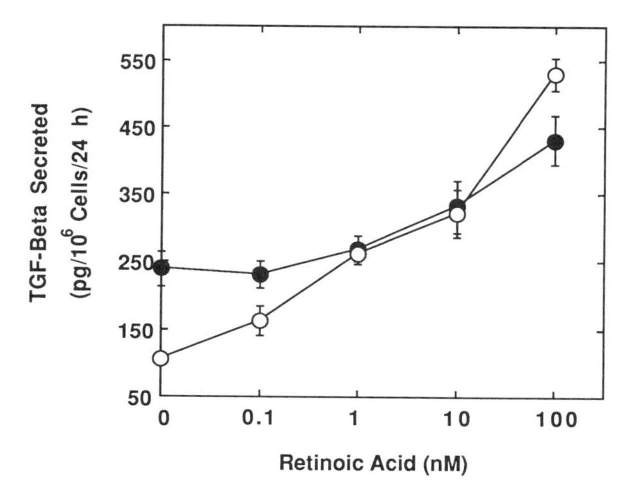

Figure 4. Dose-dependence of TGF-ß induction by retinoic acid in normal HKc and HKc/HPV16. Normal HKc (●) or HKc/HPV16d-2 (O) were treated with increasing concentrations of RA with media changes every 24 h. Conditioned media were pooled, concentrated by ultrafiltration, and acid-activated. Total TGF-ß in the conditioned media was quantified by the radioreceptor competition assay as described under "Materials and Methods". Each value is the mean ± SD of triplicate determinations. Reproduced with permission from Batova *et al.*[42]

(Figure 5). A similar experiment conducted with normal HKc also found significant induction (1.5-fold) after 24 h of RA exposure, and further induction (up to 2-fold) of TGF-ß secretion after 2 days of RA treatment (data not shown).

We next sought to determine if several independently derived HKc/HPV16 lines, as well as HKc/GFI, also displayed enhanced TGF-ß production in response to RA. As shown in Table 1, RA treatment of all HKc/HPV16 and HKc/GFI lines resulted in enhanced (1.4 to 3.2-fold) levels of total secreted TGF-ß. While the absolute levels of secreted TGF-ß varied substantially between the different lines, all HKc/HPV16 and HKc/GFI

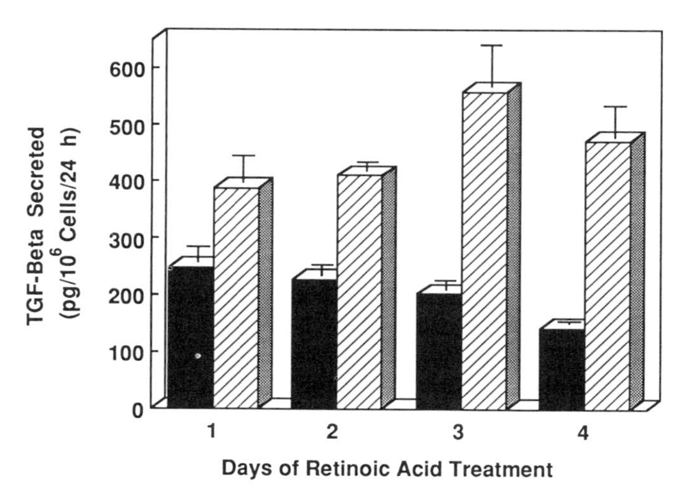

Figure 5. Time course of TGF-ß induction by retinoic acid in HKc/HPV16. HKc/HPV16d-2 were cultured in the absence (solid bars) or the presence (hatched bars) of 100 nM RA for the days indicated with media changes every 24 h. Conditioned media collected at the end of each 24 h period was concentrated by ultrafiltration and acid-activated. Total TGF-ß in the conditioned media was quantified by the radioreceptor competition assay as described under "Materials and Methods." Each value is the mean ± SD of triplicate determinations. Reproduced with permission from Batova *et al.*[42]

Table 1. Total TGF-ß secreted by normal HKc and HKc/HPV16 lines cultured in the absence and the presence of retinoic acid[a]

Cells	Retinoic Acid	Total TGF-ß Secreted[b] (pM)	Total TGF-ß Secreted[b] (pg/10^6 cells/24 h)	Ratio (+RA/-RA)
Normal HKc	–	8.1 ± 0.4	452 ± 20	—
Normal HKc	+	11.8 ± 0.7	656 ± 38	1.5
HKc/HPV16d-2	–	5.5 ± 0.2	183 ± 8	—
HKc/HPV16d-2	+	17.4 ± 0.2	579 ± 5	3.2
HKc/GFId-2	–	3.2 ± 0.3	121 ± 12	—
HKc/GFId-2	+	5.2 ± 1.4	194 ± 52	1.6
HKc/HPV16d-4	–	10.4 ± 0.5	849 ± 43	—
HKc/HPV16d-4	+	19.3 ± 1.7	1578 ± 141	1.9
HKc/GFId-4	–	5.9 ± 0.8	479 ± 64	—
HKc/GFId-4	+	13.7 ± 0.4	1117 ± 30	2.3
HKc/HPV16d-5	–	17.1 ± 1.7	1373 ± 133	—
HKc/HPV16d-5	+	24.3 ± 2.0	1953 ± 160	1.4
HKc/GFId-5	–	7.2 ± 0.1	674 ± 6	—
HKc/GFId-5	+	18.2 ± 1.5	1716 ± 145	2.6

[a]HKc were cultured in the absence or the presence of RA (100 nM) for 2 days with a media change at 24 h. Conditioned media were collected during the final 24 h of RA treatment, concentrated by ultrafiltration and acid-activated. Total TGF-ß in the conditioned media was quantified by the radioreceptor competition assay as described under "Materials and Methods." Reproduced with permission from Batova *et al.*[42]

[b] The data are the means ± SD of triplicate determinations.

lines we have examined displayed increased secreted levels of total TGF-ß when cultured in the presence of RA. It should be noted that we have compared the ability of RA to induce TGF-ß production in both low and high passage HKc/HPV16 and have found no consistent differences in either basal levels of TGF-ß or the fold induction of TGF-ß secretion by RA.

We have also used a sandwich enzyme-linked immunosorbent assay[66] to determine the levels of TGF-ß1 and TGF-ß2 isotypes secreted by normal HKc, HKc/HPV16, and HKc/GFI, cultured in the absence and the presence of RA. RA treatment enhanced both TGF-ß1 and TGF-ß2 levels in normal HKc, HKc/HPV16, and HKc/GFI. Induction of TGF-ß1 by RA averaged 2.2-fold, while TGF-ß2 levels were generally induced to a greater extent, an average of 5.3-fold.[42]

D. Induction by Retinoic Acid of TGF-ß1 and TGF-ß2 mRNA

The studies just described demonstrated that RA induced the secretion of TGF-ß1 and TGF-ß2 in normal HKc and HPV16-immortalized HKc. To determine if RA also induced the steady-state levels of TGF-ß mRNA, Northern blot analysis was performed using TGF-ß1 and TGF-ß2 specific cDNA probes. Polyadenylated RNA was isolated from HKc/HPV16 cultured in the absence or the presence of RA (100 nM) for various times (1.5-24 h). Northern blot analysis with the TGF-ß1 probe revealed a single 2.5-kb transcript, that was induced about 1.5-fold after 3 h of RA treatment and 3-fold after 12 and 24 h of RA exposure (Figure 6). Hybridization with a human ß2-microglobulin cDNA probe indicated that similar

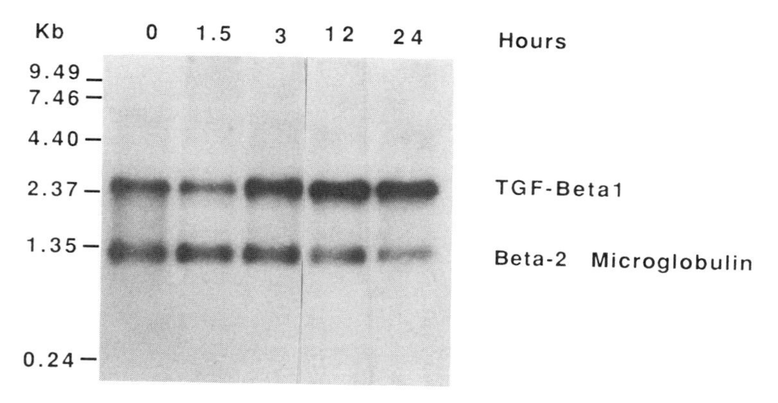

Figure 6. Retinoic acid increases TGF-ß1 mRNA in HKc/HPV16. Polyadenylated RNA was isolated from HKc/HPV16d-2 treated for the indicated times with 100 nM RA. Polyadenylated RNA (7 μg/lane) was separated by formaldehyde-agarose gel electrophoresis, transferred to Gene Screen filters, and probed with TGF-ß1 and ß2-microglobulin specific cDNA probes as described under "Materials and Methods". Exposure time for both TGF-ß1 and ß2-microglobulin was 7 h. The position of RNA molecular weight markers is on the left. Reproduced with permission from Batova *et al.*[42]

amounts of poly $(A)^+$ RNA were loaded in each lane. Northern blot analysis with a TGF-ß2 specific probe revealed four transcripts of sizes 6.0, 5.0, 3.6 and 2.7- kb, similar in size to those reported for mouse keratinocytes.[41,67] All four TGF-ß2 transcripts were induced by RA in a time-dependent manner (Figure 7). The major transcripts of sizes 5 and 6-kb were induced about 8-fold after 6 h of RA treatment and about 50-fold after 24 h of RA treatment (Figure 7).

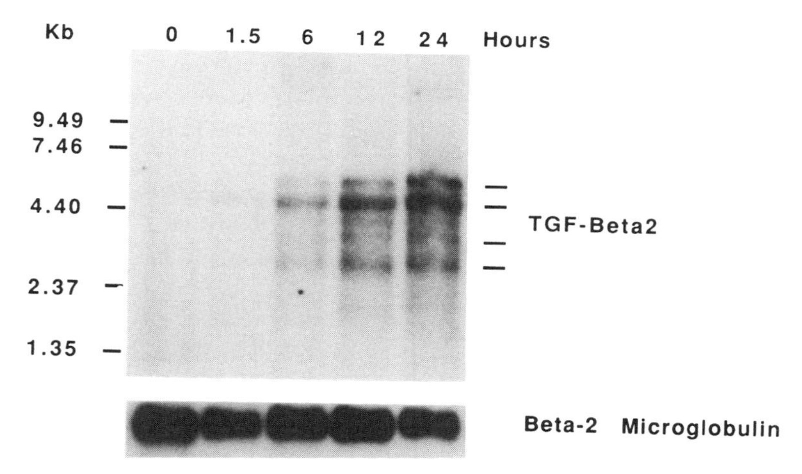

Figure 7. Retinoic acid increases TGF-ß2 mRNA in HKc/HPV16. Polyadenylated RNA was isolated from HKc/HPV16d-2 treated for the indicated times with 100 nM RA. Polyadenylated RNA (10 μg/lane) was separated by formaldehyde-agarose gel electrophoresis, transferred to Gene Screen filters, and probed with TGF-ß2 and ß2-microglobulin specific cDNA probes as described under "Materials and Methods". Exposure times were 3 days for TGF-ß2, and 4.5 h for ß2-microglobulin. The position of RNA molecular weight markers is on the left. The bars on the right indicate the position of the four TGF-ß2 transcripts. Reproduced with permission from Batova *et al.*[42]

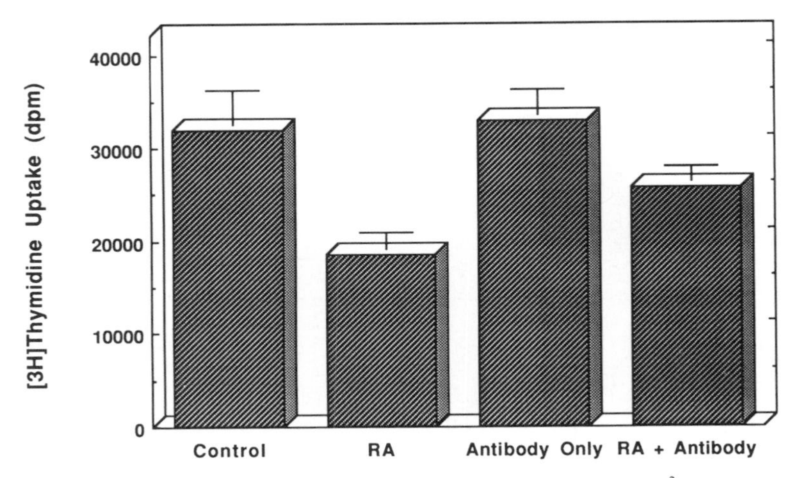

Figure 8. Antibodies to TGF-ß partially relieve retinoic acid inhibition of [^{3}H]thymidine uptake in HKc/HPV16. Low passage HKc/HPV16d-1 were plated (50,000 cells/well) into 24-well clusters in CM. The following day the cells were re-fed with 1.0 ml of CM only (control) or CM containing RA (100 nM), anti-TGF-ß antibodies only (5 µg/ml) or both RA and anti-TGF-ß antibodies and 24 h later [^{3}H]thymidine was added for 16 h. [^{3}H]Thymidine uptake was determined as described under "Materials and Methods." Each value is the mean ± SD of at least triplicate determinations.

E. Anti-TGF-ß Antibodies Partially Relieve Retinoic Acid Inhibition of [^{3}H]Thymidine Uptake by HKc/HPV16

We reasoned that if RA inhibition of growth of HPV16-immortalized cells was mediated through TGF-ß, neutralizing anti-TGF-ß antibodies should at least partially reverse inhibition of DNA synthesis by RA. The antibodies used in this experiment recognize and neutralize both TGF-ß1 and TGF-ß2. As shown is Figure 8, RA alone (100 nM) inhibited [^{3}H]thymidine uptake about 40%. However, in the presence of the anti-TGF-ß antibodies, RA inhibition of [^{3}H]thymidine uptake was 20%, about half that observed in the absence of the antibodies (Figure 8). The anti-TGF-ß antibodies alone did not influence [^{3}H]thymidine uptake by HKc/HPV16.

F. Effect of TGF-ß1 and TGF-ß2 on the Proliferation of HKc/HPV16 During *In Vitro* Progression

Previous studies by Pietenpol *et al.*[68] demonstrated that HKc immortalized with HPV16/E6-E7 under the control of a heterologous promoter (human ß-actin promoter) display resistance to the growth inhibitory effects of TGF-ß. We therefore compared the sensitivity of normal HKc, early to late passage HKc/HPV16, and HKc/DR, to inhibition of [^{3}H]thymidine uptake by TGF-ß1 and TGF-ß2. Normal HKc and very low passage HKc/HPV16 (passage 7) displayed similar dose-dependent inhibition of [^{3}H]thymidine uptake in response to increasing concentrations of TGF-ß1 (Figure 9). However, the cells became increasingly resistant to TGF-ß1 during *in vitro* progression, with the proliferation of HKc/DR being virtually unaffected by TGF-ß1 treatment (Figure 9). Similarly, TGF-ß2 inhibited [^{3}H]thymidine uptake in low passage (passage 14) but not high passage (passage 187) HKc/HPV16, HKc/GFI, or HKc/DR (Figure 10). Overall, loss of growth inhibition by RA paralleled loss of TGF-ß sensitivity.

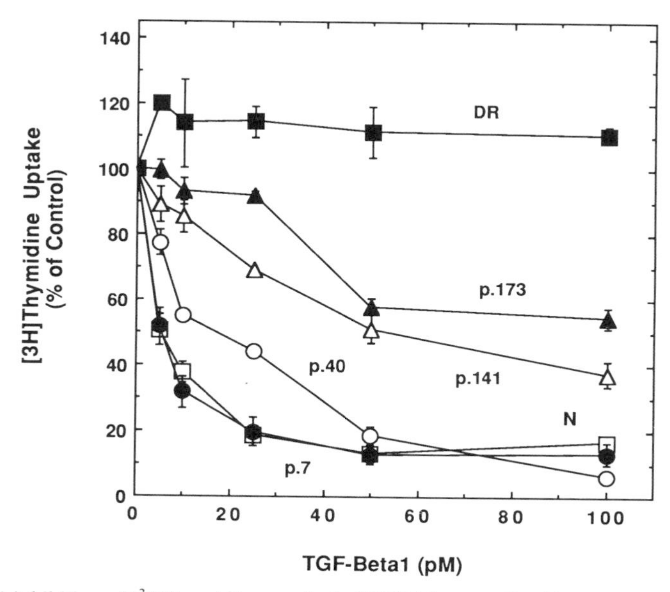

Figure 9. TGF-ß1 inhibition of [^{3}H]thymidine uptake in HPV16-immortalized human keratinocytes. Normal HKc (□), HKc/HPV16d-1 passage 7 (●), HKc/HPV16d-1 passage 40 (○), HKc/HPV16d-1 passage 141 (△), HKc/HPV16d-1 passage 173 (▲), and HKc/DR (■) were plated into 24-well clusters in CM. The following day the cells were re-fed with 1.0 ml of CM containing the indicated concentrations of TGF-ß1 and 24 h later [^{3}H]thymidine was added for 16-18 h. [^{3}H]Thymidine uptake was determined as described under "Materials and Methods." Each value is the mean ± SD of triplicate determinations.

IV. DISCUSSION

HPV16-immortalized HKc represent a unique model to study multistep human cell carcinogenesis *in vitro*. We have used this model system to study the sensitivity of HPV16-immortalized cells to the antiproliferative effects of the vitamin A metabolite, RA, and the cytokine TGF-ß, during *in vitro* progression. Overall, these studies found that as HPV16-immortalized cells progress in culture, they become increasingly resistant to the antiproliferative effects of both RA and TGF-ß. In general, loss of growth inhibition by RA paralleled loss of TGF-ß sensitivity.

We have previously shown that HKc/HPV16 are more sensitive to growth and differentiation control by RA than normal HKc and that this increased sensitivity is likely to be mediated by an inhibition of the expression of the HPV16 oncogenes E6 and E7.[13-15] E6 and E7 are the transforming genes of HPV16 and their constant expression is necessary to maintain continuous growth of cells transformed by HPV16.[69] Our results are supported by studies of Bartsch *et al.*[44] who found that RA can regulate HPV18 E6/E7 expression at the transcriptional level in HeLa cell hybrids. In addition, HPV16 immortalized human ectocervical cells, like HKc/HPV16, display an increased sensitivity to RA control of differentiation.[70]

The mechanism by which RA inhibits the expression of the HPV early genes is not known. However, recent results suggest that the effects of RA on E6 and E7 expression may be mediated by the cytokine TGF-ß. First, Woodworth *et al.*[43] reported that TGF-ß markedly inhibits the expression of HPV16 E6 and E7 in HPV16-immortalized human genital epithelial cells. In addition, as we have described in this chapter and in a recent paper,[42] RA

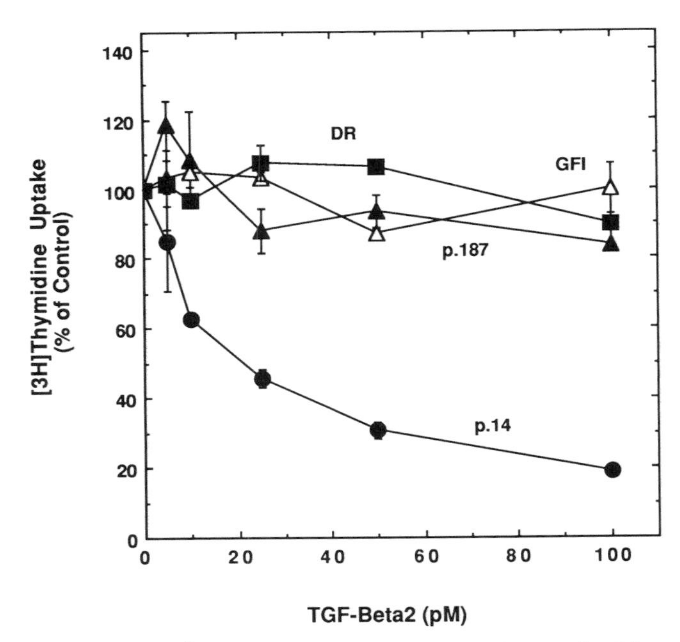

Figure 10. TGF-ß2 inhibition of [^{3}H]thymidine uptake in HPV16-immortalized human keratinocytes. HKc/HPV16d-1 passage 14 (●), HKc/HPV16d-1 passage 187 (▲), HKc/GFI (△), and HKc/DR (■) were plated into 24-well clusters in CM. The following day the cells were re-fed with 1.0 ml of CM containing the indicated concentrations of TGF-ß2 and 24 h later [^{3}H]thymidine was added for 16-18 h. [^{3}H]Thymidine uptake was determined as described under "Materials and Methods." Each value is the mean ± SD of triplicate determinations.

treatment of HPV16-immortalized cells induces TGF-ß secretion. It is thus likely that an autocrine loop exists whereby RA induces the secretion of TGF-ß, which in turn inhibits growth, through an inhibition of E6 and E7 expression.

In this chapter we have shown that as HPV16-immortalized cells progress in culture they become increasingly resistant to the growth inhibitory effects of both RA and TGF-ß. The mechanism for this acquired resistance is not yet known. However, we are actively determining whether the loss of sensitivity to RA and TGF-ß is paralleled by a loss in the ability of RA and TGF-ß to modulate E6 and E7 expression in late passage HKc/HPV16, HKc/GFI, and HKc/DR. The loss of sensitivity to TGF-ß that we observe in our HPV16 progression model is consistent with the observation that some human squamous cell carcinoma cell lines[71], as well as HPV16-positive malignant cervical carcinoma cell lines (Caski and Siha),[72] are refractory to growth inhibition by TGF-ß. Loss of sensitivity to TGF-ß clearly precedes malignant conversion since high passage HKc/HPV16, HKc/GFI, and HKc/DR, which exhibit increasing resistance to TGF-ß, are non-tumorigenic. Also, Pietenpol *et al.*[68] have reported TGF-ß resistance in HPV16, HPV18, or SV40-immortalized non-tumorigenic HKc. Loss of TGF-ß sensitivity during progression of HPV16-immortalized cells is unlikely to be associated with a loss of TGF-ß receptors, which has previously been shown to be a rare event.[46] In addition, both squamous carcinoma cells and cervical carcinoma cells which showed resistance to TGF-ß maintained the ability to bind TGF-ß.[71,72] Therefore, the molecular changes leading to TGF-ß resistance are downstream of receptor binding. Our multistep model of HPV16-immortalized HKc, in which the cells become progressively resistant to TGF-ß, is particularly well suited for future studies defining the molecular changes leading to TGF-ß resistance.

In conclusion, our earlier studies showing that the vitamin A metabolite, RA, dramatically inhibited the growth of HPV16-immortalized HKc, the expression of the HPV16 oncogenes in HKc/HPV16, and HPV16-mediated immortalization of normal HKc[13-15] provided a biochemical explanation for the epidemiological evidence which links a diet poor in vitamin A or its precursor, ß-carotene, to an increased risk of cervical cancer.[16-19] The studies presented in this report have used HPV16-immortalized HKc as an *in vitro* model of multistep human cell carcinogenesis. The results of these studies have shown that HPV16-immortalized HKc become increasingly resistant to the growth inhibitory effects of RA and TGF-ß as they progress from immortalization toward tumorigenicity. The results support a model whereby RA inhibition of growth and HPV16 early gene expression in HPV16-immortalized HKc is mediated via the cytokine TGF-ß. In addition, the loss of RA responsiveness during tumorigenic progression suggests that RA may be most effective in the chemoprevention of cervical cancer and in the treatment of HPV-induced lesions at early preneoplastic stages. In fact, this conclusion is consistent with the recent results of a randomized phase III clinical trial of cervical neoplasia which found that topical RA enhanced regression of moderate dysplasia (CINII) but had no effect on regression in cases of severe dysplasia.[32,33]

V. SUMMARY

Retinoids (vitamin A and its natural and synthetic derivatives) have shown potential as chemopreventive agents, and diets poor in vitamin A and/or its precursor ß-carotene have been linked to an increased risk of cancer at several sites including the cervix. Human papillomavirus (HPV) plays an important role in the etiology of cervical cancer. We have developed an *in vitro* model of cancer progression using human keratinocytes (HKc) immortalized by HPV16 DNA (HKc/HPV16). Although immortal, early passage HKc/HPV16, like normal HKc, require epidermal growth factor (EGF) and bovine pituitary extract (BPE) for proliferation and undergo terminal differentiation in response to serum and calcium. However, following prolonged culture, growth factor independent HKc/HPV16 lines that no longer require EGF and BPE can be selected (HKc/GFI). Further selection of HKc/GFI produces lines that are resistant to serum- and calcium- induced terminal differentiation (HKc/DR). HKc/DR, but not early passage HKc/HPV16, are susceptible to malignant conversion following transfection with viral Harvey *ras* or Herpes simplex virus type II DNA. We have investigated the sensitivity of low to high passage HKc/HPV16 and HKc/GFI to growth control by all-*trans*-retinoic acid (RA, an active metabolite of vitamin A). Early passage HKc/HPV16 are very sensitive to growth inhibition by RA, and in these cells RA decreases the expression of the HPV16 oncogenes E6 and E7. However, as the cells progress in culture they lose their sensitivity to RA. Growth inhibition by RA may be mediated through the cytokine transforming growth factor-beta (TGF-ß), a potent inhibitor of epithelial cell proliferation. RA treatment of HKc/HPV16 and HKc/GFI results in a dose- and time-dependent induction (maximal of 3-fold) in secreted levels of TGF-ß. Also, Northern blot analysis of mRNA isolated from HKc/HPV16 demonstrated that RA treatment induced TGF-ß1 and TGF-ß2 expression about 3- and 50-fold, respectively. We next studied the effect of TGF-ß1 and TGF-ß2 on the proliferation of early to late passage HKc/HPV16, HKc/GFI and HKc/DR. While early passage HKc/HPV16 were as sensitive as normal HKc to growth inhibition by TGF-ß1 and TGF-ß2, the cells became increasingly resistant to TGF-ß during *in vitro* progression, with the proliferation of HKc/DR being virtually unaffected by TGF-ß1 or TGF-ß2 treatment. Overall, loss of growth inhibition by RA parallels loss of TGF-ß sensitivity. These results support a mechanism in which RA regulates growth control by enhancing the production of TGF-ß, which, then inhibits cellular proliferation in an autocrine or paracrine manner. In addition, the loss of RA and TGF-ß sensitivity observed during progression of HKc/HPV16 suggests that retinoids may be most effective in the prevention of cervical cancer and in the treatment of HPV-induced lesions at early preneoplastic stages.

ACKNOWLEDGMENTS

This work was made possible by generous support from the American Institute for Cancer Research (Grants 88A05 and 91B38), the South Carolina Endowment for Children's Cancer Research, the South Carolina Cancer Center Challenge Award, and the Center for Cancer Treatment and Research at Richland Memorial Hospital (Columbia, South Carolina).

REFERENCES

1. H. zur Hausen, Papillomaviruses in human cancers, *Mol Carcinogenesis* 1:147 (1988).
2. H. zur Hausen, Papillomaviruses in anogenital cancer as a model to understand the role of viruses in human cancers, *Cancer Res* 49:4677 (1989).
3. K.V. Shah, and P.M. Howley, Papillomaviruses, *in*: "Virology", B.N. Fields, D.M. Knipe et al., eds. Raven Press, New York (1990).
4. L. Pirisi, S. Yasumoto, M. Feller, J. Doniger, and J.A. DiPaolo, Transformation of human fibroblasts and keratinocytes with human papillomavirus type 16 DNA, *J Virol* 61:1061 (1987).
5. L. Pirisi, K.E. Creek, J. Doniger, and J.A. DiPaolo, Continuous cell lines with altered growth and differentiation properties originate after transfection of human keratinocytes with human papillomavirus type 16 DNA, *Carcinogenesis* 9:1573 (1988).
6. C.D. Woodworth, J. Doniger, and J.A. DiPaolo, Immortalization of human foreskin keratinocytes by various human papillomavirus DNAs corresponds to their association with cervical carcinoma, *J Virol* 63:159 (1989).
7. M. Durst, R.T. Dzarlieva-Petrusevska, P. Boukamp, N.E. Fusenig, and L. Gissmann, Molecular and cytogenetic analysis of immortalized human primary keratinocytes obtained after transfection with human papillomavirus type 16 DNA, *Oncogene* 1:251 (1987).
8. P. Kaur, and J.K. McDougall, Characterization of primary human keratinocytes transformed by human papillomavirus type 18, *J Virol* 62:1917 (1988).
9. C.D. Woodworth, P.E. Bowden, J. Doniger, L. Pirisi, W. Barnes, W.D. Lancaster, and J.A. DiPaolo, Characterization of normal human exocervical epithelial cells immortalized in vitro by papillomavirus types 16 and 18 DNA, *Cancer Res* 48:4620 (1988).
10. L.L. Zyzak, L.M. MacDonald, A. Batova, R. Forand, K.E. Creek, and L. Pirisi, Increased levels and constitutive tyrosine phosphorylation of the epidermal growth factor receptor contribute to autonomous growth of human papillomavirus type 16 immortalized human keratinocytes, *Cell Growth & Differ* 5:537 (1994).
11. J.A. DiPaolo, C.D. Woodworth, N.C. Popescu, D.L. Koval, J.V. Lopez, and J. Doniger, HSV-2-induced tumorigenicity in HPV16-immortalized human genital keratinocytes, *Virology* 177:777 (1989).
12. J.A. DiPaolo, C.D. Woodworth, N. Popescu, and J. Doniger, Induction of human cervical squamous cell carcinoma by sequential transfection with human papillomavirus type 16 DNA and viral Harvey *ras*, *Oncogene* 4:394 (1989).
13. L. Pirisi, A. Batova, G.R. Jenkins, J.R. Hodam, and K.E. Creek, Increased sensitivity of human keratinocytes immortalized by human papillomavirus type 16 DNA to growth control by retinoids, *Cancer Res* 52:187 (1992)
14. K.E. Creek, G.R. Jenkins, M.A. Khan, A. Batova, J.R. Hodam, W.H. Tolleson, and L. Pirisi, Retinoic acid suppresses human papillomavirus type 16 (HPV16)-mediated transformation of human keratinocytes and inhibits the expression of the HPV16 oncogenes, *in*: "Diet and Cancer: Markers, Prevention, and Treatment," M.M. Jacobs, ed. Plenum Press, New York (1994).

15. M.A. Khan, G.R. Jenkins, W.H. Tolleson, K.E. Creek, and L. Pirisi, Retinoic acid inhibition of human papillomavirus type 16-mediated transformation of human keratinocytes, *Cancer Res* 53:905 (1993).
16. A. Bernstein, and B. Harris, The relationship of dietary and serum vitamin A to the occurrence or cervical intraepithelial neoplasia in sexually active women, *Am J Obstet Gynecol* 148:309 (1984).
17. T. Hirayama, Diet and cancer, *Nutr Cancer* 1:67 (1979).
18. J.R. Marshall, S. Graham, T. Byers, M. Swanson, and J. Brasure, Diet and smoking in the epidemiology of cancer of the cervix, *J Natl Cancer Inst* 70:847 (1983).
19. S.L. Romney, P.R. Palan, C. Duttagupta, S. Wassertheil-Smoller, J. Wylie, G. Miller, N.S. Slagle, and D. Lucido, Retinoids and the prevention of cervical dysplasias, *Am J Obstet Gynecol* 141:890 (1981).
20. R. Peto, R. Doll, J.D. Buckley, and M.B Sporn, Can dietary beta-carotene materially reduce human cancer rates? *Nature* 290:201 (1981).
21. J.S. Bertram, L.N. Kolonel, and F.L. Meyskens, Rationale and strategies for chemoprevention of cancers in humans, *Cancer Res* 47:3012 (1987).
22. S.M. Lippman, J.F. Kessler, and F.L. Meyskens, Retinoids as preventive and therapeutic anticancer agents (part I), *Cancer Treat Rep* 71:391 (1987).
23. S.M. Lippman, S.E. Benner, and W.K. Hong, Chemoprevention: Strategies for the control of cancer, *Cancer* 72:984 (1993)
24. W.K. Hong, and L.M. Itri, Retinoids and human cancer, *in*: "The Retinoids: Biology, Chemistry, and Medicine," M.B. Sporn, A.B. Roberts and D.S. Goodman, eds., p. 597, Raven Press, New York (1994).
25. S.M. Lippman, S.E. Brenner, and W.K. Hong, Cancer chemoprevention, *J Clin Oncol* 12:851 (1994).
26. V. Graham, E.A. Surwit, S. Weiner, and F.L. Meyskens, Phase II trial of ß-all-trans-retinoic acid for cervical intraepithelial neoplasia delivered via a collagen sponge and cervical cap, *West J Med* 145:192 (1986).
27. F.L. Meyskens, V. Graham, M. Chvapil, R.T. Dorr, D.S. Alberts, and E.A. Surwit, A phase I trial of ß-all-*trans*-retinoic acid delivered via a collagen sponge and a cervical cap for mild or moderate intraepithelial cervical neoplasia, *J Natl Cancer Inst* 71:921 (1983).
28. F.L. Meyskens, and E.S. Surwit, Clinical experience with topical tretinoin in the treatment of cervical dysplasia, *J Am Acad Dermatol* 15:826 (1986).
29. Y.M. Peng, D.S. Alberts, V. Graham, E.A. Surwit, S. Weiner, and F.L. Meyskens, Cervical tissue uptake of all-*trans*-retinoic acid delivered via a collagen sponge-cervical cap delivery device in patients with cervical dysplasia, *Invest New Drugs* 4:245 (1986).
30. E.A. Surwit, V. Graham, W. Droegemueller, D. Alberts, M. Chvapil, R.T. Dorr, J.R. Davis, and F.L. Meyskens, Evaluation of topically applied *trans*-retinoic acid in the treatment of cervical intraepithelial lesions, *Am J Obstet Gynecol* 143:821 (1982).
31. S.A. Weiner, E.A. Surwit, V.E. Graham, and F.L. Meyskens, A phase I trial of topically applied *trans*-retinoic acid in cervical dysplasia-clinical efficacy, *Invest New Drugs* 4:241 (1986).
32. F.L. Meyskens, E. Surwit, T.E. Moon, J.M. Childers, J.R. Davis, R.T. Dorr, C.S. Johnson, and D.S. Alberts, Enhancement of regression of cervical intraepithelial neoplasia (moderate dysplasia) with topically applied all-*trans*-retinoic acid: a randomized trial, *J Natl Cancer Inst* 86:539 (1994).
33. M.B. Sporn and A.B. Roberts, Cervical dysplasia regression induced by all-*trans*-retinoic acid, *J Natl Cancer Inst* 86:476 (1994).
34. S.M. Lippman, J.J. Kavanagh, M. Paredes-Espinoza, F. Delgadillo-Madrueno, P. Paredes-Casillas, W.K. Hong, E. Holdener, and I.H. Krakoff, 13-*cis*-Retinoic acid plus interferon alpha-2a: Highly active systemic therapy for squamous cell carcinoma of the cervix, *J Natl Cancer Inst* 84:241 (1992).
35. S.M. Lippman, J.J. Kavanagh, M. Paredes-Espinoza, F. Delgadillo-Madrueno, P. Paredes-Casillas, W.K. Hong, G. Massimini, E. E. Holdener, and I.H. Krakoff, 13-*cis*-Retinoic acid plus interferon alpha-2a in locally advanced squamous cell carcinoma of the cervix, *J Natl Cancer Inst* 85:499 (1993).
36. G.D. Shipley, M.R. Pittelkow, J.J. Wille, R.E. Scott, and H.L. Moses, Reversible inhibition of normal human prokeratinocyte proliferation by type ß transforming growth factor-growth inhibitor in serum-free medium, *Cancer Res* 46:2068 (1986).
37. R.J. Coffey, C.C. Bascom, N.J. Sipes, R. Graves-Deal, B.E. Weissman, and H.L. Moses, Selective inhibition of growth-related gene expression in murine keratinocytes by transforming growth factor ß, *Mol Cell Biol* 8:3088 (1988).
38. R.J. Coffey, N.J. Sipes, C.C. Bascom, R. Graves-Deal, C.Y. Pennington, B.E. Weissman and H.L. Moses, Growth modulation of mouse keratinocytes by transforming growth factors, *Cancer Res* 48:1596 (1988).
39. M. Reiss, and A.C. Sartorelli, Regulation of growth and differentiation of human keratinocytes by type ß transforming growth factor and epidermal growth factor, *Cancer Res* 47:6705 (1987).

40. K. Matsumoto, K. Hashimoto, M. Hashiro, H. Yoshimasa, and K. Yoshikawa, Modulation of growth and differentiation in normal human keratinocytes by transforming growth factor-ß, *J Cell Physiol* 145:95 (1990).
41. A.B. Glick, K.C. Flanders, D. Danielpour, S.H. Yuspa, and M.B. Sporn, Retinoic acid induces transforming growth factor-ß2 in cultured keratinocytes and mouse epidermis, *Cell Regulation* 1:87 (1989).
42. A. Batova, D. Danielpour, L. Pirisi, and K.E. Creek, Retinoic acid induces secretion of latent transforming growth factor ß1 and ß2 in normal and human papillomavirus type 16-immortalized human keratinocytes, *Cell Growth & Differ* 3:763 (1992).
43. C.D. Woodworth, V. Notario, and J.A. DiPaolo, Transforming growth factors beta 1 and 2 transcriptionally regulate human papillomavirus (HPV) type 16 early gene expression in HPV-immortalized human genital epithelial cells, *J Virol* 64:4767 (1990).
44. D. Bartsch, B. Boye, C. Baust, H. zur Hausen, and E. Schwarz, Retinoic acid-mediated repression of human papillomavirus 18 transcription and different ligand regulation of the retinoic acid receptor ß gene in non-tumorigenic and tumorigenic HeLa hybrids, *EMBO J* 11:2283 (1992).
45. L. Madisen, N.R. Webb, T.M. Rose, H. Marquardt, T. Ikeda, D. Twardzik, S. Seyedin, and A.F. Purchio, Transforming growth factor-ß2: cDNA cloning and sequence analysis, *DNA* 7:1 (1988).
46. L.M. Wakefield, D.M. Smith, T. Masui, C.C. Harris, and M.B. Sporn, Distribution and modulation of the cellular receptor for transforming growth factor-beta, *J Cell Biol* 105:965 (1987).
47. R. Pircher, D.A. Lawrence, and P. Jullien, Latent beta-transforming growth factor in nontransformed and Kirsten sarcoma virus-transformed normal rat kidney cells, clone 49F, *Cancer Res* 44:5538 (1984).
48. P.A. Dennis, and D.B. Rifkin, Cellular activation of latent transforming growth factor ß requires binding to the cation-independent mannose 6-phosphate/insulin-like growth factor type II receptor, *Proc Natl Acad Sci USA* 88:580 (1991).
49. C.A. Frolik, and J.E. DeLarco, Radioreceptor assay for transforming growth factors, *Methods Enzymol* 146:95 (1987).
50. L.M. Wakefield, An assay for type-ß transforming growth factor receptor, *Methods Enzymol* 146:167-173 (1987).
51. S. Yasumoto, J. Doniger, and J.A. DiPaolo, Differential early viral gene expression in two stages of human papillomavirus type 16 DNA-induced malignant transformation, *Mol Cell Biol* 7:2165 (1987).
52. E. Ohmura, M. Okada, N. Onoda, Y. Kamiya, H. Murakami, T. Tsushima, and K. Shizume, Insulin-like growth factor I and transforming growth factor α as autocrine growth factors in human pancreatic cancer cell growth, *Cancer Res* 50:103 (1990).
53. M.A. Anzano, D. Rieman, W. Prichett, D.F. Bowen-Pope, and R. Greig, Growth factor production by human colon carcinoma cell lines, *Cancer Res* 49:2898 (1989).
54. D.R. Hofer, E.R. Sherwood, W.D. Bromberg, J. Mendelsohn, C. Lee, and J.M. Kozlowski, Autonomous growth of androgen-independent human prostatic carcinoma cells: Role of transforming growth factor α, *Cancer Res* 51:2780 (1991).
55. M.-E. Forgue-Lafitte, A.-M. Coudray, B. Breant, and J. Mester, Proliferation of the human colon carcinoma cell line HT29: Autocrine growth and deregulated expression of the c-*myc* oncogene, *Cancer Res* 49:6566 (1989).
56. R. Kurokawa, S. Kyakumoto, and M. Ota, Autocrine growth factor in defined serum-free medium of human salivary gland adenocarcinoma cell line HSG, *Cancer Res* 49:5136 (1989).
57. M.B. Sporn, and A.B. Roberts, Autocrine growth factors and cancer, *Nature* 313:745 (1985).
58. M. Reiss, E.B. Stash, V.F. Vellucci, and Z.-L. Zhou, Activation of the autocrine transforming growth factor α pathway in human squamous carcinoma cells, *Cancer Res* 51:6254 (1991).
59. V.M. Macaulay, M.J. Everard, J.D. Teale, P.A. Trott, J.J. Van Wyk, I.E. Smith, and J.L. Millar, Autocrine function for insulin-like growth factor I in human small cell lung cancer cell lines and fresh tumor cells, *Cancer Res* 50:2511 (1990).
60. O.M. El-Badry, C. Minniti, E.C. Kohn, P.J. Houghton, W.H. Daughaday, and L.J. Helman, Insulin-like growth factor II acts as an autocrine growth and motility factor in human rhabdomyosarcoma tumors, *Cell Growth & Differ* 1:325 (1990).
61. R. Derynck, D.V. Goeddel, A. Ullrich, J.U. Gutterman, R.D. Williams, T.S. Bringman, and W.H. Berger, Synthesis of messenger RNAs for transforming growth factors α and ß and the epidermal growth factor receptor by human tumors, *Cancer Res* 47:707 (1987).
62. J.H. Mydlo, J. Michaeli, C. Cordon-Cardo, A.S. Goldenberg, W.D.W. Heston, and W.R. Fair, Expression of transforming growth factor α and epidermal growth factor receptor messenger RNA in neoplastic and nonneoplastic human kidney tissue, *Cancer Res* 49:3407 (1989).

63. A. Goppinger, F.M. Wittmaack, H.O. Wintzer, H. Ikenberg, and T. Bauknecht, Localization of human epidermal growth factor receptor in cervical intraepithelial neoplasias, *J Cancer Res Clin Oncol* 115:259 (1989).
64. T. Bauknecht, M. Kohler, I. Janz, and A. Pfleiderer, The occurrence of epidermal growth factor receptors and the characterization of EGF-like factors in human ovarian, endometrial, cervical and breast cancer, *J Cancer Res Clin Oncol* 115:193 (1989).
65. W. Gullick, J.J. Marsden, N. Whittle, B. Ward, L. Bobrow, and M.D. Waterfield, Expression of epidermal growth factor receptors on human cervical, ovarian, and vulval carcinomas, *Cancer Res* 46:285 (1986)
66. D. Danielpour, K.-Y. Kim, L.L. Dart, S. Watanabe, A.B. Roberts, and M.B. Sporn, Sandwich enzyme-linked immunosorbent assays (SELISAs) quantitate and distinguish two forms of transforming growth factor-beta (TGF-ß1 and TGF-ß2) in complex biological fluids, *Growth Factors* 2:61 (1989).
67. A.B. Glick, D. Danielpour, D. Morgan, M.B. Sporn, and S.H. Yuspa, Induction and autocrine receptor binding of transforming growth factor-ß2 during terminal differentiation of primary mouse keratinocytes, *Mol Endocrin* 4:46 (1990).
68. J.A. Pietenpol, R.W. Stein, E. Moran, P. Yaciuk, R. Schlegel, R.M. Lyons, M.R. Pittelkow, K. Munger, P.M. Howley, and H.L. Moses, TGF-ß1 inhibition of c-*myc* transcription and growth in keratinocytes is abrogated by viral transforming proteins with pRB binding domains. *Cell* 61:777 (1990).
69. T. Crook, J.P. Morgenstern, L. Crawford, and L. Banks, Continued expression of HPV-16 E7 protein is required for maintenance of the transformed phenotype of cells co-transformed by HPV-16 plus EJ-ras, *EMBO J* 8:513 (1989).
70. C. Agarwal, E.A. Rorke, J.C. Irwin, and R.L. Eckert, Immortalization by human papillomavirus type 16 alters retinoid regulation of human ectocervical epithelial cell differentiation, *Cancer Res* 51:3982 (1991).
71. C.D. Hebert, and L.S. Birnbaum, Lack of correlation between sensitivity to growth inhibition and receptor number for transforming growth factor ß in human squamous carcinoma cell lines, *Cancer Res* 49:3196 (1989).
72. L. Braun, M. Durst, R. Mikumo, and P. Gruppuso, Differential response of nontumorigenic and tumorigenic human papillomavirus type 16-positive epithelial cells to transforming growth factor ß1, *Cancer Res* 50:7324 (1990).

12

SHORT-CHAIN FATTY ACIDS AND MOLECULAR AND CELLULAR MECHANISMS OF COLONIC CELL DIFFERENTIATION AND TRANSFORMATION

Leonard H. Augenlicht, Anna Velcich, and Barbara G. Heerdt

Albert Einstein Cancer Center
Department of Oncology
111 East 210th Street
Bronx, New York 10467

I. INTRODUCTION

There has been enormous progress in defining structural alterations of genes common in colonic cancer. Accumulation of mutations and deletions in APC, p53, DCC, and Ki-ras are the *sine qua non* of the disease, although the roles and interactions of these genetic alterations in the biological and clinical heterogeneity of the disease are not understood. Of even greater interest from the point of view of disease prevention are inherited mutations found in the APC gene and in genes which encode mismatch repair functions, which lead to the high frequency of colonic cancer in familial polyposis (FAP) and hereditary non-polyposis colon cancer (HNPCC) families, respectively[1-6]. However, unlike inherited childhood cancers, such as retinoblastoma and Wilm's tumor, inherited colon cancer takes decades to develop. This can be partially understood in the context of what has been learned regarding genetic alterations in sporadic cancers. For example, one may assume that the inherited mutation supplies the first of a series of genetic alterations, and by so doing reduces the overall time and elevates the probability of accumulating the necessary number of alterations in genes such as those cited above. In the case of FAP, the inheritance of a mutant allele for APC provides one of the key events directly, since mutations in the gene occur somatically in over 70% of human colon tumors, and the mutation is sufficient to initiate development of intestinal tumors in mice[7,8]. In HNPCC, the situation is more complicated, in that the inherited mutations in genes responsible for DNA mismatch repair fail to repair errors which arise during DNA synthesis at tens of thousands of loci throughout the genome, and it must be presumed that among this plethora of changes there are key genes, again perhaps coincident with some of those mentioned above, which are eventual targets that lead to tumor formation. It is also not yet understood whether an inherited mutant allele for a gene necessary for DNA repair is penetrant at all, or whether significant defects in repair arise only following mutation in the second allele - that is, whether the initial mutation is dominant

Diet and Cancer, Edited under the auspices of the American Institute of Cancer Research
Plenum Press, New York, 1995

or recessive[9]. Nevertheless, it is possible to understand the early incidence and increased frequency of cancer in FAP and HNPCC in the context of the paradigm that has emerged which envisions that an accumulation of a limited number of discrete genetic alterations are necessary and sufficient for tumor development.

Similarly, it is possible to explain dietary effects on colonic cancer incidence in terms of this paradigm. For example, it can be hypothesized that dietary agents produce mutagens (eg free radicals) or elements that prevent the formation or decrease the concentration of such mutagens, and that this alters the frequency of mutational events throughout the genome, including loci which can modulate the incidence of cancer. However, it is also possible that there are more subtle mechanisms at work which account for the relatively modest shifts in cancer incidence which are associated with decades of exposure to particular diets.

The evidence that the situation is more complex than a simple accumulation of a discrete number of genetic alterations comes from the data which led to the development of the paradigm. It was shown 5 years ago that besides the relatively frequent alterations at loci encompassing DCC, p53, APC and Ki-ras, there were extensive and heterogeneous deletions throughout the genome of colonic tumors, involving in some tumors, up to 50% of the loci investigated and thus potentially many tens of thousands of genes[10]. This has been complemented more recently by the use of a sensitive method of comparative genomic hybridization, which has shown that low-level sequence amplification is much more widespread in the genome of colonic and other tumors than was appreciated by cytogenetic or standard molecular analyses[11-13]. Indeed, low-level amplification of the c-myc gene has been demonstrated in particularly aggressive colonic tumors[14], and may have prognostic value as determined from frequency of amplification in tissue from phase III studies of adjuvant therapy for colonic cancer (Augenlicht et al, unpublished). This work on structural alterations of genes in colonic tumors was preceeded by a wealth of data which showed that transformation in general was linked to alterations in expression of about 10%, or 1000, of the sequences normally expressed in a cell (eg[15]), and this has been specifically demonstrated for the development of both adenomas and carcinomas in the colon[16].

Thus, there is tremendous complexity in the events associated with tumor formation, and hence ample reason to suspect that environmental agents, such as nutritional factors, can interact with metabolic pathways in significant ways to alter the probability of tumor development and progression. Approximately 10 years ago we initiated a program to investigate this complexity, and several significant insights into the cellular and molecular events linked to tumor formation have developed from those initial experiments.

II. THE TRANSFORMATION OF COLONIC EPITHELIAL CELLS IS ASSOCIATED WITH HIGHLY PLEIOTROPIC EFFECTS ON PATTERNS OF GENE EXPRESSION IN COLONIC TUMORS AND IN THE FLAT MUCOSA AT RISK FOR TUMOR DEVELOPMENT

Utilizing methodology which permitted quantitative analysis of level of expression of each of four thousand cDNA sequences in small biopsies of human tissue, we demonstrated that 4% of the sequences assayed were modulated in expression in adenomas compared to the flat mucosa at low-risk for tumor development, and that this number increased to 7% in colonic carcinomas[16]. This degree of perturbation of gene expression in the transformed compared to the normal tissue is consistent with what is known in more simple systems of transformation from analysis of shifts in populations of RNA sequences assayed by solution hybridization techniques (reviewed in[16-22]). Most striking in these initial results was the finding that the flat mucosa from individuals at high genetic risk for development of colonic cancer (ie FAP and HNPCC) exhibited even more pronounced alterations in expression, involving up to 25% of the sequences analyzed[16,17]. Our interpretation of this observation was that the inherited mutation in these individuals generated highly pleiotropic and heterogeneous affects on gene expression in the tissue, perhaps by altering patterns of differentiation and cellular interactions in the mucosa. In fact, it has been shown recently that both rodent and human colonic tissue at elevated risk for tumor development exhibit heterogeneous enzyme altered foci[23,24]. Such foci, each expressing distinct patterns of colonic epithelial cell markers, are a direct reflection of both improper differentiation along many pathways and the great heterogeneity this generates in the mucosa at risk. The relatively narrower spectrum of alterations in the neoplasias which subsequently arise reflects the clonal expansion of an individual initiated cell in the tumor[17].

III. A PANEL OF SEQUENCES COULD BE SELECTED WHOSE PATTERN OF EXPRESSION WAS EFFECTIVE IN DISTINGUISHING THE GENETIC HIGH-RISK FLAT MUCOSA FROM LOW-RISK

From the data base on expression in tissue biopsies, we selected a panel of 30 sequences whose altered *pattern* of expression characterized the mucosa at genetic risk[17]. Figure 1 illustrates these data. For each of the biopsies of high risk tissue (FAP and HNPCC) and low-risk tissue, the presence of a shaded area in the figure reflects altered expression of a particular sequence (represented by the dots along the top of the figure) in the tissue at high-risk. The open spaces among the high-risk biopsies are false negatives, while the shaded areas among the low-risk biopsies are false positives. However, it is clear that the overall pattern of expression of the 30 selected sequences distinguishes the high-risk from the low-risk tissue. Since the two risk groups have very different genetic etiologies (mutations in APC for FAP, and in mismatch repair genes for HNPCC), the commonality of change for the two groups among the selected sequences suggests that there may be common pathways which converge on the transformed phenotype, even if the initiating mechanisms differ.

There is clearly great clinical potential in being able to identify colonic mucosa at risk for tumor development. Investigations using these sequences as intermediary markers in screening and intervention studies are underway in our laboratory.

IV. AMONG THIS PANEL OF SEQUENCES WHICH CHARACTERIZES HIGH-RISK ARE COORDINATELY REGULATED GENES ENCODED ON THE MITOCHONDRIAL H STRAND

Each of the 30 cloned sequences in this panel which characterizes risk has been at least partially sequenced. The genes fall into two groups. Three of the clones which decrease in expression are mitochondrial genes for COX (cytochrome oxidase) subunit II and III, and ND3[25,26]. It is interesting that all three decrease in expression since all genes on the mitochondrial H strand are coordinately regulated through a common promotor, generating a contiguous transcript which is post-transcriptionally processed into individual mRNA molecules[27,28]. Thus, the independent identification of three dif-

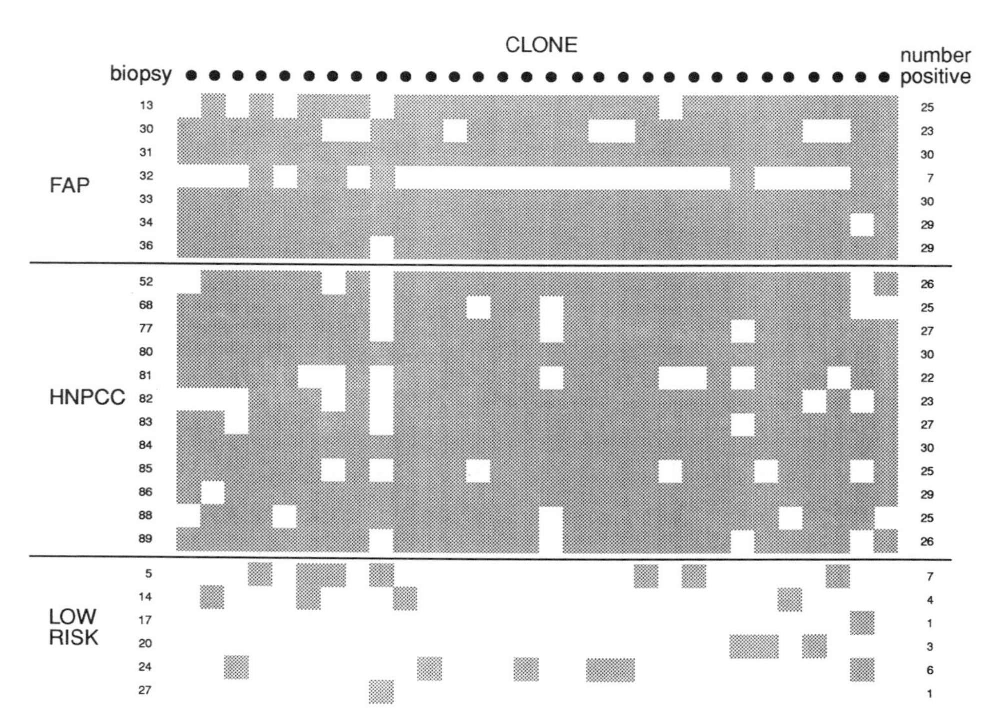

Figure 1. .Pattern of expression of selected cloned sequences in biopsies from the flat colonic mucosa of high-risk (FAP and HNPCC) and low-risk individuals. Biopsy samples are identified by the number at left; the selected clones are represented by the solid circles along the top. A shaded area is inserted if the data for a particular clone and biopsy characterized the biopsy as tissue at high-risk. The number at the right is the total number of sequences of the 30 which scored the biopsy as high risk. Open areas for the FAP and HNPCC biopsies are false positives, while shaded areas for the low-risk biopsies are false negatives. Reprinted from[17].

ferent mitochondrial genes, all of which decrease similarly in expression, is a validation of the scanning and image processing methodology used in the initial analysis. The second group of sequences, which are elevated in expression in high-risk tissue, contain different members of the repetitive Alu family[26,29]. These sequences will be returned to in the final section.

V. IN VITRO EXPERIMENTS DEMONSTRATED THAT SHORT-CHAIN FATTY ACIDS, GENERATED IN THE COLONIC MUCOSA BY FERMENTATION OF FIBER, COULD INDUCE BOTH DIFFERENTIATION AND A RETURN OF EXPRESSION OF THE MITOCHONDRIAL GENES BACK TOWARDS LEVELS WHICH CHARACTERIZE LOW-RISK

Since we envision risk for, and progression of, tumor development in the target tissue as dependent upon defective differentiation of the colonic epithelial cells, we investigated whether induction of differentiation in culture could induce these mitochondrial genes back towards normal levels of expression. Moreover, since short-chain fatty

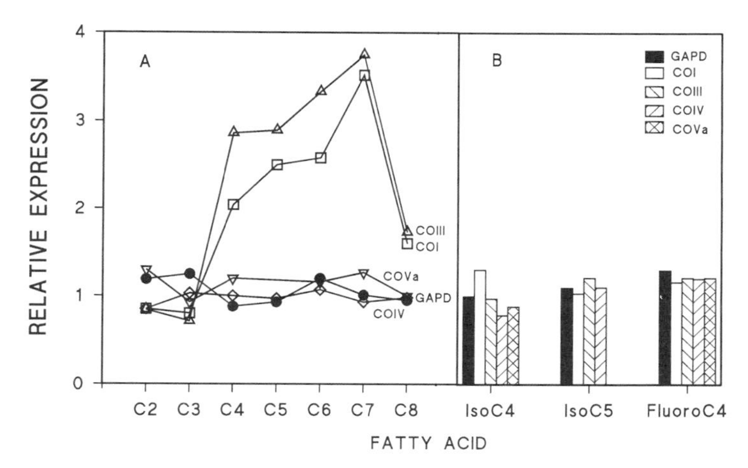

Figure 2. Effects of fatty acids on gene expression in HT29 cells. Relative expression of genes which encode COX (cytochrome oxidase) subunits I, III, IV, and Va and glyceraldehyde dehydrogenase is shown in HT29 cells treated with unbranched short-chain fatty acids of different lengths (A), or in cells treated with branched fatty acids or a non-metabolizable fluorinated derivative of butyrate. Reprinted from[30].

acids are effective inducers of colonic epithelial cell differentiation in culture and *in vivo*, and are the principal energy source for colonic epithelial cells *in vivo* (reviewed in[25,30,31]), we focused on their ability to induce both differentiation and mitochondrial gene expression. Figure 2 summarizes the results of these investigations. Unbranched fatty acids from 3 to 7 carbon atoms in length were effective inducers of COX I and III, both encoded on the mitochondrial H strand, but not of COX IV and Va, two subunits of this multisubunit mitochondrial enzyme which are encoded in the nucleus, synthesized in the cytoplasm, and transported into the mitochondria (fig 2A, from[30]). Thus, the effect of fatty acids is specific for subunits encoded by the mitochondrial genome. Figure 2B illustrates that branched fatty acids (isoC4 and isoC5), and a non-metabolizable derivative (fluoroC4) do not induce either mitochondrial or nuclear encoded genes. Therefore, the specificity for induction of mitochondrial genes may reside in the metabolism of the short-chain fatty acids by beta oxidation within the mitochondria[30]. This induction of mitochondrial gene expression also results in elevated cytochrome oxidase activity in partially purified mitochondria, suggesting that synthesis of the mitochondrially encoded subunits is rate limiting for assembly of functional enzyme[30].

These experiments led to the general hypothesis that in tissue at elevated risk and in colonic tumors, deficient mitochondrial gene expression leads to abnormalities in mitochondrial function and defects in cellular differentiation. Interestingly, structural and functional defects in mitochondria have been well-documented in colonic carcinoma cells[32-34]. We have carefully studied the mitochondrial H and L strand promotors in the D-loop region of colonic tumors and have found no evidence thus far for mutations or instability which could account for the decreased expression of the mitochondrial genome in the tumor cells[28]. Although we have not systematically sequenced the entire 16KB mitochondrial genome from a number of tumors, we have ruled out the presence of major deletions[35]. We therefore do not yet understand the biochemical mechanism responsible for the decreased expression, or for the elevation upon induction of differentiation.

VI. SPONTANEOUS DIFFERENTIATION CAN BE DEMONSTRATED IN COLONIC EPITHELIAL CELLS IN CULTURE, AND THESE CELLS PROCEDE TO APOPTOSIS, THE END STAGE OF DIFFERENTIATION. SHORT-CHAIN FATTY ACIDS WHICH INDUCE MITOCHONDRIAL GENE EXPRESSION AND DIFFERENTIATION ALSO INDUCE APOPTOSIS, AND THIS MAY BE A MECHANISM BY WHICH FIBER INFLUENCES INCIDENCE OF COLONIC CANCER

The question then arose as to the nature of the link between defective mitochondrial gene expression and differentiation, and elevated risk, tumor growth and progression. In the normal colonic mucosa, cells differentiate and cease proliferating as they migrate from a stem cell population in the lower third of the crypt to the lumenal surface. Near the top of the crypt, cells undergo apoptosis - programmed cell death - and are shed into the lumen. The link between extrusion into the lumen and apoptotic death is not clear, but we have recently demonstrated that in tissue culture, colonic epithelial cells which are shed into the medium exhibit a more differentiated phenotype and also undergo apoptosis[31]. The data shown in figure 3 are for HT29 cells. Cells which accumulate in the medium over 72 hours

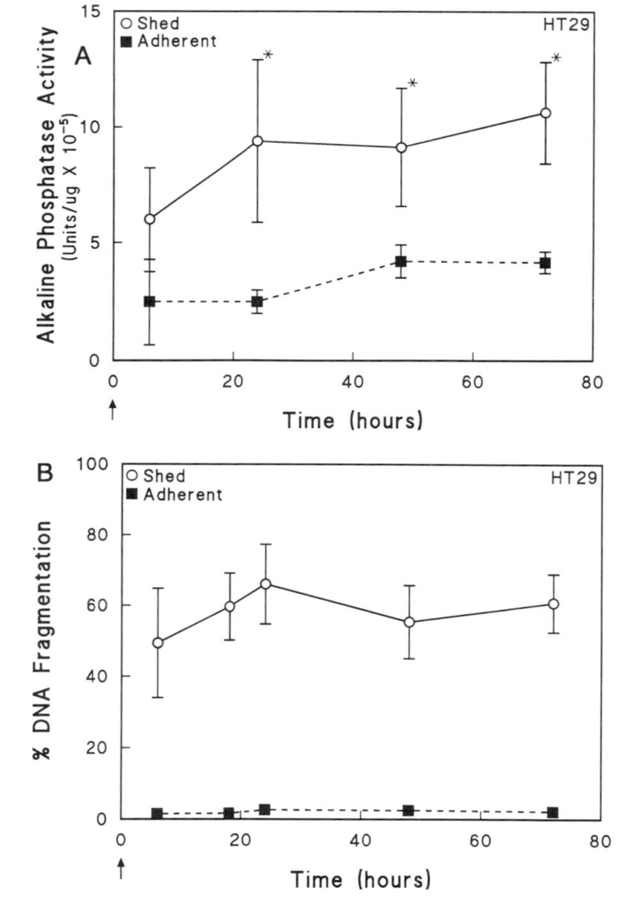

Figure 3. Differentiation and apoptosis in HT29 cells. Alkaline phosphatase (A), a marker of colonic cell differentiation, and non-random DNA cleavage (B), were quantitatively assayed in the shed and adherent cell populations during un-induced culture of HT29 cells from 6 to 72 hours. Reprinted from[31].

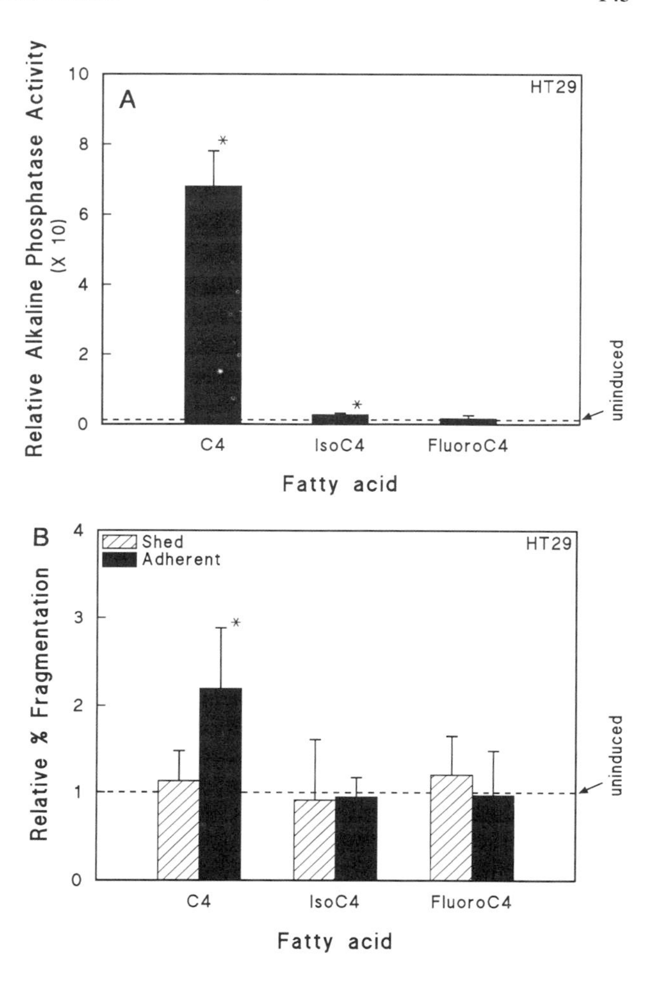

Figure 4. Induction of differentiation and apoptosis in HT29 cells. Alkaline phosphatase (A) and non-random DNA cleavage (B), were quantitatively assayed in HT29 cells treated for 72 hours with butyrate (C4), branched C4, or fluorinated C4. Reprinted from[31].

exhibit higher levels of alkaline phosphastase activity (3A), and higher levels of non-random DNA cleavage (into nucleosomal size fragments, which is characteristic of apoptotic cells, not shown), than their adherent counterparts[31].

Since short-chain fatty acids stimulate differentiation, the question was asked whether such stimulation would also result in increased apoptosis. For this experiment, it was necessary to study the adherent population in which both differentiation and apoptosis are initiated. Figure 4 A and B illustrate that both alkaline phosphatase and DNA cleavage were stimulated in the adherent population by the unbranched 4 carbon atom butyrate, but not by either isoC4 or fluoroC4[31]. Thus, only those short-chain fatty acids which stimulate mitochondrial gene expression are effective in inducing both differentiation and apoptosis of HT29 cells.

Our working model, therefore, is that defective mitochondrial gene expression, and hence impaired mitochondrial function, is incompatible with terminal differentiation of some lineages of colonic epithelial cell differentiation. Therefore, cells in the colonic mucosa at risk, in which mitochondrial gene expression is decreased, may not undergo apoptosis, the normal end-stage of terminal cell differentiation. These cells may instead persist in the crypt where they continue to accumulate mutations leading to neoplastic growth. Short-chain fatty acids, derived from fiber fermentation, partially overcome this block and thereby reduce the probability of cells in the population progressing and producing tumors[31].

The data also illustrate another important point. Although the affects of butyrate on differentiation and apoptosis are significant, it is clear that in both the induced and uninduced cell populations, only a small proportion of the cells undergo apoptosis and are shed into the medium. It is likely that this is biologically and clinically relevant. Small shifts in the rate of apoptosis can dramatically affect cell populations even in a few days[36], and alterations of cancer incidence in relation to nutritional factors are generally only a few fold with decades of exposure. Thus, an important principal is that the affects of nutritional and chemopreventive agents on the biochemistry and physiology of cells are likely to be subtle. Indeed, agents which stimulate extensive apoptosis, or other dramatic changes in the colonic mucosa, would likely be poor candidates for prevention in healthy individuals over extended periods of time.

VII. FUTURE DIRECTIONS

a. How Do Short-Chain Fatty Acids and Other Inducers of Differentiation Interact with Mutational Events Involving Genes Such as c-myc, APC and DCC in Affecting the Balance among Cell Proliferation, Cell Differentiation and Apoptosis?

One of the strengths of the working model we have described is that it suggests experiments on interactions among inducers of differentiation in the colon and genetic alterations which characterize the disease. First, the c-myc gene, which is both overexpressed and amplified in colonic tumors[14,37-39], can stimulate either cell proliferation or apoptosis[40]. Thus, the question can be asked whether the balance between these two cellular processes can be shifted by the effects of inducers of differentiation. This is made interesting by the suggestion that c-myc deregulation may be an early event in tumorigenesis, linked to mutations in the APC gene[41-43], and our observation that low-level amplification of c-myc can be seen in the an area of atypia - a premalignant lesion (Augenlicht et al, unpublished). Therefore, one can envision that modulating the activity of c-myc, either directly or through its molecular partners such as max and mad, in the presence of different signals for differentiation, may shift the liklihood of cells progressing down one pathway or another.

This leads directly to a second series of experiments. It has recently been demonstrated that the APC protein associates with beta-catenins[44,45], cytoplasmic proteins found bound to transmembrane cadherin proteins in Adherence Junctions. These proteins may function in cell adhesion and cytoskeletal integrity[46], and indeed the APC protein has been found associated with cytoskeleton microtubules and their assembly[47,48]. B-catenin exhibits homology to a Drosophila protein called *Armadillo* involved in pattern formation in the embryonic cuticle, and has also been implicated in axis formation during Xenopus development[46]. Thus, as we have speculated, APC protein is likely involved in the normal cell-cell and cell-matrix interactions which govern differentiation of colonic epithelial cells as they migrate up the crypt axis. It will therefore be of interest to determine how the wild-type protein and mutants interact with pathways of differentiation and subsequent apoptosis stimulated by different inducing agents.

Finally, it is of interest that expression of the DCC protein is also associated with differentiation, in this case, specifically with cell lineages which secrete mucin[49]. Since the protein is related to cell surface adhesion molecules[49,50], studies of the interaction of expression of this gene with inducers of differentiation and apoptosis could cast light on the relationship of cell shedding into the lumen and apoptosis of mucin secreting cells.

Again, however, there are additional levels of complexity that must be kept in mind. Inducers of colonic cell differentiation fall into different classes and should not be thought of generically. For example, butyrate does not induce the expression of the colonic mucin gene MUC2 in HT-29 cells, a gene whose expression is highly stimulated by forskolin and TPA, inducers of protein kinase A and C, respectively[51]. Moreover, we have recently found that butyrate, forskolin, TPA and other inducers are highly specific as regards the biochemical concomitants of differentiation that they induce[52] (Velcich et al, submitted), although a tight link between induction of mitochondrial gene activity and eventual progression to apoptosis continues to be supported by the data[53]. Therefore, many signals from nutritional, differentiation and growth factors are integrated within the cell in determining the final phenotype. While it may be straightforward to demonstrate affects of these various agents in culture, determining how they interact in the colonic mucosa with each other and with the many genetic alterations which arise in the colonic mucosa to determine relative risk for tumor development may be a formidable task.

b. Can Functions Encoded by Other Genes Which Are Altered in Expression in Colonic Mucosa at Risk Be Modulated by Interaction with Dietary Factors?

As discussed earlier, elevated risk is brought about in Hereditary Non-polyposis Colon Cancer by inherited mutations in genes which encode DNA mismatch repair functions. We have recently reported that the genes which are elevated in *expression* in colonic mucosa at risk[17] contain oligo A regions which are particularly susceptible to errors in DNA replication, and indeed we have demonstrated that instability at these loci exists in colonic tumors[29]. Of particular interest in the context of this discussion is that we have also shown quantitative variability in the extent of instability at different loci and in different patients. At present, nothing is known of the relationship among the variety of mutations which may be found in different mismatch repair genes, or of modifying loci, to the penetrance of the phenotype. However, the fact that different loci and patients exhibit quantitative variation raises the possiblity that mismatch repair function may also be modulated by dietary factors. While speculative, this too may become a fruitful area for investigation in the future.

ACKNOWLEDGMENTS

This work was supported by grant 92A05 from the American Institute for Cancer Research, and grants CA56858, CA59932, and P30-CA-1330 from the National Cancer Institute.

REFERENCES

1. J. Groden, A. Thliveris, W. Samowitz, M. Carlson, L. Gelbert, H. Albertson, G. Joslyn, J. Stevens, L. Spiro, M. Robertson, L. Sargeant, K. Krapcho, E. Wolff, R. Burt, J.P. Hughes, J. Warrington, J. McPherson, J. Wasmuth, D. Le Paslier, H. Abderrahim, D. Cohen, M. Leppert, and R. White, Identification and characterization of the familial adenomatous polyposis coli gene. Cell. 66:589 (1991).
2. I. Nishisho, Y. Nakamura, Y. Miyoshi, Y. Miki, H. Ando, A. Horii, K. Koyama, J. Utsunomiya, S. Baba, P. Hedge, A. Markham, A.J. Krush, G. Petersen, S.R. Hamilton, M.C. Nilbert, D.B. Levy, T.M. Bryan, A.C. Preisinger, K.J. Smith, L.-K. Su, K.W. Kinzler, and B. Vogelstein, Mutations of chromosome 5q21 genes in FAP and colorectal cancer patients. Science. 253:665 (1991).

3. R. Fishel, M.K. Lescoe, M.R.S. Rao, N.G. Copeland, N.A. Jenkins, J. Garber, M. Kane, and R. Kolodner, The human mutator gene homolog MSH2 and its association with hereditary nonpolyposis colon cancer. Cell. 75:1027 (1993).
4. C.E. Bronner, S.M. Baker, P.T. Morrison, G. Warren, L.G. Smith, M.K. Lescoe, M. Kane, C. Earabino, J. Lipford, A. Lindblom, P. Tannergard, R.J. Bollag, A.R. Godwin, D.C. Ward, M. Nordenskjold, R. Fishel, R. Kolodner, and R.M. Liskay, Mutation in the DNA mismatch repair gene homoloque hMLH 1 is associated with hereditary non-polyposis colon cancer. Nature. 368:258 (1994).
5. F.S. Leach, N.C. Nicolaides, N. Papadopoulos, B. Liu, J. Jen, R. Parsons, P. Peltomaki, P. Sistonen, L.A. Aaltonen, M. Nystrom-Lahti, X.-Y. Guan, J. Zhang, P.S. Meltzer, J. Yu, F. Kao, D.J. Chen, K.M. Cerosaletti, R.E.K. Fournier, S. Todd, T. Lewis, R.J. Leach, S.L. Naylor, J. Weissenbach, J. Mecklin, H. Jarvinen, G.M. Petersen, S.R. Hamilton, J. Green, J. Jass, P. Watson, H.T. Lynch, J.M. Trent, A. de la Chapelle, K.W. Kinzler, and B. Vogelstein, Mutations of a mutS homolog in hereditary nonpolyposis colorectal cancer. Cell. 75:1215 (1993).
6. N. Papadopoulos, N.C. Nicolaides, Y.-F. Wei, S.M. Ruben, K.C. Carter, C.A. Rosen, W.A. Haseltine, R.D. Fleischmann, C.M. Fraser, M.D. Adams, J.C. Venter, S.R. Hamilton, G.M. Petersen, P. Watson, H.T. Lynch, P. Peltomaki, J.-P. Mecklin, A. de la Chapelle, K.W. Kinzler, and B. Vogelstein, Mutation of a mutL homolog in hereditary colon cancer. Science. 263:1625 (1994).
7. L.-K. Su, K.W. Kinzler, B. Vogelstein, A.C. Preisinger, A.P. Moser, C. Luongo, K.A. Gould, and W.F. Dove, Multiple intestinal neoplasia caused by a mutation in the murine homolog of the APC gene. Science. 256:668 (1992).
8. R. Fodde, W. Edelmann, K. Yang, C. van Leeuwen, C. Carlson, B. Renault, C. Breukel, E. Alt, M. Lipkin, P.M. Khan, and R. Kucherlapati, A targeted chain-termination mutation in the mouse Apc gene results in multiple intestinal tumors. Proc. Nat. Acad. Sci. USA. 91: (1994).
9. R. Parsons, G. Li, M.J. Longley, W. Fang, N. Papadopoulos, J. Jen, A. de la Chapelle, K.W. Kinzler, B. Vogelstein, and P. Modrich, Hypermutability and mismatch repair deficiency in RER+ tumor cells. Cell. 75:1227 (1993).
10. B. Vogelstein, E.R. Fearon, S.E. Kern, S.R. Hamilton, A.C. Preisinger, Y. Nakamura, and R. White, Allelotype of colorectal carcinomas. Science. 244:207 (1989).
11. A. Kallioniemi, O.-P. Kallioniemi, D. Sudar, D. Rutovitz, J.W. Gray, F. Waldman, and D. Pinkel, Comparative genomic hybridization for molecular cytogenetic analysis of solid tumors. Science. 258:818 (1992).
12. A. Kallioniemi, O.-P. Kallioniemi, J. Piper, M. Tanner, T. Stokke, L. Chen, H.S. Smith, D. Pinkel, J.W. Gray, and F.M. Waldman, Detection and mapping of amplified DNA sequences in breast cancer by comparative genomic hybridization. Proc. Nat. Acad. Sci. USA. 91:2156 (1994).
13. T. Ried, I. Petersen, H. Holtgreve-Grez, M.R. Speicher, E. Schrock, S.d Manoir, and T. Cremer, Mapping of multiple DNA gains and losses in primary small cell lung carcinomas by compararative genomic hybridization. Cancer Res. 54:1801 (1994).
14. B.G. Heerdt, S. Molinas, D. Deitch, and L.H. Augenlicht, Aggressive subtypes of human colorectal tumors frequently exhibit amplification of the c-myc gene. Oncogene. 6:125 (1991).
15. M. Groudine and H. Weintraub, Activation of cellular genes by avian RNA tumor viruses. Proc.Natl.Acad.Sci.USA. 77:5351 (1980).
16. L.H. Augenlicht, M.Z. Wahrman, H. Halsey, L. Anderson, J. Taylor, and M. Lipkin, Expression of cloned sequences in biopsies of human colonic tissue and in colonic carcinoma cells induced to differentiate in vitro. Cancer Res. 47:6017 (1987).
17. L.H. Augenlicht, J. Taylor, L. Anderson, and M. Lipkin, Patterns of gene expression that characterize the colonic mucosa in patients at genetic risk for colonic cancer. Proc.Natl.Acad.Sci.USA. 88:3286 (1991).
18. L.H. Augenlicht, B. Heerdt, S. Molinas, and M. Lipkin,Patterns of gene expression that characterize the mucosa at risk for development of colorectal cancer,Calcium, vitamin D, and prevention of colon cancer,,eds.,M. Lipkin, H.L. Newmark, and G. Kelloff,1991,CRC Press,Boca Raton,pp. 283-300
19. L.H. Augenlicht and B.G. Heerdt, Modulation of gene expression as a biomarker in colon. Journal of Cellular Biochemistry. 16G:151 (1992).
20. L.H. Augenlicht, G. Corner, S. Molinas, and B.G. Heerdt,Genetic biomarkers,Cancer Chemoprevention,,eds.,L. Wattenberg, M. Lipkin, C.W. Boone, and G.J. Kelloff,1992,CRC Press,Boca Raton,Fla,pp. 559-569
21. L.H. Augenlicht and B.G. Heerdt,Colonic Carcinoma: a common tumor with multiple genomic abnormalities,Biochemical and Molecular Aspects of Selected Cancers, vol 2,,eds.,T. Pretlow and T. Pretlow,1994,Academic Press,NY,pp. 47-91
22. L.H. Augenlicht,Gene structure and expression in colon cancer,Cell and Molecular Biology of Colon Cancer,,eds.,L.H. Augenlicht,1989,CRC Press,Boca Raton, Florida,pp. 165-186

23. B.J. Barrow, M.A. O'Riordan, T.A. Stellato, B.M. Calkins, and T.P. Pretlow, Enzyme-altered foci in colons of carcinogen-treated rats. Cancer Res. 50:1911 (1990).
24. N.R. Hughes, R.S. Walls, R.C. Newland, and J.E. Payne, Gland to gland heterogeneity in histologically normal mucosa of colon cancer patients demonstrated by monoclonal antibodies to tissue-specific antigens. Cancer Res. 46:5993 (1986).
25. B.G. Heerdt, H.K. Halsey, M. Lipkin, and L.H. Augenlicht, Expression of mitochondrial cytochrome c oxidase in human colonic cell differentiation, transformation, and risk for colonic cancer. Cancer Res. 50:1596 (1990).
26. J.S. Chen and L.H. Augenlicht. Sequence analysis of clones altered in expression in risk for colonic cancer. in preparation (1994).
27. D.A. Clayton, Nuclear gadgets in mitochondrial DNA replication and transcription. Trends Biochem. Sci. 16:107 (1991).
28. B.G. Heerdt, J.S. Chen, L.R. Stewart, and L.H. Augenlicht, Polymorphisms, but lack of mutations or instability, in the promotor region of the mitochondrial genome in human colonic tumors. Cancer Res. 54:3912 (1994).
29. J.S. Chen, B.G. Heerdt, and L.H. Augenlicht, Presence and instability of repetitive elements in sequences the altered expression of which characterizes risk for colonic cancer. Cancer Res. *55*, 174 (1995).
30. B.G. Heerdt and L.H. Augenlicht, Effects of fatty acids on expression of genes encoding subunits of cytochrome c oxidase and cytochrome c oxidase activity in HT29 human colonic adenocarcinoma cells. J.Biol.Chem. 266:19120 (1991).
31. B.G. Heerdt, M.A. Houston, and L.H. Augenlicht, Potentiation by specific short-chain fatty acids of differentiation and apoptosis in human colonic carcinoma cell lines. Cancer Res. 54:3288 (1994).
32. J.R. Wong and L.B. Chen,Recent advances in the study of mitochondria in living cells,Advances in Cell Biology,2,,,1988,JAI Press, Inc.,,pp. 263-290
33. I.C. Summerhayes, T.J. Lampidis, S.D. Bernal, J.J. Nadakavukaren, K.K. Nadakavukaren, E.L. Shepherd, and L.B. Chen, Unusual retention of rhodamine 123 by mitochondria in muscle and carcinoma cells. Proc. Nat. Acad. Sci. USA. 79:5292 (1982).
34. J.S. Modica-Napolitano, G.D. Steele, and L.B. Chen, Aberrant mitochondria in two human colon carcinoma cell lines. Cancer Res. 49:3369 (1989).
35. B.G. Heerdt and L.H. Augenlicht, Absence of detectable deletions in the mitochondrial genome of human colon tumors. Cancer Commun. 2:109 (1990).
36. W. Bursch, S. Paffe, B. Putz, G. Barthel, and R. Schulte-Hermann, Determination of the length of the histological stages of apoptosis in normal liver and in altered hepatic foci of rats. Carcinogenesis. 11:847 (1990).
37. G. Yander, H. Halsey, M. Kenna, and L.H. Augenlicht, Amplification and elevated expression of c-myc in a chemically induced mouse colon tumor. Cancer Res. 45:4433 (1985).
38. M.D. Erisman, J.K. Scott, R.A. Watt, and S.M. Astrin, The c-myc protein is constitutively expressed at elevated levels in colorectal carcinoma cell lines. Oncogene. 2,:367 (1988).
39. M.D. Erisman, P.G. Rothberg, R.E. Diehl, C.C. Morse, J.M. Spandorfer, and S.M. Astrin, Deregulation of c-myc gene expression in human colon carcinoma is not accompanied by amplification or rearrangement of the gene. Mol. Cell. Biol. 5:1969 (1985).
40. G.I. Evan, A.H. Wyllie, C.S. Gilbert, T.D. Littlewood, H. Land, M. Brooks, C.M. Waters, L.Z. Penn, and D.C. Hancock, Induction of apoptosis in fibroblasts by c-myc protein. Cell. 69:119 (1992).
41. M.D. Erisman, J.K. Scott, and S.M. Astrin, Evidence that the familial adenomatous polyposis gene is involved in a subset of colon cancers with a complementable defect in c-myc regulation. Proc. Nat. Acad. Sci. USA. 86:4264 (1989).
42. P.G. Rothberg, J.M. Spandorfer, M.D. Erisman, R.N. Staroscik, H.F. Sears, R.O. Petersen, and S.M. Astrin, Evidence that c-myc expression defines two genetically distinct forms of colorectal adenocarcinoma. Br. J. Cancer. 52:629 (1985).
43. C. Rodriguez-Alfageme, E.J. Stanbridge, and S.M. Astrin, Suppression of deregulated c-MYC expression in human colon carcinoma cells by chromosome 5 transfer. Proc. Nat. Acad. Sci. USA. 89:1482 (1992).
44. B. Rubinfeld, B. Souza, I. Albert, O. Muller, S.H. Chamberlain, F.R. Masiarz, S. Munemitsu, and P. Polakis, Association of the APC gene product with b-catenin. Science. 262:1731 (1993).
45. L.-K. Su, B. Vogelstein, and K. Kinzler, Association of the APC tumor suppressor protein with catenins. Science. 262:1734 (1993).
46. Peifer, Cancer, catenins, and cuticle pattern: a complex connection. Science. 262:1667 (1993).
47. K.J. Smith, D.B. Levy, P. Maupin, T.D. Pollard, B. Vogelstein, and K.W. Kinzler, Wild-type but not mutant APC associates with the microtubule cytoskeleton. Cancer Res. 54:3672 (1994).

48. S. Munemitsu, B. Souza, O. Muller, I. Albert, B. Rubinfeld, and P. Polakis, The APC gene product associates witht microtubules in vivo and promotes their assemby in vitro. Cancer Res. 54:3676 (1994).
49. L. Hedrick, K.R. Cho, E.R. Fearon, T.-C. Wu, K.W. Kinzler, and B. Vogelstein, The DCC gene product in cellular differentiation and colorectal tumorigenesis. Genes Dev. 8:1174 (1994).
50. E.R. Fearon, K.R. Cho, J.M. Nigro, S.E. Kern, J.W. Simons, J.M. Ruppert, S.R. Hamilton, A.C. Preisinger, G. Thomas, K.W. Kinzler, and B. Vogelstein, Identification of a chromosome 18q gene that is altered in colorectal cancers. Science. 247:49 (1990).
51. A. Velcich and L.H. Augenlicht, Regulated expression of an intestinal mucin gene in HT29 colonic carcinoma cells. J.Biol.Chem. 268:13956 (1993).
52. A. Velcich, L. Palumbo, A. Jarry, C. Laboisse, J. Rachevskis, and L. Augenlicht, Patterns of expression of lineage specific markers during the in vitro differentiation of HT29 colon carcinoma cells. submitted (1994).
53. B.G. Heerdt, M.A. Houston, J.J. Rediske, and L.H. Augenlicht. Relationships among differentiation enhanced mitochondrial activity and apoptosis in human colonic carcinoma cells.

13

FISH OIL AND CELL PROLIFERATION KINETICS IN A MAMMARY CARCINOMA TUMOR MODEL

Nawfal W. Istfan,* Jennifer Wan, and Zhi-Yi Chen

Clinical Nutrition Unit
Boston University Medical Center Hospital
88 East Newton St.
Boston, Massachusetts 02118

ABSTRACT

In vivo bromodeoxyuridine (BrdUrd) labelling and bivariate BrdUrd/DNA analysis was used to evaluate cell cycle kinetics in a rat tumor model known to be sensitive to dietary fatty acid manipulation. Fish oil supplementation significantly reduced the rate of BrdUrd movement relative to DNA content, indicating prolongation of the DNA replication time. This finding, which accounted for most of the decrease in tumor growth rate in the fish oil-fed group, represents the first description of an alteration in S phase duration by an extrinsic factor. The significance of this finding is discussed in relation to current understanding of cell cycle regulation.

Fish oil feeding is associated with slower growth rate in certain tumors (1,2). According to current concepts of cellular proliferation (3), regulation of growth by extrinsic factors is thought to precede the S phase. This statement is based on the notion that, within a given cell type, DNA replication time (S phase duration) is constant (4-6). Extensive evidence also supports an on/off mechanism of cell cycle regulation at the level of entry into the S phase (3).

In this report, we present evidence showing, for the first time, that the S phase duration of fat-responsive tumor cells can be altered by dietary manipulation of fatty acids. Furthermore, these differences in S phase duration appear to account for all the *in vivo* variation in tumor growth resulting from fish oil feeding. Although the mechanism of this phenomenon remains unclear, our observations support increasing evidence for a regulatory

* Correspondence: Nawfal W Istfan M.D., Ph.D. , Clinical Nutrition Unit, Boston University Medical Center, 88 East Newton Street, Boston, MA 02118. Tel: (617) 638-8557; Fax: (617)-638-5977.

Diet and Cancer, Edited under the auspices of the American Institute of Cancer Research
Plenum Press, New York, 1995

step at the level of the nucleus. They are also important for understanding the relationship between dietary fat and tumor growth.

METHODS

We used an implantable mammary tumor cell line (13762 MAT) to generate a fat-responsive solid tumor model in the Fischer rat. We and others (1) have previously demonstrated reduction in the rate of tumor growth in this model after several weeks of feeding a diet rich in w-3 fatty acids (fish oil). All experimental procedures with animals were reviewed and approved by the Institutional Animal Care and Use Committee.

Female Fischer rats were first pair-fed on two diets, safflower and fish oil enriched, designed to provide 15% of calories as fat. Calories, protein, and micronutrient content were identical for the two diets. After six weeks of feeding, the rats were inoculated subcutaneously with 10^7 tumor cells. The solid phase of the tumor was generally detectable by the fifth day after inoculation and subsequently followed an exponential growth phase over the next 10 days. Changes in tumor dimensions were measured daily in all rats.

Cell cycle kinetics were determined on the tenth day following tumor implantation. The thymidine analog bromodeoxyuridine (BrdUrd) was administered by intraperitoneal injection to pulse-label cells actively synthesizing DNA, as previously described (8,9). Rats from both dietary groups were then killed at one and six hours after injection. The tumors were dissected, and the cells dissociated after treatment with collagenase and hyaluronidase. The dissociated cells were stained with FITC-labelled anti-BrdUrd monoclonal antibody and propidium iodide for detection of DNA synthesis activity and measurement of cellular DNA content, respectively, by flow cytometry.

To estimate the rate of DNA synthesis time, we measured the change in the position of BrdUrd-labelled cells relative to the cellular DNA content. This is done by measuring the DNA content of undivided BrdUrd-labelled cells as a function of time. One useful method for expressing the position of BrdUrd-labelled cells relative to G1 and G2 cells is by use of the relative movement (RM), which is defined as follows:

$$RM = \frac{(Fl_U - Fl_1)}{(Fl_2 - Fl_1)}$$

where Fl_U, Fl_1, and Fl_2 are the mean red fluorescence (DNA content) of labelled undivided cells, G_1, and G_2 cells, respectively (8).

RESULTS

Tumor volumes are summarized in Figure 1, which documents the exponential growth pattern of the tumor as well as the growth-retarding characteristic of fish oil feeding. Expressed as a first-order rate constant, tumor growth was approximately 30% slower in the fish oil group relative to the safflower oil group.

The results of cell cycle distribution and labelling index (percent of cells taking up BrdUrd) are summarized in Table 1. Differences noted in these parameters were not statistically significant between the two dietary groups. Also not significant were the differences in labelling indices (BrdUrd uptake at 1 hour) which are summarized in Figure 2.

Examples of the change in RM between one and six hours are shown in Figure 3 for the two diet groups. DNA content of BrdUrd-labelled cells was similar for the two groups after one hour of pulse labelling with the thymidine analogue. In contrast, six hours after

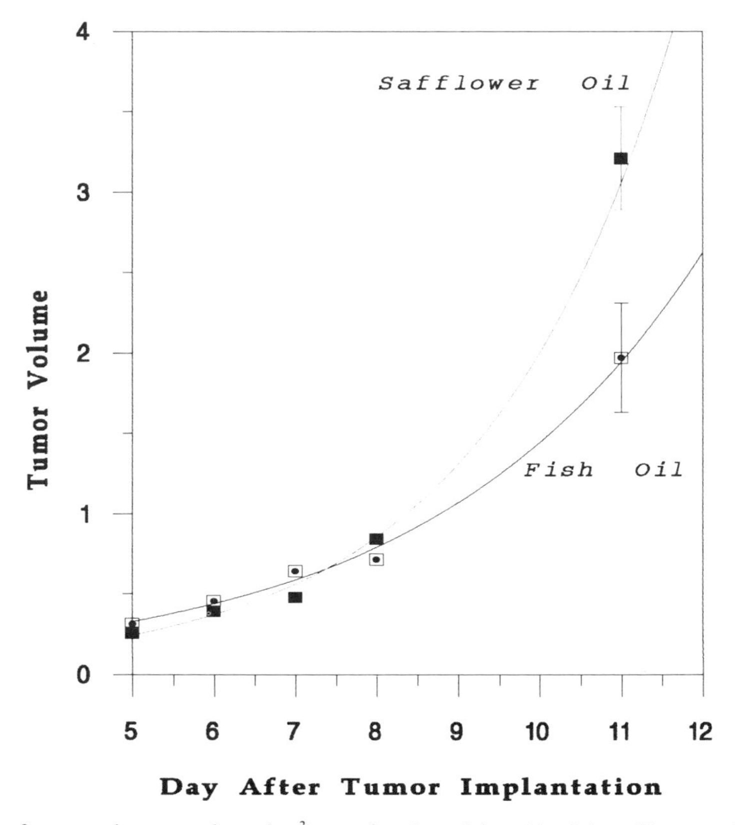

Figure 1. Plot of measured tumor volume (cm^3) as a function of time (days) in safflower and fish oil fed tumor-bearing rats during exponential growth. Tumor volumes were estimated from 3-dimensional measurements with calipers. Lines represent least-square computer-generated regression curves. The first-order growth rate constants (k_g) derived from these regression analyses were 0.47 day^{-1} in the safflower oil group and 0.30 day^{-1} in the fish oil group. The tumor volume doubling times (Td) , 35.4 and 55.5 hours, respectively, are derived from the equation $Td = Ln\ 2/k_g$.

pulsing, DNA content in the BrdUrd-labelled cells was 0.78 relative to the cells in G2 in the fish oil group, and 0.92 relative to G2 in the safflower oil group. The mean values of BrdUrd relative movement are summarized in Table 2, which indicates significantly slower tumor DNA replication in the fish oil group ($p < .001$).

Since RM is approximately linear in time (10), it is possible to estimate mean DNA synthesis time as the time at which the mean red fluorescence of the BrdUrd-labelled cells is equal to the mean red fluorescence of G2 cells (*i.e.*, RM equal 1.0). It is also possible to

Table 1. Summary of cell-cycle phase distribution and labelling index

Group	N	BrdUrd labelling time (h)	Go/G1-phase (%)	S-phase (%)	G2/M-phase (%)	Labelling Index (%)
FO I	8	1	63.2 ±0.7	29.6 ±1.1	7.2 ±0.6	27.1 ±2.2
SO I	7	1	62.2 ±1.1	31.4 ±1.3	6.4 ±0.5	32.1 ±1.0

Values are mean ± SEM.

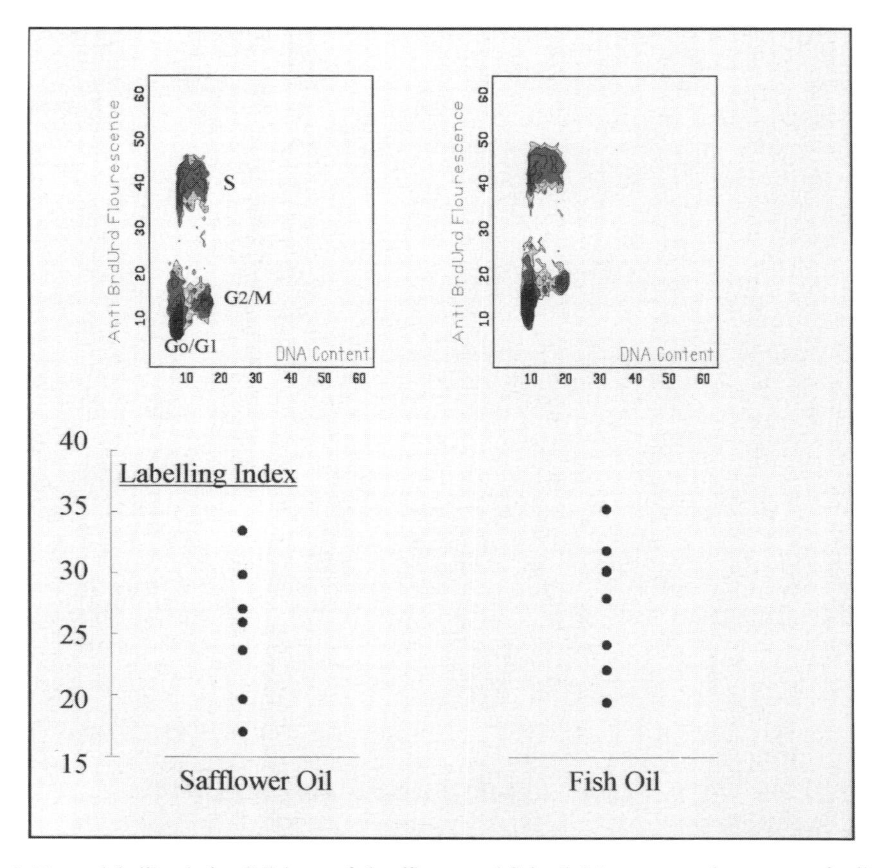

Figure 2. Tumor labelling index (LI) in rats fed safflower and fish oil. LI represents the percent of cells with BrdUrd uptake as determined by bivariate DNA/BrdUrd analysis. Cells were dissociated from the solid phase by enzymatic treatment with a mixture of collagenase/hyaluronidase and analyzed by flow cytometry 1 hour after intraperitoneal BrdUrd injection in the tumor-bearing rats. Differences in LI were not statistically significant (n=7). Top panel shows an example contour plots of bivariate BrdUrd/DNA distribution in tumor cells representing the two diet ary groups.

obtain estimates of cell cycle times based on established kinetic principles applicable to exponential growth (11). The results of such analysis are summarized in Table 3, showing a significant prolongation of the S phase in the tumors of rats fed the fish oil diet ($p < .001$). Although the estimate of time spent in G1 was longer in the fish oil group, these differences were not statistically significant. Neither were the differences in growth fraction significant between the two groups.

DISCUSSION

Control of cellular proliferation can be viewed within the perspective of two key questions: what events initiate the proliferative process; and how, within cells committed to divide, is DNA replication regulated? The first question pertains to the activation of cells prior to their entry into the S phase. It is at this level that signal transduction pathways that

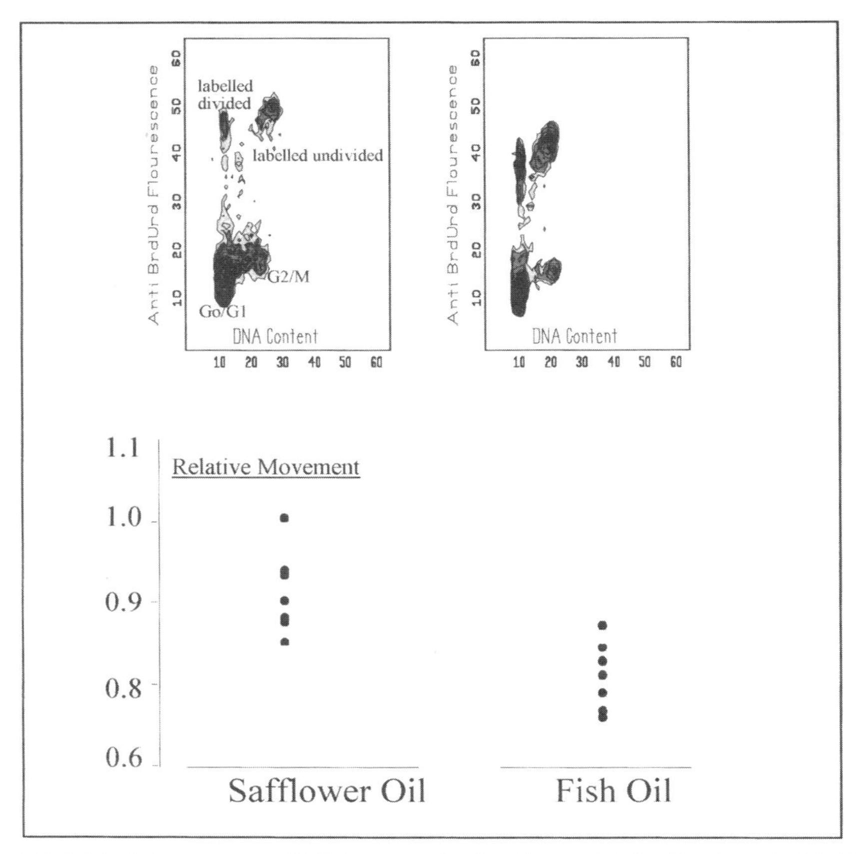

Figure 3. BrdUrd relative movement (RM) in tumor cells after 6 hours of BrdUrd in vivo pulse labelling. RM is calculated according to the following equation:

$$Rm = \frac{(Fl_U - Fl_1)}{(Fl_2 - Fl_1)}$$

where Fl_U, Fl_1, and Fl_2 are the mean red fluorescence (DNA content) of labelled undivided cells, G_1, and G_2 cells, respectively. Undivided BrdUrd-labelled cells from fish oil fed rats had a significantly lower increase in DNA content after 6 hours of pulsing (mean RM ± SE 0.78±0.01 in fish oil group, and 0.91±0.02 in safflower oil group, $p<0.001$). Top panel shows an example of 6-hour RM in safflower and fish oil fed rats.

link external stimuli to cellular proliferation mechanisms have been extensively characterized in the recent literature (3). In contrast, little is known about the regulation of DNA synthesis, especially in higher eukaryotic cells. In kinetic terms, the phase of cellular activation is reflected in the parameter growth fraction (Table 3) and in the duration of G1 phase, neither of which were significantly changed by fish oil feeding. Although we cannot completely exclude the possibility of a small prolongation in G1 phase, the major impact of fish oil feeding was expressed in the estimates of mean DNA synthesis time. By considering the magnitude of change in tumor volume growth (Figure 1) and the kinetic parameters of cellular proliferation (Table 3), it is possible to conclude that differences in the cell cycle times, predominantly explained by the prolongation of the S phase in the fish oil group, accounted for most of the variation in tumor growth between the two groups.

Table 2. Summary of Labelled Divided and Undivided Cells, and Relative Movement of BrdUrd

Group	N	BrdUrd labelling time (h)	Labelled-undivided cells(flu) (%)	Labelled-divided Cells (fld) (%)	V	Relative Movement
FO I	8	1	27.1 ±2.2	—	0.24 ±0.02	0.47 ±0.01
FO II	7	6	25.4 ±2.2	12.9 ±1.3	0.29 ±0.02	0.78+++ ±0.01
SO I	7	1	32.1 ±0.9	—	0.28 ±0.01	0.46 ±0.02
SO II	8	6	17.3+++ ±2.7	17.3 ±2.7	0.24 ±0.02	0.91*** ±0.02+++

Values are Mean ± SEM.
Relative movement is the DNA content of BrdUrd labelled cells relative to G1/Go and G2/M cells. The parameter v is defined as ln (1 + flu)/(1 – fld/2) where fld and flu are the fractions of labelled divided and undivided cells, respectively (8).
*** p< 0.001 vs. Fish oil group for the same BrdUrd labelling time.
+++ p< 0.001 comparing measurements at 1 and 6 hours after pulse labelling within the same dietary group.

The mechanism of changes in S phase duration remains conjectural. Total DNA synthesis time is theoretically a function of the rate of the polymerase reaction and of the frequency of initiation of replication forks and their spatial distribution within the replicating genome. Although both factors may contribute to temporal variations in the S phase, almost all available evidence strongly implicates the latter explanation (12-15). In fact, the dynamics of the polymerase reaction are not completely understood. As recently noted by Kornberg (16), the kinetics of binding as well as the rates of movement along the DNA strands continue to be largely unexplored and unknown.

On the other hand, initiation of replication at origin DNA is a highly ordered process. Replication foci, defined as the sites at which DNA synthesis occurs, begin to appear in the interphase nucleus. Although details of how these foci are spatially arranged into clusters within the nucleus are largely unknown, it is generally accepted that the duration of DNA synthesis within each cluster remains constant (3). Consequently, the total DNA synthesis phase depends on the frequency with which the replication foci are organized (13,14). Apart

Table 3. Summary of Cell-Cycle Kinetic Estimates

Treatment		$T_{Go/G1}$ (h)	T_s (h)	$T_{G2/M}$ (h)	T_c (h)	Potential Doubling Time (Tpot, h)	Actual Doubling Time (Td, h)	Growth Fraction
FO	7	16.0 ±4.9	15.0 ±1.0	2.2 ±0.5	33.2 ±3.0	41.5 ±4.1	55.5 ±3.7	0.87 ±0.04
SO	8	13.7 ±1.8	9.2*** ±0.5	2.2 ±0.6	25.2 ±2.7	22.9*** ±1.2	35.4*** ±3.0	0.81 ±0.04

Values are Mean ± SEM.
*** p< 0.001 vs. Fish oil group by ANOVA.
T_c = Total cell-cycling time.$T_{Go/G1}$ = Go/G1 time; T_s = mean DNA synthesis time or S-phase time; $T_{G2/M}$ = G2/M time. In calculating these kinetic estimates, we assume an exponentially growing cell population with the possibility of cell loss as previously defined by Steel (11). T_s is estimated from the change in relative BrdUrd movement as explained in the Text. Tpot is calculated from the equation ln 2 (Ts/v), where v is defined and summarized in Table 2. To calculate Tc, TG1/Go and T G2/M we denote by GF (growth fraction) the ratio of proliferating to total tumor cells. During exponential growth, the following three relationships hold: log (GF + 1) = (log2) (Tc/Tpot); TG2/M = Tc [log (1 + PG2M)/log 2]; and (Ts+TG2M) = Tc [log (1 + Ps + PG2M)/log 2] where Ps and PG2M are the proportion of tumor cells in the S and G2/M phases, respectively. Finally, TGo/G1 is estimated from the relationship Tc= Ts+TG2/M+TG1/Go.

from changes occurring early in embryonal development, as described in Drosophila (17), the organization of replicating foci into clusters appears to follow a uniform pattern thus accounting for the observation that S phase duration is characteristic of a given cell type. The prolongation of S phase duration described in this report, as a result of fish oil feeding, is a novel finding which points to the possibility of changes in the the temporal organization of replication clusters by dietary fatty acids.

Several studies have previously documented growth retardation in similar rat tumor models. However, most of the emphasis has centered around the ability of fish oil to reduce synthesis of prostaglandins and leukotrienes derived from arachidonic acid. Extensive literature implicates these molecules in signal transduction pathways for a number of cellular functions such as smooth muscle contractility, platelet aggregation, and mitogenesis. Furthermore, prostaglandins have been shown to be increased in tumor tissue and to correlate with proliferative activity. Based on this background and our finding of prolonged S phase duration, it is possible that prostaglandins mediate the effect of fish oil at the level of the nucleus through a signal transduction pathway akin to that described at the cell surface. Alternatively, fatty acid incorporation into the nuclear membrane may influence DNA replication through generation of phospholipid domains that can attract specific proteins and affect the creation and migration of structures within the nuclear envelope. This hypothesis is consistent with a growing body of evidence that implicates the nuclear membrane in the regulation of DNA replication (19-21). For example, in cell-free systems derived from Xenopus egg extract, DNA replication commences after the formation of a nuclear envelope (19). Recent experiments have clearly shown that this nuclear envelope is essential for replication (20,21) and that, unlike isolated nuclei, nuclei encapsulated within the same membrane undergo synchronous replication. Based on this evidence, Leno and Laskey and have concluded that the nuclear membrane defines the DNA replication unit (21). Although their hypothesis favors a role for the nuclear membrane pore and transport processes, it is still possible that changes in membrane structure, as by manipulation of dietary fat, may also affect the binding properties of membrane components to chromatin in a manner that influences the organization of DNA replication clusters.

In conclusion, this is the first report documenting the modulation by extrinsic factors of S phase duration in a tumor model. While providing insight into the mechanism of interactions between diet and tumor growth, the experimental model described in this report may prove to be helpful for understanding the role of the nucleus and possibly the nuclear membrane in regulating DNA replication in eukaryotic cells.

REFERENCES

1. Wan, J. M. F. , Istfan, N. W. , Chu, C. C. , *Metabolism* **40,** 577-584 (1991).
2. Abou-El-Ela, S. A., Prasse, K.W, Carroll, R. , *Lipids* **23,** 948-954 (1988).
3. Pardee, A. B., *Science,* **246** 603-608 (1989).
4. O'Farrell, P. H., Edgar, B. A, Lakich, D., *Science* **246,** 635-640 (1989).
5. Denhardt, D.T., Edwards, D. R, Parfett, C. L.J., *Biochim. Biophys. Acta* **865**, 83-125 (1986).
6. Cross, F., Roberts, J., Weintraub, H., *Annu Rev Cell Biol* **5,** 341-396 (1989).
7. Baserga, R., The Biology of Cell Reproduction (Harvard University Press, Cambridge, MA, 1985).
8. Dolbeare, F., Gratzner, H., Pallavicini, M. G., *Proc. Natl. Acad. Sci. USA* **80,** 5573-5577 (1983).
9. Fogt, F., Wan, J., O'Hara, C., *Cytometry* **12**,33-41 (1991).
10. White, R. W. and Meistrich, M. L., *Cytometry* **7,** 486-490 (1986).
11. Steel, G. G., *Growth Kinetics of Tumors: Cell Population Kinetics in Relation to the Growth and Treatment of Cancer* (Clarendon, Oxford, U.K., 1977).
12. Edenberg, H. J., and Huberman, J. A., *Annu. Rev. Genet.* **9,** 245-284 (1975).
13. Hand, R., *Cell* **15** 317-325 (1978).
14. Houseman, D. and Huberman, J.A., *J. Mol. Biol.* **94,** 173-181 (1975).

15. Jackson, D.A., *Bio Essays* **12**, 87-89 (1990).
16. Kornberg, A., *J. Biol. Chem.* **263,** 1-4 (1988);
17. Blumenthal, A. B., Kriegstein, H. J., Hogness, D. S., *Cold Spring Harbor Symp. Quant. Biol.* **38**, 205-223 (1974).
18. Dingwall, C. and Laskey, R., *Science* **258,** 942-947 (1992).
19. Blow, J. J., and Watson, J. V., *EMBO J* **6,** 1997-2002 (1987).
20. Cox, L. S., *J Cell Sci* **101,** 43-53 (1992).
21. Leno, G. H., and Laskey, R. A., *J Cell Biol* **112,** 557-566 (1991).

ABSTRACTS

RETINOIC ACID INDUCED EFFECTS ON RB TUMOR SUPPRESSOR EXPRESSION AND CELL DIFFERENTIATION DEPENDENT OF RAF

Andrew Yen, Megan Williams, Joseph D. Platko, Channing Der and Mark Hisaka
Cancer Biology Laboratory, Department of Pathology
Cornell University, Ithaca, New York

Expression of an activated raf transgene accelerated the terminal myeloid differentiation of HL-60 human promyelocytic leukemia cells induced by retinoic acid. A similar result was obtained when 1,25-dihydroxy vitamin D3 was used to induce monocytic differentiation. The stable transfectants were derived by transfecting HL-60 cells with DNA encoding an N-terminal truncated raf-1 protein. In normal HL-60 cells retinoic acid is known to induce a CSF-1 dependent metabolic cascade culminating in GO arrest and phenotypic conversion. Early in this cascade, expression of the RB tumor suppressor gene product is down regulated. A progressive redistribution of the form of the protein from largely hyper-phosphorylated protein to the hypo-phosphorylated form begins later with GO arrest and differentiation. In the activated raf transfected cells, RB down regulation occurred more rapidly, consistent with accelerated differentiation. But the conversion to the hypo-phosphorylated form was not accelerated and occurred after GO arrest and phenotypic conversion to myeloid differentiated cells. Thus raf activation appears to be a component of the retinoic acid induced metabolic cascade culminating in terminal differentiation. In this cascade raf activation promotes RB down regulation. The data are consistent with a model in which raf is an effector of this CSF-1 dependent metabolic cascade which culminates in terminal cell differentiation, and RB down regulation is one of the downstream consequences of raf action. Furthermore they indicate that RB down regulation may be an essential component of the cellular processes causing GO arrest and differentiation, but RB hypo-phosphorylation is more likely a consequence thereof and not a cause.

High cholesterol diet and modulation of sterol composition of the target membranes: A new approach to antihepatoma chemotherapy.

Alexander M. Feigin
Monell Chemical Senses Center, Philadelphia, PA 19104

A selective increase in the sensitivity of liver tumors to some antitumor agents may be achieved by taking advantage of the very specific metabolic imbalances that are typical for all known hepatomas: 1) a high rate of cholesterol biosynthesis (even higher than that in normal liver), and 2) an absence of feed-back regulation of this process. To increase the sensitivity of primary hepatomas to some antitumor agents we suggest the use a combination of a special diet with different inhibitors of cholesterol biosynthesis (AY-9944, ketoconazole, etc.) to induce accumulation of sterols of different structures selectively in hepatoma cells. Since the structure of membrane sterols strongly influences the ability of these sterols to modulate the barrier properties of membranes, as well as the affinity of sterols for certain membrane active agents such as polyene antibiotics, some polypeptide antibiotics and toxins, etc., the changes of structure and content of membrane sterol can increase: i) the permeability of membranes for antitumor agents that have targets inside the cell; ii) the sensitivity to antitumor agents that bind with membranes and eliminate their barrier properties, like polyene antibiotics or some peptide antibiotics; and iii) the sensitivity of tumor cells to physical factors (e.g. temperature or radiation).

Regulation of platelet-derived growth factor receptor by retinoic acid and cyclic AMP in differentiating cells. C. Wang* and C.D. Stiles#. * Center of Molecular Biology for Oral Diseases, University of Illinois at Chicago, Chicago, IL; # Division of Molecular and Cellular Biology, Dana-Farber Cancer Institute, Harvard Medical School, Boston, MA.

Retinods have a wide spectrum of biological effects, ranging from controlling embryonic development to preventing or reversing cancer, in both cultured cells and experimental animals. Although the precise mechanisms by which retinoids control various biological processes remain elusive, they are known to have direct effects on expression of cellular genes, of which many encode growth factors, oncogenes and extracellular matrix proteins. Since some of these responsive molecules are involved in both normal and tumorogenic development, it is reasonable to suggest that interruption to such network might be responsible for uncontrolled cell behavior.

In this study, we examined the effect of retinoids on the expression of growth factor ligand-receptor system in embryonic carcinoma (EC) cells (line F9). F9 cells are fast growing embryonic stem-like cells which can form tumors when re-introduced into mouse blastocyst. Yet differentiated F9 cells (by RA or in combination with cyclic AMP) fail to do so under the same condition. This suggests that anti-cancer nature of retinoids may be partly attributed to its ability to lead abnormal cells into a "differentiation" pathway instead of a "proliferation" pathway. One possible way for retinoids to do so is by controlling the expression of various growth factors and growth factor receptors. Here, we reported that retinoids can modulate the expression of platelet-derived growth factor (PDGF) ligand-receptor system during cell differentiation process. Our studies showed that PDGFA was down regulated by retinoic acid whereas PDGF alpha receptor was up regulated in F9 cells. The increase in PDGF alpha receptor transcripts were resulted from an increase level of gene transcription. Here, we present data on the identification and characterization of this cis-element.

TRANSITION TEMPERATURE OF MAMMARY MITOCHONDRIAL ATPase IN DMBA-TREATED RATS FED DIFFERENT LIPID DIETS

T.C.Cavalcanti[1], F. Guimarães[1], H.F. Gumerato[2] & Q.S. Tahin[3]
[1] Laboratório de Pesquisas Bioquímicas. CAISM/UNICAMP.CP 6151- Campinas, São Paulo - Brasil, 13081-970. [2] Departamento de Ciência de Alimentos FEA/UNICAMP. [3] Tumor Markers Laboratory. Fox Chase Cancer Center. Philadelphia/USA.

Membrane-dependent properties are modulated by dietary lipids. The mitochondria of tumoral cells usually differ in number and lipid membrane composition when compared with normal cells. In this report we described the effect of DMBA treatment on transition temperature (T_t) of mitochondrial ATPase on the mammary Sprague-Dawley rats fed different lipid diets.

Female rats 45-days-old were i.g. inoculated with 0.5 mg DMBA/0.5 ml soybean oil. Controls were non-treated DMBA rats. We used 4 isocaloric semi-synthetic lipid diets: commercial chow (Purina, CCP), 10% sunflower oil (SFO), 10% coconut fat diet (CO) and reference diet (5.0% soybean oil + 2.5% coconut fat + 2.5% olive oil, RF). DMBA-treated and control animals were fed one these diets during 4,6,8 and 10 months.

We observed an increase in T_t of mitochondrial ATPase in DMBA-treated rats fed CCP, SFO or RF diets, and a decrease in this enzyme activity in CO dietary group in comparison to control rats. This result was found even mammary tissues had ductal adenocarcinoma grade II. Fibroadenomas with increased T_t were observed in controls animals fed CCP diet.

Increase on T_t of mammary mitochondrial ATPase in rats feeding CCP, SFO or RF diets may be due mitochondrial membrane incorporation of the dietary mono and polyunsaturated fatty acids. By other hand, the decrease in T_t of this enzyme in CO rat groups can be due the dietary incorporation of saturated fatty acids by these membranes.

Because mammary DMBA carcinogenesis is fatty acid-dependent and mitochondrial ATPase is also a membrane-dependent enzyme, whose activity can be alterated by dietary fatty acids manipulation, our results suggest this enzyme as an important tool for the study of mammary-induced cancer progression.

T.C.Cavalcanti is CNPq researcher.

SQUAMOUS CELL CARCINOMA LINES FAIL TO RESPOND TO THE PRODIFFERENTIATING ACTIONS OF 1,25 DIHYDROXY VITAMIN D AND CALCIUM DESPITE NORMAL LEVELS OF THE VITAMIN D RECEPTOR.

ANITA V.RATNAM, MEI JHY SU, DANIEL D.BIKLE
Department of Medicine, University of California,
San Francisco, CA 94121

Normal Human Keratinocytes (NHK) differentiate from a highly proliferative basal cell to a terminally differentiated cornified cell in culture in the presence of physiological levels of extracellular calcium. In contrast,squamous carcinoma cells(SCC) differentiate less well than NHK under identical conditions. Our studies with Northern blot analysis elucidate that the differentiation process is associated with a concomitant increase in the expression of genes for involucrin (I) and transglutaminase (TG)- the natural substrate and enzyme involved with cornified envelope formation. The mRNA for both I and TG were not detected in all the SCC lines studied *viz.*,SCC4,12B2, 12F2,A431 and HACAT,when they were grown in MEM/KGM without serum. Addition of 5% FCS for 48 hours triggered the expression of these genes which could then be maintained subsequently in KGM in the absence of serum, suggesting that serum factors play a key role in inducing these genes but only as an initial trigger. Serum was not required for the expression of these genes in NHK. In NHK,extracellular calcium (CaO)and 1,25 dihydroxy vitamin D3 (1,25 D) stimulate the expression of I and TG in a dose dependent manner, while the SCC lines failed to respond to these factors regardless of whether these cells had been pre-exposed to serum. An important factor that mediates 1,25 D stimulated gene expression is the Vitamin D Receptor (VDR).

However, VDR mRNA levels in all the SCC lines examined were comparable to those in NHK. Furthermore, the VDR protein levels and affinity in SCC lines as assessed by ligand binding analysis were comparable to those of NHK. These studies indicate that serum, but not calcium or 1,25 D stimulate expression of I and TG genes in SCC lines, suggesting that a serum response element rather than calcium or Vitamin D response element may specifically potentiate transcription in the SCC lines. Secondly, the mediators facilitating 1,25 D and calcium action on gene expression other than the VDR may be missing or defective in SCC lines. Thus, SCC lines appear to lack some but not all the mediators by which NHK respond to the prodifferentiating actions of calcium and 1,25 D. Further experiments are intended to elucidate the defective pathway.

DIFFERENTIAL EFFECTS OF BUTYRATE AND ACETATE ON CACO-2 PHENOTYPE. Marc D. Basson, MD, PhD, Fu Hong, MD

Colorectal cancer is the second leading cause of cancer death. Epidemiologic data suggests that the intake of dietary fiber is associated with a decreased risk of colonic carcinogenesis It has been theorized that the promotion of colonocyte differentiation by short chain fatty acids produced by bacterial fermentation of fiber contributes to its protective action. However, the fermentation of different dietary fibers yields different ratios of different short chain fatty acids. We therefore sought to determine whether the two common short chain fatty acids butyrate and acetate had similar effects on the phenotype of the human colonic Caco-2 cell line, focusing on the expression of the brush border digestive enzyme dipeptidyl peptidase (DPDD) and the ability of the cells to adhere to type I collagen as markers of differentiation and calculating the effects of these agents on Caco-2 doubling time. After treatment with 10 mM butyrate or acetate for 48 hours, cell doubling time was quantitated by logarithmic transformation of serial cell counts, DPDD specific activity was quantitated in protein-matched aliquots of cell lysates and adhesion to collagen I was measured one hour after plating a single cell suspension across a collagen I substrate. Results are expressed as mean ± SE and were analyzed by ANOVA and Bonferroni's modified t tests.

Both butyrate and acetate modulated Caco-2 differentiation, adhesion to matrix, and doubling time. However, butyrate proved markedly more potent in this regard.

	DPDD (u/ng protein)	Adhesion (cells/HPF)	Doubling time (hours)
Control	3.43 ± 0.12	56.9 ± 0.6	23.7± 0.7
Acetate (10 mM)	5.78 ± 0.11*	27.25 ± 0.5*	34.9 ± 1.1*
Butyrate (10 mM)	8.53 ± 0.18*	2.6 ± 0.2*	46.2±1.0*

* $p<0.001$ by ANOVA

In particular, the acetate appeared substantially less potent than butyrate in equimolar concentrations.

Although extrapolation from cell culture data to the in vivo condition must always be cautious, these results suggest that different short chain fatty acids may exhibit different potencies with regard to the promotion of colonocyte differentiation and the inhibition of colonocyte proliferation. Thus, if the fermentation of dietary fiber to short chain fatty acids is desired to promote colonocyte differentiation and inhibit proliferation, it may be relevant to consider the molar ratios of butyrate and acetate produced by digestion of different fibers.

STORAGE OF INTACT BETA-CAROTENE (BC) AND VITAMIN A IN RAT ORGANS FOLLOWING BC CARRIER MEDIATED SUPPLEMENTATION [1]
The lung as a BC target organ. Relevance to cancer BC-Vit.A chemoprevention.

Amalia BIANCHI-SANTAMARIA and Leonida SANTAMARIA
"C. Golgi" Institute of General Pathology, Centro Tumori, University of Pavia, 27100 Pavia, Italy.

Objective. To assess beta-carotene (BC) and retinyl esters distribution and storage in different organs from rats BC-supplemented with different carrier mediations, namely, standard rat diet ± arachidic oil (a.o.) or cristalline BC in arachidic oil only, by gavage, twice a week.
Materials and Methods. 25 adult female Wistar rat, housed one per cage, were divided in 5 groups and given diets as follows. *Group A*: standard diet (s.d.), for 12 weeks. *Group B*: s.d. + 1ml arachidic oil (a.o.) by gavage twice a week, for 12 weeks. *Group C*: s.d. added a.o., for 12 weeks. *Group D*: s.d. supplemented with BC beadlets added cristalline BC-a.o.; each rat was given 350mg/week of BC, for 12 weeks. *Group E*: s.d. and cristalline BC-a.o. only, by gavage, twice a week; each rat was given the same 350mg/week of BC, for 12 weeks. Then, the animals were sacrificed and BC levels together with retinyl palmitate presence (from BC metabolism) were assessed by HPLC analysis in liver, lung, kidney, small intestine, mesenteric fat, brain, spleen, stomach and blood plasma.
Results. In *Group A*, no BC or retinyl palmitate was detected in any organ, except for blood plasma, where BC was found in traces. In *Group B*, BC and retinyl palmitate were surprisingly detected in lung and in liver respectively and in traces in blood plasma. Here, the origin of both compounds could be from arachidic oil and/or from s.d. contents through a.o. as a carrier. In *Group C*, no BC or retinyl esters were detected in any tissue but traces were found in blood plasma. In *Group D*, supplemented BC beadlets added cristalline BC- a.o., BC storage was highest in liver, small intestine and spleen (from 4.2 to 45.2 nmols/g wet tissue) lesser in lung, kidney, stomach, blood serum; retinyl palmitate was present in liver and lung. In *Group E*, supplemented with cristalline BC-a.o. only, by gavage, twice a week, BC was distributed and stored in all the selected organs except for brain and stomach. Remarkably, the highest storage was detected in lung (from 116.30 to 5,763.62 nmols/g wet tissue) up to about 25 times more with reference to the storage in the small intestine. In this group the retinyl palmitate assay was positive only in liver. After discontinuing BC supplementation by gavage, this high lung BC storage lasted even after 10 days.
Conclusions. 1) a.o. is a relevant BC carrier; 2) rats like humans have the capacity to absorbe intact BC and to store it in tissue; 3) the use of both beadlets BC in the diet and cristalline BC-a.o. by gavage had to facilitate the pioneering clearcut demonstration of BC chemoprevention against skin benzo(a)pyrene photocarcinogenesis in mice[2]; 4) the data of lung as a BC target organ, as pointed out also in piglets[3], might explain epidemiological findings on inverse correlation between BC blood serum levels and lung cancer as well as current promising BC-Vit.A human interventions to increase survival in patients with non-surgically treatable lung cancer occurences.

[1] Bianchi Santamaria A., Dell'Orti M., Frigoli G., Gobbi M., Arnaboldi A. and Santamaria L. *Internat. J. Vit. Nutr. Res.*, **64**, 15-20, 1994. [2] Santamaria L. *et al.*. *Boll Chim Farmaceutico*, 119, 745-748, 1980; *Experientia*, **39**, 1043-1045, 1983. [3] Schweigert F.J. *et al.* 10th Int. Symp. Carotenoids, Trondheim, Norway, Book of Abstract CL 3-4, 1993.

ACTIVITY OF BETA-CAROTENE IN PATIENTS WITH THE AIDS RELATED COMPLEX (ARC)
Updated report

Amalia BIANCHI-SANTAMARIA, *Sergio FEDELI and Leonida SANTAMARIA
Camillo Golgi Institute of General Pathology, Centro Tumori, University of Pavia, 27100 Pavia, Italy;
* Delmati Hospital, Division of Internal Medicine, 20079 Sant'Angelo Lodigiano, Milano, Italy.

Objective. To update a pilot single blind study on the activity of beta-carotene (BC) supplementation in patients with AIDS related complex (ARC) under current medical treatment.[1]

Method. Since 1989, 84 patients suffering from ARC, CD4 count at baseline not less than 400 cells/mm^3 (including asymptomatic immunodeficiency, multiple district lymphoadenopathy syndrome (LAS) with no opportunistic infections), were recruited for a single blind study. These patients belonged to the "Mondo X" drug addicts' communities, led by a Franciscan monk Father Eligio, in Italy. Thus, they were controlled for compliance in being supplemented, in a blind manner, with 60 mg BC daily, for cycles of 20 days per month, as adjuvant to their current clinical treatments. The control group consisted of data from near 200 ARC patients from "Mondo X" who had been treated for more than a decade at the above division of Delmati Hospital. In all patients the following immunological parameters were checked: counts of CD4 and CD8 cells/mm^3, responses to HIV (Elisa and Western Blot test), multitest, toxotest, Hbs, HCV, Delta and CMV antibodies.

Results. Of the 84 patients so far recruited as above, 31 proved to comply well with this current ongoing single blind study. The others dropped out from follow up due to the well known unreliability of drug-addicts. Nevertheless, they had been followed up for at least 6-12 months giving enough information for conclusive data. This BC treatment produced a non significant increase in CD4 count, as well as in inverted CD4/CD8 ratio, no improvement in multiple district lymphoadenopathy, but complete recovery from objective and subjective symptoms such as asthenia, feverishness, nocturnal sweating, loss in weight, working efficiency impairment, leading as a result to restoration of general health conditions and, remarkably, working efficiency. Of the above 31 ARC well complying patients, 7 developed into AIDS terminal phase (increase of β_2 microglobulin associated with p24 antibody reduction, occurrence of PcP, toxoplasmosis, oral candidosis, Kaposi sarcoma, HIV encephalitis) with dropping of CD4 counts to 100 or less and with CD4/CD8 = 0.06 (0.02-0.12); they, after few months worsened in their conditions and subsequently died.

Conclusions. The above data of this ongoing study are relevant in pointing out BC supplementation as a remarkable adjuvant intervention in restoring the general health conditions and working efficiency. As far as mechanism of action is concerned, BC should be active in scavenging oxygen free radicals that are likely to be continuously formed in AIDS molecular pathology, thus producing immunorecovery. The latter should occur at the level of white monocyte system rather than lymphocytes.

[1]Bianchi-Santamaria A. *et al. Med Oncol & Tumor Pharmacother.*, **9**, 151-153, 1992.

Genistein Programs Against Mammary Cancer

Coral A. Lamartiniere, P.I.
Department of Pharmacology and Toxicology
University of Alabama at Birmingham

It is estimated that there will 182,000 new cases of breast cancer in the United States in 1994. Chemically induced mammary carcinogenesis in rats provides a good model of this human pathology. We have investigated the potential of genistein, a phytoestrogen component of soybeans to protect against dimethylbenz(a)anthracene (DMBA) induced mammary adenocarcinomas. We have used a novel approach to exert a chemopreventive effect i.e. treatment of female rats during a limited critical period of development. Newborn female rats were treated with geinistein or the vehicle, dimethylsulfoxide (DMSO) on days 2,4, and 6 postpartum only. On day 50 postpartum, the two groups were exposed to DMBA to induce mammary tumors. From this study, we found that female rats treated neonatally with genistein had increased latency and decreased incidence and multiplicity of palpable tumors as compared to female rats receiving DMSO during the neonatal period. Investigations into mammary cell differentiation revealed that neonatal genistein treatment resulted in fewer terminal end buds and terminal ducts (undifferentiated and more susceptible ductal structures) and more lobules (more differentiated structures) in 50 day old animals (day of DMBA exposure). Cell proliferation as evaluated by immunohistochemical staining for proliferating cell nuclear antigen was decreased in mammaries of genistein treated animals. It appears that early exposure to genistein can program cell differentiation and proliferation in the young adult to result in protection against mammary cancer.

NON-PROMOTING EFFECTS OF LEAN BEEF IN THE RAT COLON CARCINOGENESIS MODEL. **B.C. Pence, M.J. Butler, D.M. Dunn and M.F. Miller, Texas Tech University Health Sciences Center, Lubbock, TX 79430**

Recent epidemiologic studies have implicated red meat consumption as a risk factor for colon cancer in both men and women. The goal of the present study was to compare, in an experimental animal model, the effects of beef with casein as a protein source on promotion of colon carcinogenesis. These factors were also examined within the context of a low and high fat diet containing either corn oil or beef tallow. Tumors were induced in Sprague-Dawley rats with 1,2-dimethylhydrazine (20 mg/kg body weight for 10 weeks). 280 male weanling rats were randomized to eight dietary treatment groups of a 2 X 2 X 2 factorial design with fat source (corn oil vs. beef tallow), fat level (5% vs. 20%), and protein source (very lean beef vs. casein) as the factors. Diets were fed *ad libitum* before, during and after carcinogen treatment for a total of 27 weeks. At termination of the study, animals were examined for location, size and type of colon or extracolonic lesion. All lesions were verified microscopically. Results demonstrated that the total number of intestinal tumors and the total incidence of tumors were significantly lower in the groups fed beef rather than casein as the protein source. High fat levels, regardless of source, increased the number of colon adenomas. Fat type had no effect on tumor numbers. These results demonstrate that when lean beef is used as the protein source in the context of a low fat diet, fewer intestinal tumors develop. These data do not support that belief that red meat consumption increases the risk for colon carcinogenesis, but underscores the importance of fat level in dietary context. Supported in part by the American Institute for Cancer Research, grant #:92B30-REV.

Stearate Inhibits Breast Cancer Cell Growth Via A 41 kDa Gi Protein Associated With the Epidermal Growth Factor Receptor

N.S.M.D. Wickramasinghe, H. Jo, J.M. McDonald, and R.W. Hardy.
Department of Pathology, University of Alabama at Birmingham, Birmingham, AL 35294

Epidemiological and animal studies have linked dietary fat to the development of breast cancer. However, little is known about the biochemical processes responsible for this effect. In order to understand how dietary fat modulates the signal transduction mechanisms involved in breast cancer cell growth, we investigated the effects of long chain saturated fatty acids (LCSFA) most commonly found in the western diet on epidermal growth factor (EGF) induced cell growth in the breast cancer cell line HS-578T. We find that LCSFA inhibit EGF-induced cell growth in a chain length dependent manner. Pretreatment (0.15 mM, 6 hrs.) with stearate (C:18) completely inhibits the EGF-induced ^{3}H thymidine incorporation while palmitate (C:16) inhibits by 67±8% and myristate (C:14) has no effect. A pertussis toxin (PTX) sensitive, 41 kDa, G-protein specifically co-immunoprecipitates with the EGF receptor (EGFR) in this cell line. We have shown that this G-protein is involved in the growth of this cell line. Time course studies indicate that the maximum ADP-ribosylation of the EGFR associated Gi-protein occurs at 1 minute of stimulation with EGF. Pretreatment of cells with PTX (0.1 μg/ml, 24 hrs.) inhibits the EGF-induced ^{3}H thymidine incorporation by 50±8% and the EGF-induced ADP-ribosylation of the 41 kDa Gi-protein is completely inhibited. In a similar manner, pretreatment of cells with stearate (0.15 mM, 6 hrs.) inhibits EGF-induced ^{3}H thymidine incorporation and EGFR associated 41 kDa G-protein ADP-ribosylation. Furthermore, pretreatment with PTX or stearate does not affect EGFR tyrosine phosphorylation. We also find that a G-protein co-immunoprecipitates with PLC-γ in EGF stimulated cells and the maximum ADP-ribosylation of this G-protein occurs at 1 minute of stimulation with EGF (1 nM). These studies establish an EGFR-Gi signaling pathway in cell growth and demonstrate that it is specifically inhibited by stearate.

Modulation of gene expression and mammary tumor incidence by calorie restriction in MMTV/v-Ha-*ras* transgenic mice.

G. Fernandes, B. Chandrasekar and JT Venkatraman
Medicine/Clinical Immunology, Uni. of Texas HSC at San Antonio, Texas

Carcinogenesis is a multistep process involving either loss of tumor suppressor genes and/or overexpression of oncogenes, cytokines and growth factors in the target tissue. Also, uninhibited production of free radicals as a result of overexpression of proinfla-mmatory cytokines and certain dietary components, may contribute to the alterations in the expression of genes associated with initiation and progression to malignancy. We have reported earlier that calories and/or source and level of dietary lipids alter the tumor incidence in C3H/Bi mice that spontaneously develop breast tumors. The present study is designed to elucidate the role of source (corn oil, CO; fish oil, FO) and level of lipid (5 and 20%), and amount of calories on mammary tumor incidence and gene expression in tumors and livers of MMTV/v-Ha-*ras* transgenic mice (Dupont) over a period of three years. The nutritionally adequate semipurified diets supplemented with equal levels of antioxidants consisted of (1) CO-5% (*ad libitum*, AL) (2) CO-20%-AL, (3) FO-20%-AL, and (4) Calorie restriction (CR, 40% less calories than CO-5% AL). The results indicate that, CR significantly inhibited tumor incidence [CR, 28%; CO-5%-AL, 83%; CO-20%, 89%, and FO-20%, 71%], tumor progression and extended life span. Northern blot analysis revealed a significant inhibition in transgene (v-Ha-ras), *c-erb* B2, IL-6, TGFβ1, whereas *p53* and antioxidant enzymes (catalase and SOD) genes expression were significantly elevated in tumors of CR mice as compared to the tumors in AL fed mice. Further livers from tumor bearing mice when analyzed, showed higher thiobarbituric acid-reactive substances (TBARS) and lower CAT, GSH-Px and SOD enzymes mRNA in AL-fed groups, whereas, CR significantly inhibited hepatic TBARS generation and enhanced antioxidant enzymes mRNA. Increasing the levels of CO from 5 to 20% or supplementing FO (20%) did not influence significantly any of the parameters studied, indicating that CR is far more effective than quantity and source of lipids in modulating breast tumor incidence in this transgenic mouse model. (Supported by NIH RO1-AG 03417 and AG 10531, and The Kleberg Foundation, San Antonio, Texas).

Kuratko, CN and Pence, BC. Dietary lipid and iron modify normal colonic mucosa without affecting phospholipase A_2 activity. Texas Tech University Health Sciences Center, Department of Pathology, Lubbock, Texas. Phospholipase A_2 (PLA_2) is the lipolytic enzyme that hydrolyses the 2-acyl position of membrane glycerophospholipids. It is elevated in some human tumors and may be involved with mechanisms of tumor promotion. PLA_2 functions as the rate-limiting step in arachidonic acid metabolism and in the removal of damaged or peroxidized membrane lipids. It interacts with phospholipase C/protein kinase C and has been proposed to function on the same signal transduction pathway as *ras* . In vitro systems have shown PLA_2 activity to be altered by variations in fatty acid and antioxidant components. First, this study examined PLA_2 activity in colon tumors produced by the AOM model using Fischer-344 rats. Secondly, the study tested the effect of iron supplementation as a potential pro-oxidant in diets of varied fatty acid composition. Diets included 35 mg/kg of iron (CD, CO, MO, or BT) or 140 mg/kg (CDFe, COFe, MOFe, BTFe) in the AIN-76A diet or diets high in either corn oil, menhaden oil, or beef tallow. Effects of dietary treatment on PLA_2 activity were examined in normal colon mucosa. Results showed tumor PLA_2 activity to be significantly higher than normal mucosa primarily as the result of increased activity within the particulate fraction. Fatty acid composition of colon mucosa was altered by both dietary fat and iron. BT-fed animals had the highest level of oleic acid and CO-fed animals had the highest level of linoleic acid. MO-fed animals had the lowest level of arachidonic acid and highest level of α-linolenic, eicosapentaenoic, docosapentaenoic, and docosahexaenoic acids. Iron supplementation in diets high in corn oil resulted in decreased membrane composition of palmitoleic and δ-linolenic acid. In spite of these changes in membrane composition, there were no changes in PLA_2 activity. These results show that PLA_2 activity is increased in AOM-induced tumors but that diet alone does not influence PLA_2 activity in this model.

Supported by NIH CA52006 and the Audrey B. Jones Cancer Fund of Texas Tech University Health Sciences Center.

REGULATION OF MONOCYTE/MACROPHAGE DIFFERENTIATION IN ACUTE PROMYELOCYTIC NB4 CELLS BY 1,25 DIHYDROXYVITAMIN D3 AND 12-O-TETRADECANOYLPHORBOL-13-ACETATE Mickie Bhatia, James B. Kirkland and **Kelly A. Meckling-Gill,** Department of Nutritional Sciences, University of Guelph, Guelph, Ontario, Canada, N1G 2W1.

NB4 cells are the first *bonafide* acute promyelocytic leukemia cell line to be described. These cells differentiate into neutrophils when treated with all-*trans* retinoic acid (ATRA). When used clinically, ATRA will also induce neutrophilic differentiation in patients with M3 and some other leukemias. An advantage of differentiation therapy in general, is that it usually avoids that debilitating side-effects associated with classic chemotherapeutic drugs. However, remissions are usually short-lived because of resistance that develops as a result of increased production of retinoic acid-binding proteins. Here we show that NB4 cells in culture are capable not only of neutrophilic differentiation but when treated with combinations of 1,25 dihydroxyvitamin D_3 (1,25 D_3) and phorbol ester, are also able to differentiate along the monocyte/macrophage pathway. This may provide an alternate route for differentiation therapy when ATRA fails or relapse occurs in leukemia treatment. Monocyte/macrophage differentiation occurs in two-steps with 1,25 D_3 providing the initial signal and phorbol ester promoting the full acquisition of the mature phenotype. Using inhibitors and activators of PKC and tyrosine phosphorylation cascades, it appears that 1,25 D_3 may signal through routes that are independent of the vitamin D receptor/vitamin D response element. Although phorbol esters could not be used clinically, an understanding of the mechanisms by which 1,25 D_3 and phorbol ester induce leukemia cells to differentiate may allow the development of drugs or other nutritional agents that could be used in cancer treatment. As well, some understanding of the role 1,25 D_3 plays in normal hematopoietic differentiation and possibly cancer prevention, may also be achieved.

Calorie and Fat Restricted Diets Affect the Growth of Transplanted RIII/Sa Mouse Mammary Tumors but not the Expression of MMTV and *Wnt*-1

Nurul H. **Sarkar**, HuiWu **Li**, and Wei **Zhao**, Department of Immunology and Microbiology, Medical College of Georgia, Augusta, GA 30912

The most important initial step in mouse mammary tumorigenesis is the activation by MMTV of a family of *int* protooncogenes. Several studies have shown that the mammary cells of mice fed a low calorie (LC) diet expresses lower levels of MMTV than the mammary cells of mice fed a high calorie (HC) diet. However, the effect of a LC diet containing high or low amounts of dietary fat on the expression of MMTV in actively growing tumor tissues has not been determined. In addition, the effect of the LC diet on the expression of *int* genes, such as *Wnt*-1, is not known.

We compared the effect of LC and HC diets on the growth of transplanted mammary tumors, and the levels of their MMTV and *Wnt*-1 expression using RIII/Sa mice as a model. Starting at 3-4 weeks of age, groups of female mice were fed either high isocaloric diets (16 kcal/day/mouse) containing 25% or 5% corn or fish oil or low isocaloric (10 kcal/day/mouse) diets containing 25% or 5% fish or corn oil. After one week, small pieces (2x2x2 mm) of tumor tissue were inserted into the fourth pair of the mammary glands, and the mice were maintained on their respective diets until sacrifice. In a separate experiment a tumor cell line (RIII/Sa-*Wnt*/1) expressing *Wnt*-1 was used for transplantation. After 12 weeks mice were sacrificed, tumor weights determined, and the level of the expression of MMTV and *Wnt*-1 in the tumors evaluated. Our results show that LC diets, regardless of the type or amount of fat, inhibited tumor growth by at least 60% in comparison to HC diets. However, mice fed a LC diet containing fish oil were also found to produce smaller tumors (20-40%) as opposed to those mice fed similar, but corn oil containing diets. Surprisingly, the levels of MMTV expression were not found to be affected by either caloric or fat content in the diet. Similarly, no significant changes in the levels of *Wnt*-1 were observed in RIII/Sa-*Wnt*/1 cell line-derived tumors from mice fed LC and HC diets.

OMEGA-3 POLYUNSATURATED FATTY ACIDS MODULATE NUCLEOSIDE TRANSPORT AND CHEMOTHERAPEUTIC DRUG TOXICITY IN TUMOR BUT NOT NORMAL CELLS. Trevor G. Atkinson, Danielle Martin, Helen de Salis and **Kelly A. Meckling-Gill.** Department of Nutritional Sciences, University of Guelph, Guelph, Ontario, Canada, N1G 2W1.

Nucleosides are cellular nutrients for hematopoietic, intestinal epithelial, and brain cells. Because *de novo* synthesis is inadequate, nucleoside needs are met by salvage via membrane-bound transporter systems. These same transporters also facilitate the entry of nucleoside drugs used in the treatment of neoplastic disease. Unfortunately, the uptake of nucleoside drugs into normal tissues and their consequent toxicity, limits the dose/efficacy of drug treatment. We have previously shown that transformation and growth factor stimulation can modulate nucleoside transport. We now report that transport may limit drug toxicity in some neoplastic cells and that drug toxicity can be modulated in cell culture by supplementing with omega-3 polyunsaturated fatty acids. Docosahexaenoic acid (22:6n-3, DHA) and eicosapentaenoic acid (20:5n-3, EPA) both result in increased adenosine transport rates in L1210 leukemia cells and increase the toxicity of cytarabine (ara-C) and adriamycin in these cells. No increase in toxicity was found in normal macrophages despite substantial changes in membrane fatty acid composition. Results obtained in a rat-2 fibroblast model showed that DHA can protect normal cells from ara-C while not substantially altering the sensitivity of tumor cells. In both these model systems then, the difference between IC_{50} values in normal and tumor cells was enhanced such that preferential killing of tumor cells or selective protection of normal cells was achieved. DHA also appeared to enhance toxicity of the anthracycline drug, adriamycin toward tumor cells without affecting normal macrophages or normal fibroblasts. This suggests that omega-3 polyunsaturated fatty acids, particularly DHA may have chemoprotective activity for at least some tumor/host-tissue sites.

RETINOIC ACID INHIBITION OF BREAST CANCER CELL GROWTH IS ASSOCIATED WITH INHIBITION OF CDK2 GENE EXPRESSION

Christine Teixeira and Christine Pratt, Department of Pharmacology, University of Ottawa, Ottawa, Ont., Canada K1H 8M5

Retinoic acid (RA) a metabolite of vitamin A, inhibits the growth of estrogen dependent breast cancer cells. The molecular mechanisms underlying this growth inhibition are unknown. Since the regulation of cellular proliferation by steroids in human breast cancer cells occurs by cell cycle specific actions in G1 phase we have investigated the effects of RA on cell cycle progression of MCF-7 estrogen receptor positive breast cancer cells. Our results indicate that RA treatment results in accumulation of MCF-7 cells in G1. Accordingly, we have also assessed the effects of RA on the regulation of the G1 cyclins and their catalytic partners the cyclin dependent kinases and the accompanying effects on the phosphorylation status of the retinoblastoma protein (Rb). Western blot analysis showed that unphosphorylated Rb protein accumulates in both MCF-7 and T47-D estrogen receptor positive breast cancer cells treated with RA for 24 hours and the effect is maximal by 48 hours. Northern blot analysis showed that the mRNA for the G1 cyclin D1 is transiently increased then decreased following RA treatment. The expression of its catalytic partner cdk4 remains constant while the expression of cdk2 is profoundly downregulated within 8 hours following exposure to RA. The mRNA levels of the G1 partner for cdk2, cyclin E are marginally upregulated by RA while those of its S phase partner cyclin A remain unchanged. Our results demonstrate that RA inhibition of breast cancer cell growth involves accumulation of underphosphorylated Rb associated with the RA mediated decrease in cdk 2 levels.

INHIBITION OF ESTROGEN RECEPTOR MEDIATED TRANSCRIPTION BY RETINOIC ACID

Christine Pratt, Department of Pharmacology, University of Ottawa, Ottawa, Ont., Canada K1H 8M5

Since retinoic acid (RA) has been shown to inhibit the growth of estrogen (E2) dependent breast cancer cells I have performed experiments to determine if a component of the antiproliferative effect is due to the ability of RA and RA receptors (RARs) to antagonize estrogen-induced transcription. I have transfected MCF-7 E2-dependent breast cancer cells with an estrogen-responsive reporter gene construct consisting of the pS2 gene promoter driving chloramphenicol acetyltransferase (CAT). Treatment of cells with E2 results in an increase in promoter activity and this induction is inhibited by the simultaneous addition of RA. Basal promoter activity was also reduced by the addition of RA. Inhibition was dependent on ligand addition since cotransfection of an RARα expression plasmid alone did not effect estrogen inducible transcription. RA treatment also inhibited E2 induction of the endogenous pS2 gene as assessed by northern blot analysis. Cotransfection of the MCF-7 cells with the reporter gene and a C-terminally truncated RARα (RARα′) alleviated the RA inhibition of transcription. Gel shift experiments with in vitro translated proteins showed that the RARα alone or in combination with the RXRß binds only weakly to the estrogen response element. In contrast the ER and the thyroid hormone receptor bind with high affinity to the ERE. The results are consistent with an inhibitory mechanism involving transcriptional "squelching" of estrogen-inducible transcription by retinoid receptor-ligand complexes.

RETINOIDS AS INHIBITORS OF GLUTATHIONE S-TRANSFERASE (GST).

A. A. Kulkarni, M. Sajan, K. Datta, P. Roy and A. P. Kulkarni
Florida Toxicological Research Center, College of Public Health
University of South Florida, Tampa, FL-33612

A hypothesis that retinoids inhibit mammalian tissue GSTs was tested. GSTs are responsible for the activation or detoxification of carcinogens. Cytosolic GSTs from female rat livers (RL-GST), adult human livers (HL-GST) and human term placentas from non-smokers (HP-GST) were purified by affinity chromatography and used in all the experiments. GST activity was assayed using incubation media containing 100 mM phosphate buffer, pH 6.5, 1.0 mM 1-chloro-2,4-dinitrobenzene (CDNB), 2.5 mM GSH and rate limiting amount of GST. An increase in the absorbance at 340 nm for CDNB conjugate formation was monitored at room temperature. The enzyme preparations used exhibited specific activity of 28$\pm$3, 54$\pm$7 and 48$\pm$12 μmoles/min/mg protein for RL-GST, HL-GST and HP-GST respectively. All trans-retinoic acid (t-RA) caused a significant decrease in RL-GST, HL-GST and HP-GST activity suggesting susceptibility of different isozymes of GST to the inhibition. Relatively, HL-GST was the most susceptible to the inhibition while RL-GST was affected the least. HP-GST was used in all the subsequent experiments. The analysis of data yielded Ki values of 20 and 41 μM for t-RA in the presence of different concentrations of CDNB and GSH, respectively. The Lineweaver-Burk plots of the data suggested noncompetitive nature of HP-GST inhibition with respect to both CDNB and GSH. The preliminary data gathered for other retinoids indicated that the magnitude of HP-GST inhibition depends upon the chemical structure of the inhibitor. IC_{50} (μM) values noted for 7 different retinoids were as follows: 143 for cis-retinoic acid, 33 for all-trans retinol, 37 for 13-cis retinol, 168 for all-trans retinyl palmitate, 158 for all-trans retinol acetate, 100 for all-trans retinal and 248 for 13-cis retinal. Taken together, the data suggest that GST inhibition may represent yet another mechanism of retinoid action. Further _in vivo_ studies are needed to confirm or refute these findings.

Selenium interacts with cysteine-rich regions of protein kinase C and induces the enzyme inactivation: its role in cancer chemoprevention

Rayudu Gopalakrishna, Zhen-hai Chen, and Usha Gundimeda
Department of Cell and Neurobiology, School of Medicine,
University of Southern California, Los Angeles, CA 90033

The mechanism by which selenium inhibits tumorigenesis is not known. Since protein kinase C (PKC) serves as a receptor for certain tumor promoters, it is possible that selenium may elicit its antitumor promoter action at this site. Selenium interaction with critical cysteinyl residues present in the proteins has been suggested to be one of its modes of action. Since PKC has cysteine-rich zinc-finger motifs which are required for phorbol ester binding and also other critical cysteine residues in the catalytic domain, we have determined whether selenium can influence PKC activity by interacting with these regions. When the rabbit brain purified PKC (a mixture of α, β, and γ isoenzymes) was incubated with low concentrations (0.05 to 5 μM) of selenite, PKC lost rapidly its kinase activity. However, the phorbol ester binding was not affected at these concentrations. Mercapto compounds inhibited this selenite-mediated inactivation of PKC. Furthermore, the kinase activity of selenite-modified PKC was regenerated by incubation with higher concentrations (>5 mM) of dithiothreitol. These studies suggested that the vicinal thiols present within the catalytic domain may be involved in the reduction of selenite. Selenocystine also inactivated PKC with an IC_{50} of 6 μM. However, selenate and selenomethionine had no effect on the PKC activity. The treatment of intact LL/2 lung carcinoma, B16 melanoma, and MCF-7 breast carcinoma cells with either selenite (0.5 to 10 μM) or selenocystine (10 μM) also resulted in an inactivation of PKC activity. Unlike with purified enzyme, in intact cells the phorbol ester binding was also affected by selenite. Protein phosphatase 2A, although it has one critical cysteine residue, was not effected by selenite. Similarly, protein kinase A activity was not effected by selenite either in the isolated form or in the intact cells. Given the fact that PKC serves as a receptor for not only phorbol ester tumor promoters but also to other structurally unrelated tumor promoters such as oxidants, selenium-mediated inactivation of PKC may have an important role in the antitumor promoting activity of selenium.
Supported by American Institute for Cancer Research grant 93B43.

THE MODULATING EFFECTS OF VITAMIN A ON ENDOTHELIOMA CELL GROWTH AND PHENOTYPE

S.J. Braunhut, D. Medeiros, M.R. Freeman, and M.A. Moses
Departments of Ophthalmology, Urology and Surgery at Children's Hospital and the Harvard Medical School, Boston, MA 02115

Normally, endothelial cell (EC) growth is stringently regulated. We have investigated the role of vitamin A, which circulates in the blood, in controlling the growth and phenotype of normal ECs. Our studies have shown that retinoids are potent inhibitors of normal EC growth and act, in part, by decreasing the activity of matrix metalloproteinases (MMPs), the enzymes that specifically degrade the extracellular matrix (ECM). We have also found that retinoid-treatment of EC changes the phenotype of the cells causing them to resemble their in vivo counterparts more closely. These findings strongly suggest that circulating vitamin A could contribute to the growth control of normal EC.

Abnormal EC proliferation, however, does occur in association with certain disease states. For example, neovascularization is associated with tumors, diabetes and retinopathy of prematurity. Aberrant EC growth is also a prominent feature of tumors of endothelial cell origin, or hemangiomas. Hemangiomas are the most common tumor of infancy and are often life-threatening. We hypothesized that retinoids might prove effective in reducing this form of endothelial cell tumor growth, whose etiology is unknown. To investigate this we examined the response of polyoma virus middle T-transfected brain endothelial cells (bENDO) and embryonic stem cells (eENDO) to retinoids. bENDO and eENDO are two cell lines which express EC markers, exhibit aberrant growth in vitro and form hemangioma-like tumors in vivo. We have treated ENDO cells with various retinoids and studied changes in their morphology, growth and MMP production, in vitro. Furthermore, we have studied the tumorigenicity of untreated and retinoid-treated ENDO cells using the chick chorioallantoic membrane (CAM) assay.

Our studies have shown that retinoid treatment of ENDO cells causes a dramatic change in their morphology. Prior to treatment, ENDO cells exhibit long, overlapping processes; the cells are not contact-inhibited at confluency and grow in multiply layers. Following a 6 day exposure to retinoids, ENDO cells form monolayers in a cobblestone pattern, similar to that of normal ECs. Proliferation studies demonstrated that retinol inhibited ENDO cell growth by at least 50%. Retinoic acid inhibited eENDOs by 30% and bENDOs to a lesser extent. Substrate gel electrophoresis revealed that untreated ENDO cells produce MMPs of 92, 72, and 68 kDa. After retinol treatment, both cell lines produced only the 92 kDa gelatinase. Retinoic acid treatment led to a reduction of the 72 kDa gelatinase in bENDOs, but not in eENDOs. Finally, 6 day pretreatment of ENDO cells with retinol reduced the tumorigenicity of both cell types in the CAM assay. In contrast, retinoic acid treatment of ENDOs caused an increase in the number and the size of the tumors. In conclusion, retinol and retinoic acid appear to differentially effect ENDO cell growth and phenotype. Furthermore, it appears that retinol-treatment reduces the production of MMPs and this in turn may reduce the invasiveness of the cells and their potential to form tumors in vivo.

ANTI-TUMOR ACTION AND MODULATION OF THE TUMOR MARKER Gal-GalNAc BY INOSITOL HEXAPHOSPHATE ($InsP_6$) IN HT-29 HUMAN COLON CARCINOMA CELLS in vitro

Yang G.Y., and Shamsuddin A.M. Department of Pathology, University of Maryland School of Medicine, Baltimore MD 21201

Phosphorylation and dephosphorylation of inositol phosphates are important intracellular events during signal transduction in mammalian system, wherein inositol 1,4,5 triphosphate acts as a second messenger. However, higher Inositol phosphates ($InsP_5$, $InsP_6$) have not been well studied eventhough they comprise the bulk of the inositol phosphate content of mammalian cells. Inositol hexaphosphate ($InsP_6$) ubiquitous in plants and animals is also a natural antioxidant by virtue of its chelating property.

A novel anti-tumor action of $InsP_6$ was demonstrated in models of experimental colon and mammary carcinogenesis in vivo and human cancer cells in vitro. We now present its effects on growth and differentiation of HT-29 human colon carcinoma cells in vitro. A dose- and time-dependent (0.33-20 mM InsP6 and 1-6 days treatment) growth inhibition was observed as tested by MTT-incorporation assay. The inhibition was statistically significant ($p<0.05$) at 1 mM concentration as early as 24 hours after treatment; IC_{50} was 0.66-3.3mM. DNA-synthesis was also suppressed by $InsP_6$ as early as 3 hours and continued upto 48 hours as determined by ^3H-thymidine incorporation assay. The expression of proliferation marker PCNA was down-regulated ($p<0.05$) by $InsP_6$ (1 and 5 mM) after 48 hours of treatment. Alkaline phosphatase activity (brush border enzyme, associated with absorptive cell differentiation) expressed both cytochemically and biochemically, increased following 1 and 5 mM $InsP_6$ treatment for 1-6 days. The expression of a mucin antigen associated with goblet cell differentiation and defined by the monoclonal antibody CMU_{10} was augmented by $InsP_6$. The mucin tumor marker Gal-GalNAc, expressed by precancer and cancer of colon but not by the normal cells and identified by galactose oxidase-Schiff sequence showed a time-dependent biphasic quantitative and qualitative change by $InsP_6$; an increased expression after 1 and 2 days of treatment followed by suppression after 3 days suggest progression of mucin synthesis and differentiation of cancer cells with reversion to normal phenotype.

Since $InsP_6$, a natural dietary ingredient of cereals and legumes (not necessarily high in fiber), inhibits growth and induces terminal differentiation of HT-29 cancer cells, it is an excellent candidate for adjuvant chemotherapy and prevention of cancer. Furthermore, because the tumor marker Gal-GalNAc is a) easily detected in rectal mucin of patients with colonic cancer and precancer with high sensitivity and specificity, and b) suppressed by $InsP_6$ treatment, it can be used to monitor the efficacy of chemoprevention by $InsP_6$ or other such agents. (Supported in part by AICR Grant #MG92B01)

Inositol hexaphosphate ($InsP_6$) inhibits mammary cancer *in vivo* and *in vitro*

Ivana Vucenik*, Guang-yu Yang, Abulkalam M. Shamsuddin
*Departmant of Medical and Research Technology, Department of Pathology
University of Maryland School of Medicine, Baltimore, Maryland

Because inositol hexaphosphate ($InsP_6$) and inositol (Ins), contained in plants and most mammalian cells, have been demonstrated to have anti-cancer and anti-cell proliferative action in several experimental models of carcinogenesis, we have examined the effect of $InsP_6 \pm$ Ins on DMBA-induced rat mammary tumor model and tested whether $InsP_6$ will be effective on human mammary cancer cell line.

Starting two weeks prior to induction with DMBA, the drinking water of female Sprague-Dawley rats was supplemented with either: 15 mM $InsP_6$, 15 mM Ins, or 15 mM $InsP_6$ + 15 mM Ins; a control group received no inositol compounds. Animals (49-day-old) were given a single dose of DMBA (5 mg/rat) in 1 ml of corn oil by oral intubation. After 45 weeks of treatment the results are as follows:

Treatment	Number of Rats	Number of Rats with Tumor	Tumor Incidence (%)	Total Number of Tumors	Mean Number of Tumors per Rat	Number of Tumors per Tumor-Bearing Rat	Tumor Burden per Rat (g)	Rats with ≥5 Tumors (%)
DMBA	40	37	92.50	113	2.83±0.36*	3.05±0.35*	31.03	17.5
DMBA + 15mM $InsP_6$	38	27	71.05‡	69	1.82±0.25†	2.46±0.21	31.86	5.3
DMBA + 15mM Ins	40	30	75.00‡	64	1.60±0.22†	2.13±0.19†	19.73	2.5‡
DMBA + 15mM $InsP_6$ 15mM Ins	38	29	76.32‡	51	1.34±1.07†	1.76±0.14†	17.58†	0.0‡

*denotes mean ± standard error
†statistically different from DMBA-only group ($p<0.05$) (student *t*-test)
‡statistically different from DMBA-only group ($p<0.05$) (chi-square test)

Our data show that $InsP_6$ significantly ($p<0.05$) inhibits DMBA-induced rat mammary carcinogenesis and Ins potentiates this action. The best results were obtained by $InsP_6$+Ins. Four additional groups not receiving DMBA, but drinking tap water, $InsP_6$, Ins, or $InsP_6$ + Ins of the same molarity as experimental groups were observed for the duration of the study to monitor for any toxicity following this long-term treatment; no significant toxicity, as evaluated by body weight gain, serum and bone mineral levels was seen.

In vitro, $InsP_6$ was a potent inhibitor of proliferation and inducer of differentiation of MCF-7 human mammary cancer cells. A 50% inhibition (IC_{50}-MTT assay) was found with $\geq$1.0 $InsP_6$. Malignant cells treated with $InsP_6$ became smaller and differentiated, as demonstrated by an increased expression of lactoalbumin.

These experiments demonstrate that $InsP_6$ is not only protective against mammary carcinoma, but may also be an important adjuvant therapeutic agent. Further studies of the molecular and cellular mechanisms of action, as well as further extensive toxicological evaluation, with the goal of introducing this agent into a clinical chemoprevention trial are needed.

(Supported by the American Institute for Cancer Research grant 92B18-REV (I.V.)

LEVEL OF DIETARY FAT AND REGULATION OF GENE EXPRESSION IN BALB/c MICE. E. A. Paisley, James Kaput, H. J. Mangian and W. J. Visek. University of Illinois College of Medicine, Urbana, IL., 61801

Our hypothesis is that certain genes regulated by dietary fat promote disease. Differential display of mRNA in liver and mammary glands (MG) of virgin female Balb/c mice fed semipurified diets containing 3 or 20% corn oil (CØ) for 2 wks revealed 20 differentially expressed bands in liver, 10 were cloned and partially sequenced. ApoE and cytochrome P450-15-alpha-hydroxylase, initially chosen for characterization and Northern analyses indicated that fat or other components in CØ regulate their expression differently between genetically distinct mouse strains. Nine of 18 genes differentially expressed in MG were sequenced and mRNA levels of several determined. We have identified genes for hnRNP G protein, rat angiotensin II, Mus musculus Balb-12 gene fragment, and mouse b2 repetetive sequence mRNA in mouse MG. Dr. Lee Romancyzck of M&M/MARS, Hackettstown, NJ, has shown that the purified CO fed in these studies contained tocopherols, plant sterols and fatty acid sterol esters in sufficient amounts to alter lipid metabolism. Constant exposure to small amounts of these substances or their microbial metabolites may lead to significant levels in the circulation. Sequences similar to sterol or fatty acid response elements (RE) occur in promoters of ApoE and P450-15-alpha-hydroxylase. The latter also contains corticosteroid RE. The RE are recognized by transcription factors that regulate gene expression as a function of physiological state and in response to different dietary components.

MODULATION OF *RAS* EXPRESSION IN HUMAN MALIGNANT CELLS BY POTENTIAL DIETARY SUPPLEMENTS. R.J. Hohl and K. Lewis., University of Iowa College of Medicine, Iowa City, IA.

The function of the growth-promoting RAS protein is dependent upon its attachment to farnesyl pyrophosphate derived from the product of hydroxymethylglutaryl coenzyme A (HMG CoA) reductase. We have observed that inhibition of HMG CoA reductase by lovastatin both impairs malignant cell proliferation and decreases RAS farnesylation. D-Limonene is the major monoterpine found in a wide variety of fruits and has been studied in animal models as an anticarcinogen. In the human-derived leukemia cell lines, THP-1 and RPMI-8402, proliferation was measured as DNA synthesis and RAS levels were measured in Western analysis using an anti-RAS antibody (NCC-RAS-004). In these cell types, lovastatin and limonene both impair DNA synthesis and decrease farnesylated RAS levels in concentration dependent manners. DNA synthesis is impaired by fifty percent with 24 hour exposure to either 10 uM lovastatin or 2 mM limonene. With similar exposure to 25 uM lovastatin there is a gradual accumulation of unfarnesylated RAS protein so that it equals lowered farnesylated RAS levels. Since 10 mM mevalonic acid prevents this effect then lovastatin likely results in decreased farnesyl pyrophosphate availability for RAS processing. When cells are exposed to 5 mM limonene there is no accumulation of unfarnesylated RAS protein despite lowered farnesylated RAS levels. Furthermore, perillyl alcohol, an *in vivo* limonene metabolite, more dramatically depresses levels of farnesylated RAS protein without resulting in the accumulation of unfarnesylated RAS protein. Previous reports, measuring radiolabeled mevalonate incorporation into RAS protein, have concluded that limonene, like lovastatin, lowers cellular RAS levels by interfering with RAS farnesylation. In contrast, our current studies show that limonene and perillyl alcohol interfere with *RAS* expression at a pretranslational or different posttranslational level than does lovastatin. These findings are suggestive of a mechanism for the anticarcinogen effects of dietary monoterpines.

Comparison of the Effect of Dietary Restriction on the Hepatic DNA-Carcinogen Adduct Formation in F344 Rats and B6C3F1 Mice

King-Thom Chung[1], Wen Chen[2], Yonggui Zhou[1], Peter P. Fu[2], Ronald W. Hart[2] and Ming W. Chou[2]. Department of Biology, The University of Memphis[1], Memphis, TN 38152, and National Center for Toxicological Research[2], Jefferson, AR 72079

Dietary restriction (DR) has been shown to result in a significant extension of maximum achievable lifespan and a reduction in the incidence of spontaneous and chemically-induced cancers in laboratory animals. However, the molecular mechanisms by which DR extends life-span and reduces tumor incidences are still not clear. Previously we used male F344 rats as the animal model to study the effect of DR on the metabolic activation of AFB_1. In this presentation, a species comparison study was initiated to study the effect of DR on AFB_1 activation in F344 rats and B6C3F1 mice.

Acute DR (60% of the food consumption of ad libitum-fed rats for 6 weeks) reduced the metabolic activation of AFB_1 in both male rats and mice. The formation of AFB_1-DNA adducts in rat liver was 5- to 8-fold more than that in mouse liver. DR reduced AFB_1-DNA adduct formation in both animal systems by about 60%. The reduction may be attributed to decreased cytochrome P-450 mediated AFB_1- epoxidation and/or the increased detoxification of AFB_1 catalyzed by hepatic glutathione S-transferase (GST) and other phase II detoxification enzymes. DR did not alter the rat liver GST activity when 1-chloro-2,4-dinitrobenzene or 2,4-dichloronitrobenzene was used as the substrate; however, a significant increase of the GST activity was found when synthesized AFB_1-8,9-epoxide was used. The activity of GST specific toward the AFB_1-epoxide was increased more than 2-fold in both DR-mouse and DR-rat liver cytosolic fractions. The mouse liver GST activity toward AFB_1-8,9-epoxide was 14.7-fold greater than the GST activity of rats. Benzo[a]pyrene was also included in the species comparison study.

Results obtained from this and previous studies indicate that the effect of DR on metabolic activation of AFB_1 in both mouse and rat is dependent upon activities of enzymes responsible for metabolic activation and detoxification of AFB_1.

Redox-cycling of etoposide (VP-16) in K562 cells by glutathione and protein sulfhydryls

Valerian E. Kagan, Jack C. Yalowich, Julia Y. Tyurina, Vladimir A. Tyurin

Departments of Environmental and Occupational Health and Pharmacology, University of Pittsburgh and Pittsburgh Cancer Institute, Pittsburgh, PA 15238

Phenoxyl radicals generated from an anticancer drug, etoposide (VP-16) (a semisynthetic phenolic derivative of the natural product podophyllotoxin), can directly oxidize thiols in model chemical systems, liposomes and homogenates of K562 cells. The VP-16 phenoxyl radical-driven oxidation of critical protein thiols may contribute significantly to its cytotoxic effects. Because low molecular weight thiols can also reduce the VP-16 phenoxyl radical thus preventing its interaction with protein thiols, in the present work we used ESR to study interactions of the VP-16 phenoxyl radicals in K562 cells with different endogenous levels of GSH. Tyrosinase-catalyzed oxidation of VP-16 results in the generation of its phenoxyl radicals with typical features in the ESR spectrum. K562 cell homogenates cause a transient disappearance of the ESR signal due to the reduction of the VP-16 phenoxyl radicals by endogenous reductants. The duration of the lag period for the reappearance of the ESR signal was linearly dependent on the content of thiols in cell homogenates. Depletion of thiols by treatment with a thiol reagent, mersalyl acid, eliminated the ability of K562 cell homogenates to reduce the VP-16 phenoxyl radical. To evaluate the role of GSH in the overall ability of intracellular thiols to reduce the VP-16 phenoxyl radical, we depleted GSH in K562 cells using buthionine-S,R-sulfoximine, BSO. This resulted in a 35-40% decrease of reduction of the VP-16 phenoxyl radical by cell homogenates. Similar effect was obtained when K562 cell homogenates were treated with GSH peroxidase + cumene hydroperoxide. In contrast, increased levels of endogenous thiols and GSH caused by teatment of cells with Cd^{2+} resulted in an increased ability of K562 cell homogenates to reduce the VP-16 phenoxyl radicals. Thus, metabolism of VP-16 by oxidative enzymes (peroxidases, cytochromes P450, tyrosinases) may trigger a redox-cycling mechanism in which VP-16 phenoxyl radicals will deplete endogenous GSH and oxidatively modify protein sulfhydryls.

Phenoxyl radicals of VP-16 oxidize sulfhydryls and inhibit Ca^{2+}-ATPase: prevention by ascorbate and low molecular weight thiols

V.B. Ritov, R. Goldman, D.A. Stoyanovsky, E.V. Menshikova, V.E. Kagan
Department of Environmental and Occupational Health, University of Pittsburgh and Pittsburgh Cancer Institute, PA 15238

Etoposide (VP-16) is one of the most widely used clinical antitumor drugs. VP-16 is a hindered phenol and the presence of the phenolic 4'-hydroxy group in the benzene ring is crucial for its antitumor activity. Several recent reports have suggested that VP-16 antitumor efficiency is dependent on the metabolic activation of the hydroxy group to consequently form phenoxyl and semiquinone radicals, quinone methide and o-quinone derivatives. Recently, we demonstrated that while VP-16 possesses a high antioxidant activity against lipid peroxidation, its phenoxyl radical is highly reactive toward low molecular weight thiols and protein thiols (GSH, dihydrolipoic acid, metallothioneins). The role of the VP-16 phenoxyl radical in its cytotoxicity is not clear. The study of interaction VP-16 phenoxyl radical with membrane can be useful for understanding mechanisms of its antitumor activity. In the present study, sarcoplasmic reticulum (SR) membranes were used to compare antioxidant effects of VP-16 in protecting membrane lipids with prooxidant effects of the VP-16 phenoxyl radical toward membrane thiols and activity of Ca^{2+}-ATPase. VP-16 acted as a potent antioxidant in protecting SR membrane lipids and *cis*-parinaric acid against peroxidation induced by an azo-initiator, 2,2'-azobis(2-amidinopropane)-dihydrochloride. Interaction of SR Ca^{2+}-ATPase with tyrosinase-generated VP-16 phenoxyl radical resulted in oxidation of about 5 moles of the Ca^{2+}-ATPase sulfhydryls per 1 mol of VP-16 suggesting that redox-cycling of VP-16 phenoxyl radicals by Ca^{2+}-ATPase thiols occured. Interaction of the VP-16 phenoxyl radicals with SR thiols was directly confirmed by our ESR studies. Vitamin C (but not vitamin E) was able to reduce the VP-16 phenoxyl radicals and to protect Ca^{2+}-ATPase against inactivation by the phenoxyl radical. Low molecular weight thiols (e.g. GSH) also prevented the VP-16 phenoxyl radical-induced inhibition of Ca^{2+}-ATPase. Interactions of the VP-16 phenoxyl radicals with SH-groups in critical biomolecules and their prevention by ascorbate and low molecular weight thiols may be important for developing new strategies to enhance the antitumor efficiency of VP-16.

Growth inhibitory effects of vitamin-D analogues and retinoids on human pancreatic cancer cells

Gerhard Zugmaier, Robert Jäger, Marco Gottardis, Klaus Havemann and Cornelius Knabbe

Abstract

Retinoids and vitamin-D are important factors that regulate cellular growth and differentiation. An additive growth inhibitory effect of retinoids and vitamin-D analogues has been demonstrated for human myeloma and myeloid leukemic cells. We set out to study the effects of the vitamin-D analog EB1089 and the retinoids all-trans-, 9-cis-retinoic acid on the human pancreatic cancer cell lines Capan 1, Capan 2 and Hs766T. All-trans retinoic acid at a concentration of 10nM inhibited the growth of Capan 1 and Capan 2 cells by 40% relative to controls. 9-cis retinoic acid was less effective. Neither all-trans retinoic acid nor 9-cis retinoic acid affected the growth of Hs766T cells. EB1089 induced a maximal growth inhibition of 25% in the three cell lines. This effect was reached at a concentration of 1nM. The combination of 1nM EB1089 with 10nM all-trans retinoic acid induced a growth inhibitory effect of 90% in Capan 1 cells and of 70% in Capan 2 cells. The transcript levels of retinoic acid receptor RAR-gamma correlated with the growth inhibitory effects of retinoids. Our data suggest combining vitamin-D analogues and retinoids as a new therapeutic concept of pancreatic cancer.

Clinical trials are under way.

Department of Medical Oncology, Marburg University Medical Center, Baldinger Street, 35033 Marburg, Germany (G.Z., R.J., K.H.).
Department of Clinical Chemistry, Hamburg University Medical Center, Germany (C.K.).
Ligand, Department Pharmacology, La Jolla, CA 92037 (M.G.).

Fatty Acid Modulation of Keratinocyte Differentiation

Ruth A. Hagerman, Susan M. Fischer and Mary F. Locniskar
Division of Nutritional Sciences, University of Texas, Austin TX 78712
UTSCC, Science Park, Smithville TX 78957

Mouse epidermal cells produce the biologically active compounds 12(S)-hydroxyeicosatetraenoic acid (HETE) from arachidonic acid, and 13(S)-hydroxyoctadecadienoic acid (HODE) from linoleic acid. The experiments described here were designed to investigate the effects of these compounds on keratinocyte differentiation, a process believed to be regulated by protein kinase C (PKC). In keratinocytes cultured from newborn Sencar mice, differentiation can be induced by increasing the calcium concentration of the culture medium from 0.05 mM to 0.12 mM. Under these conditions, an epidermal marker of differentiation, keratin 1 (K1), was expressed within 24 hours of the increase in calcium concentration. Cells treated at the time of calcium increase with 50 nmol of the phorbol ester 12-0-tetradecanoylphorbol-13-acetate (TPA) or 0.02 μmol of the membrane permeable diacylglycerol, 1-oleoyl-2-acetyl-sn-glycerol, demonstrated an inhibition of K1 expression at 24 and 48 hours. However, K1 protein returned to control levels at 72 hours. Keratinocytes treated with 10^{-8} M 12(S)-HETE at the time of calcium increase showed an inhibition of K1 expression at 24 hours, with a return to control levels by 48 hours. Treatment with 13(S)-HODE alone did not down-regulate K1 expression, and pre-treatment with 13(S)-HODE inhibited the decrease in K1 protein caused by 12(S)-HETE. Cells were treated with 100 μM of the PKC inhibitor H7 to investigate the involvement of PKC in the expression of K1. Neither TPA, 12(S)-HETE nor 13(S)-HODE/12(S)-HETE had an effect on K1 expression after PKC inhibition. These data support the hypothesis that processes of keratinocyte differentiation, as indicated by the expression of K1 protein, are controlled by a PKC mediated pathway. In addition, these results suggest that keratinocyte differentiation can be modulated by 12(S)-HETE, a metabolite of arachidonic acid.

Supported by American Institute for Cancer Research and NIH (CA46886).

EXACERBATION OF UV-CARCINOGENESIS BY HIGH DIETARY LIPID INTAKE IS PARTIALLY MEDIATED VIA SUPPRESSION OF IMMUNE FUNCTION

H.S. Black, G. Okotie-Eboh, J. Gerguis, J.I. Urban, and J.I. Thornby.

Baylor College of Medicine and Veterans Affairs Medical Center, Houston, Texas 77030

Previous studies have shown that a high level of dietary lipid (corn oil) exacerbates UV-carcinogenic expression. Using a carcinogenic protocol that allows inferences to be made regarding the segment of the carcinogenic continuum at which dietary lipid exerts its principal effect upon UV-carcinogenesis, we have demonstrated that high dietary fat produces its exacerbating effect during the post-initiation stage. As high dietary lipid significantly suppresses delayed type hypersensitivity, we sought to determine whether dietary lipid influences UV-carcinogenic expression through modulation of immune function. To address this question, tumor transplantation studies were conducted within the temporal constraints of the UV-carcinogenic protocol. UV-induced tumor tissue was transplanted to the flanks of recipient animals (HRA.HRII-c/+/Skh hairless mice) receiving various periods (0, 6, 11 wks) of UV radiation (0.4 SBUs/day) and either high (12%, by wt.) or low (0.75%) corn oil containing diets. There were no significant differences in median tumor rejection times between dietary groups at zero time and six weeks of UV. However, after eleven weeks of UV the low fat diet reflected a tumor rejection time that was comparable to that of no UV, i.e., 21 days. The high fat exhibited a tumor rejection time of greater than 63 days, significantly ($P<.01$) longer than that of the low fat group. Furthermore, suppression of the tumor rejection response by high fat occurred at a time when we had demonstrated that high fat exacerbates carcinogenic expression. When tumor multiplicity of animals on the respective diets were determined, this parameter was about 4-times greater in high fat fed animals. These data are compatible with the concept that tumor rejection is similar or analogous with natural tumor surveillance expected to occur during carcinogenesis. In conclusion, the exacerbation of UV-carcinogenic expression by high dietary lipid levels is mediated in large part through suppression of immune response. Supported by the American Institute for Cancer Research grant 90BW03.

FUMONISINS: Carcinogenic mycotoxins that appear to act via disruption of sphingolipid metabolism. A. H. Merrill, Jr., Ph.D., Biochemistry Dept., Emory University School of Medicine, Rollins Research Center, Atlanta, GA 30322-3050

Fumonisins are produced by *Fusarium moniliforme* and related fungi that are common contaminants of corn, millet, sorghum, and other agricultural products. Consumption of these mycotoxins is responsible for two diseases of agricultural importance (equine leukoencephalomalacia and porcine pulmonary edema), and have been implicated in human cancers of the esophagus, liver, and stomach. These mycotoxins are potent inhibitors of ceramide synthase (Wang et al., J. Biol. Chem. 266:14486-14490, 1991) and disruption of sphingolipid metabolism has been proposed to account for both the toxicity and the carcinogenicity of fumonisins. To explore the possible mechanisms for the carcinogenicity, we have examined two hypotheses: 1) fumonisins may serve as a promoter by being mitogenic for some cell types; and 2) fumonisins may contribute to carcinogenesis by suppression of the immune system. To study the first hypothesis, growth-arrested Swiss 3T3 cells were treated with fumonisin B_1 and DNA synthesis was measured. Both fumonisin B_1 and sphinganine, which accumulates in cells treated by fumonisins, increased [^{3}H]thymidine incorporation into DNA as well as cell number and the % of cells in S phase, as measured by FACS. Therefore, fumonisins appear to be mitogenic. To explore this further and our second hypothesis, fumonisin B_1 was administered to BALB/c mice i.p. (1 to 100 ug) and the ability of the mice to produce plaque-forming cells (PFC) against sheep red blood cells (SRBC) was determined. When administered the same day as the SRBC, a lower number of PFC were produced; however, when administered daily, there was a 4 to 12-fold increase in the number of PFC on days 3 and 4 after SRBC injection; by day 5, the number of PFC returned to near control levels. These finding suggests that fumonisins both inhibit and enhance the immune response of BALB/c mice to foreign antigens. To test the possibility that this may be due to a mitogenic effect, B and T cells were treated with fumonisin B_1 *in vitro*; however, this did not stimulate [^{3}H]thymidine incorporation into DNA. Thus, the mechanisms whereby fumonisins affect cell growth and the immune system are not necessarily the same. Supported by grants from the AICR and the USDA.

Plasma Essential Fatty Acid Concentrations, Tumor Growth and Diet-Restriction

Leonard A. Sauer and Robert T. Dauchy, Laboratory for Cancer Research, Medical Research Institute, The Mary Imogene Bassett Hospital, Cooperstown, New York 13326

Nutritional studies in laboratory animals have shown that increased ingestion of linoleic acid stimulates growth of carcinogen-induced, spontaneous and transplanted tumors. How dietary linoleic acid affects tumor growth, however, is not yet understood. Previous research performed in this laboratory suggested that increased plasma concentrations of the essential fatty acids (EFA), linoleic plus arachidonic, increased growth of transplanted tumors *in vivo* and increased [^{3}H]thymidine incorporation via a dose-dependent relationship in tumors perfused *in situ*. We proposed that the ambient plasma EFA concentrations in arterial blood were major determinants of tumor growth rates *in vivo*. This hypothesis was tested in this study by manipulating the plasma EFA concentrations in tumor-bearing rats.

Weanling male Buffalo rats (n = 15-20/group) were fed semi-purified diets either *ad lib* or 30% restricted. The diets fed *ad lib* contained either 0.0, 0.03%, 0.7% or 2.2% (w/w) linoleic acid. Restricted rats were fed once daily (at 1500 hours) a portion of either a 0.7% or 1% linoleic acid diet that equaled 70% of the amount consumed by rats fed the 0.7% linoleic acid diet *ad lib*. Restricted rats fed the 1% diet ingested the same amount of linoleic acid as the rats fed the 0.7% diet *ad lib*. After two months blood samples were collected (by heart puncture) through a 24 hour feeding period. Consecutive blood collections were every 4 hours and were spaced 3 days apart. Plasma lipids were extracted and saponified, and the fatty acids were methylated and measured by gas chromatography. Each rat was then implanted with hepatoma 7288CTC as a subcutaneous, tissue-isolated tumor. The latent periods and rates of tumor growth were estimated for an additional 20 days. The animals were killed and the tumors excised, weighed and frozen in liquid N_2 for lipid analysis.

Peak plasma fatty acid concentrations coincided with the periods of active feeding, i.e., the dark period for rats fed *ad lib* and 1500 to 2200 hours for the diet-restricted rats. Mean daily EFA concentrations (means ± SEM) were 0.0, 0.34±0.04, 0.87±0.05 and 1.2±0.1 mg/ml plasma for rats fed *ad lib* diets containing 0.0, 0.03%, 0.7% and 2.2% linoleic acid, respectively. Mean tumor growth rates in these *ad lib* fed groups were 0.0, 0.85±0.08, 1.5±0.09, and 2.1±0.09 g/day, respectively. Mean daily EFA concentrations in restricted rats fed the diets containing 0.7 or 1% linoleic acid were 0.59±0.02 and 0.67±0.02 mg/ml plasma, respectively; mean tumor growth rates in these rat groups were 0.6±0.06 and 1.32±0.07 g/day, respectively. Comparison of these data by regression analysis showed that mean daily plasma EFA concentrations and mean tumor growth rates were significantly correlated ($r = 0.94$, $P < 0.01$). The results suggest that the ambient plasma EFA concentrations regulate tumor growth *in vivo* in both *ad lib* fed and diet-restricted animals. (Supported by AICR Grant No. 90A42.)

EFFECTS OF DIETARY LINOLEIC ACID (LA) ON THE GROWTH, INVASION AND METASTASIS OF TWO HUMAN BREAST CANCER CELL LINES

Jeanne M. Connolly and David P. Rose, American Health Foundation, Valhalla, NY, 10595.

We have compared the effects of dietary LA intake on the growth and metastasis of the human breast cancer cell lines MDA-MB-435 and MDA-MB-231 in nude mice, and their invasive capacities *in vitro*. Each tumor cell line was injected into a thoracic mammary fat pad of 60 mice, with equal numbers assigned to isocaloric diets containing 23% (w/w) total fat and 2 or 12% (w/w) LA (30 mice/group). The growth of the primary tumors was monitored for a 12 week period, and lung metastasis evaluated both macroscopically and microscopically after necropsy at 12 weeks. The growth rate of primary tumors from both cell lines was higher in mice fed the high-LA diet than those fed the low-LA diet. The initial growth rate of all MDA-MB-231 tumors was slower than that of MDA-MB-435 tumors, suggesting that their angiogenic mechanisms are different, but by 6 weeks the MDA-MB-231 tumors exhibited an acceleration of growth which was enhanced by the high-LA diet. Final primary tumor weights of MDA-MB-231 tumors exceeded those of MDA-MB-435 tumors ($p<0.0001$), and in both cell lines were higher in mice fed high-LA than low-LA diets (10.2 ± 1.4 vs. 6.7 ± 1.4, and 3.6 ± 0.1 vs. 3.3 ± 0.1 respectively; each $p<0.001$). Macroscopic lung metastasis was significantly higher in the MDA-MB-435 mice fed the high-LA diet (67% vs. 33%; $p<0.02$), reflecting *in vitro* enhancement of invasion by the addition of LA to culture medium. Few macroscopic lung metasteses were observed in the MDA-MB-231 mice. Upon microscopic examination of lungs from MDA-MB-231 tumor-bearing mice, a high incidence of diffusely infiltrative micrometastasis was observed, which was enhanced by high-LA diet (68% high-LA vs. 42% low-LA). We conclude that dietary LA enhances breast cancer tumor growth and metastasis, probably by mechanisms including both angiogenesis and invasion.

The Form of Vitamin E Administered Topically Can Determine Whether Prevention or Enhancement of Photocarcinogenesis Ensues. Gensler, H.L., Gerrish, K., Peng, Y-M. and Xu, M-J. Departments of Radiation Oncology and Internal Medicine, Cancer Center, University of Arizona, Tucson, AZ 85724.

Previous studies in our laboratory demonstrated that the dl-α-tocopherol form of vitamin E prevents skin cancer and the immunosuppression induced by UVB-irradiation in mice (Nutr. Cancer 15:97-106, 1991). These studies concern the capacity of α-tocopheryl acetate or α-tocopheryl succinate, thermostable esters of vitamin E, to reduce photo-carcinogenesis. Female BALB/c mice were treated with dl-α-tocopheryl acetate (12.5, 25, or 50 mg) or α-tocopheryl succinate (2.5, 12.5, or 25 mg) applied topically three times weekly for 3 weeks before UV treatments began, and throughout the experiment. Mice received approximately 1 x 10^6 Jm^{-2}, delivered intermittently over an 18 week period by FS40 Westinghouse sunlamps. Rather than a reduction in tumor incidence, treatment with α-tocopheryl acetate or succinate resulted in a 30 or 40% increase, respectively, in tumor incidence. Since there was no significant difference in tumor incidence between dose groups, tumor incidence data was pooled for different doses of treatment with α-tocopheryl acetate or α-tocopheryl succinate and yielded a significant enhancement of photocarcinogenesis (p=0.0114, p=0.0262, respectively, log rank test). Measurement of skin concentrations after 16 or 17 weeks of vitamin E treatments demonstrated that there were dose dependent increases in the levels of the esters.

However, there was only 10-20% as many μg of α-tocopherol as of α-tocopheryl acetate or succinate per g wet weight of skin, indicating that little conversion of these esters to free α-tocopherol occurs in the skin. Only the free tocopherol acts as an antioxidant. The lack of chemoprevention by these esters, and the limited conversion of the esters to free α-tocopherol in the skin suggest that the free form of α-tocopherol is necessary for tumor prevention. The increase in skin tumors in animals treated with α-tocopheryl acetate suggests that caution is warranted concerning the current trend for inclusion of α-tocopheryl acetate in commercially available sunscreen lotions.

(Supported by NIH grants CA 27502 and 44504)

Biochemical, biophysical, and immunologic analyses of plasma membranes and exfoliated vesicles from leukemia cells modified with omega-3 fatty acids *in vivo* and *in vitro*. Laura J. Jenski, Mustapha Zerouga, Lian Zhang, and William Stillwell. Department of Biology, Indiana University - Purdue University at Indianapolis, IN 46202-5132.

Docosahexaenoic acid (DHA, 22:6), a long chain polyunsaturated omega-3 fatty acid, perturbs membrane structure and affects membrane protein expression. Thus, DHA incorporation into tumor cells may affect the tumor's immunogenicity by altering tumor antigen expression. One mechanism by which DHA may exert this effect is by inducing membrane domains rich or poor in DHA. Membrane proteins would, in turn, segregate into the most favorable domains. Because membrane domains may be exfoliated from the tumor cell surface, DHA may alter tumor cell immunogenicity either by retaining tumor antigens in the plasma membrane or exporting them into exfoliated domains. In this study, DHA was incorporated into murine tumor membranes from dietary fish oil or by culture with free fatty acid. Phospholipid classes were separated by solid phase extraction on aminopropyl bonded silica minicolumns, and fatty acid methyl esters were analyzed by gas liquid chromatography. Membrane structure was studied with fluorescent membrane probes (MC540 and pyrene). Total proteins in exfoliated and plasma membranes were compared by SDS-PAGE, and major histocompatibility complex (MHC) proteins (an estimate of tumor antigen expression) were quantified by ELISA. Our results thus far suggest that DHA, when present at high concentration in diet or culture medium, becomes incorporated into the plasma membrane and preferentially exfoliated. The protein content of exfoliated and plasma membranes was not identical, with MHC molecules showing some preference for exfoliated membrane. The presence of DHA altered the protein distribution in the two membrane fractions, and enhanced the proportion of MHC protein in the exfoliated vesicles. Although preliminary, these results are consistent with an effect of DHA on membrane domain formation and the selective exfoliation of distinct protein-containing domains. The significance of these findings for the immunogenicity of the tumor cells remains to be tested, but they imply an effect of diet on tumor antigen expression.

COMBINATION OF TRANS-RETINOIC ACID AND TAMOXIFEN MODULATE PLASMA LEVELS OF TOTAL AND FREE IGF-I AND IGF-I BINDING PROTEINS IN PATIENTS WITH ADVANCED BREAST CANCER

Homayoun P., Gupta MK., Van Lente F., Tuason LJ. and Budd T.
The Cleveland Clinic Foundation, 9500 Euclid Ave., Cleveland, OH 44195

Retinoids have shown tumor growth inhibition and a synergistic activity with hormonal manipulation in human breast cancer cell lines and rat mammary carcinoma. Tamoxifen and retinoids when used separately in cancer patients with advanced disease, caused a decrease in plasma insulin-like growth factor-I (IGF-I). We have investigated the effect of combination therapy (tamoxifen dose 20 mg/day and increasing doses of all-trans-retinoic acid (tRA); 70, 110 and 150 mg/day given orally) on the serum levels of total and free IGF-I IGF binding protein-3 (IGFBP-3) in patients with advanced breast cancer (ER+ PR+). Total IGF-I was measured by RIA after acid-ethanol extraction using a truncated IGF- (des-IGF-I) as radioligand. Free IGF-I was measured by RIA after HPLC separation and IGFBP-3 was measured by immunoradiometric assay.

Pretreatment total IGF-I levels correlated better with free IGF-I levels than post-reatment levels ($r=0.73$ vs. $r=0.4$). We observed a reduction in total IGF-I in 8/11 patients after five weeks of treatment, regardless of the tRA dose(mean +/-SE: pre= 122 +/- 17, post= 91 +/- 12; $p=0.102$). By the 11 th week only three have demonstrated a further decrease in total IGF-I; the other three had different degrees of increase and two were not tested. The changes in the levels of free IGF-I did not reach significant level ($p<0.05$). The levels of IGFBP-3 were not affected.

Our data show that regardless of tRA dose, the combination therapy causes a decrease in total IGF-I levels in most patients. Because the free IGF-I levels were not correlated with total IGF-I and the levels of IGFBP-3 were similar, we suggest further investigation on IGFBP-1

TOXIC EFFECT OF FARNESOL ON CELL VIABILITY IS PRECEDED BY INHIBITION OF PROTEIN KINASE C (PKC) ACTIVITY: DIFFERENCES BETWEEN NEOPLASTIC HELA S3K CELLS AND NON-NEOPLASTIC CF-3 CELLS

Eugenia M. Yazlovitskaya and George Melnykovych

Department of Microbiology, Molecular Genetics and Immunology, University of Kansas Medical Center, 3901 Rainbow Blvd., Kansas City, KS 66103 and Department of Veterans Affair Medical Center, 4801 Linwood Blvd., Kansas City, MO 64128.

Recently we have shown that isoprenoid farnesol caused apoptotic cell death (J. Haug et al, BBA, in press). Farnesol toxicity was significantly higher in neoplastically derived cells compare to cells of non-neoplastic origin (I. Adany et al, Cancer Lett,. in press). Here we follow up these earlier findings to study further the mechanism of farnesol toxicity. Farnesol in the concentration range from 20 μM to 40 μM decreased viability of neoplastically derived HeLa S3K cells. Viability of non-neoplastic CF-3 cells was not affected by farnesol in the same concentration range. In HeLa S3K cells 20 μM farnesol caused translocation of protein kinase C (PKC) from membrane to cytosolic fraction. This translocation was evident after 1 h of incubation with farnesol and occurred in time dependent manner for up to 14 h of incubation. In these cells farnesol also prevented induction of PKC transfer from cytosol to membranes by phorbol 12-myristate 13-acetate (PMA), known activator of PKC. In CF-3 cells intracellular localization of PKC was not affected by farnesol. These data suggest that the mechanism of induction of cell death by farnesol involves inhibition of PKC activity. The same mechanism may be responsible for different sensitivity of neoplastic and non-neoplastic cells to farnesol toxicity.

INFLUENCE OF DIET AND OTHER PARAMETERS ON COLORECTAL MUCOSAL CELL PROLIFERATION IN PATIENTS WITH ADENOMATOUS POLYPS

Jenny A Matthew[1], Steve Middleton, Alison Prior[2], Hugh J Kennedy[2], Ian W Fellows[2] and Ian T Johnson[1]

[1]AFRC Institute of Food Research, Norwich Research Park, Colney, Norwich
[2]Norfolk and Norwich Hospital, Brunswick Road, Norwich

Patients at risk of colorectal neoplasia are reported to have an abnormally high rate of crypt cell proliferation throughout the colon, and a displacement of mitotic activity towards the gut lumen (Lipkin, 1988). We compared the relative importance of diet, age and sex as determinants of the crypt cell proliferation rate in patients with adenomatous polyps.

Biopsies were obtained from flat mucosa in the mid-rectum of 37 patients undergoing diagnostic colonoscopy, who had either an adenomatous polyp or a history of polyps, but no other evidence of disease. Biopsies were fixed and whole crypts were microdissected and analyzed for crypt mitotic rate (CMR; total numbers of mitoses per crypt), and spatial distribution of mitoses, as described elsewhere (Matthew et al, 1994). Patients also completed a quantitative food questionnaire containing photographs to facilitate estimation of portion size. The data were analyzed by multiple regression using CMR as the dependent variable and a backward stepwise procedure to eliminate variables making no significant contribution to the model.

Preliminary analysis showed a significantly lower intake of energy and fat in women. In the multiple regression model there was no evidence of significant effects of age, sex or total fibre intake on CMR. However there was a positive association with the degree of displacement of the mitotic zone and total energy intake, and a weak negative relationship with total fat intake ($p = 0.002$; Adjusted $R^2 = 31\%$). Few previous studies have examined the effects of demographic variables and habitual diet on crypt cell proliferation in free-living subjects. Fireman et al (1989) observed no effects of age or sex on CMR but there were differences between urban and rural males that may have been due to differences of diet. The present study suggests that dietary energy intake promotes colorectal cell proliferation independently of fat.

Fireman Z, Rozen P, Fine N, Chetrit A
Influence of demographic parameters on rectal epithelial proliferation
Cancer Letters 1989; 47: 133-140

Lipkin M.
Biomarkers of increased susceptibility to gastrointestinal cancer: New application to studies of cancer prevention in human subjects.
Cancer Research 1988; 48: 235-245.

Matthew JA, Pell JD, Prior A, Kennedy HJ, Fellows IW, Gee JM, Burton J, Johnson IT.
Validation of a simple technique for the detection of abnormal mucosal cell replication in humans.
European Journal of Cancer Prevention (in press)

RETINOID REGULATION OF MAMMARY EPITHELIAL CELL PROLIFERATION AND DIFFERENTIATION. **Ping-Ping Lee and Margot M. Ip, Roswell Park Cancer Institute, Buffalo, NY 14263**

The role of retinoids in the proliferation and differentiation of normal mammary epithelial cells (MEC) was investigated using a primary culture model in which MEC from virgin rats were cultured within a reconstituted basement membrane using defined serum-free medium. The retinobenzoic acid derivative RE80, and the natural retinoid all trans retinoic acid (RA), each inhibited proliferation with IC50's of 10^{-10} and 10^{-8} M, respectively. Moreover, both retinoids stimulated end bud colonies to differentiate into lobular alveolar colonies, and suppressed the development of squamous colonies, an observation that was even more striking under conditions in which morphological development was inhibited (e.g. in the absence of EGF). Unexpectedly, at later times in culture (after day 12-14 of a 21-day culture period), RE80 and RA both stimulated cell death within the most differentiated (multilobular-alveolar) population, an effect that was shown to require hydrocortisone. This suggests that retinoids may be accelerating morphological differentiation, culminating in a retinoid-induced apoptotic cell death resembling the normal involution of the mammary gland *in vivo*. In addition to effects on proliferation and morphogenesis, both RE80 and RA markedly stimulated functional differentiation of the MEC, as assessed by the accumulation of the casein family of milk proteins. This stimulation of casein was observed both under optimal medium conditions, as well as in the absence of EGF or hydrocortisone; of note was the observation that RE80 could completely replace the requirement for EGF, and could partially replace the requirement for hydrocortisone, for casein accumulation in this model. In contrast, casein accumulation by MEC exhibited an absolute requirement for prolactin; RE80 could not substitute. The mechanism by which retinoids stimulate casein is not yet known, but could involve a direct activation of casein transcription by binding of the appropriate retinoid receptor to the casein promoter, or an indirect effect by modulation of the activity of other hormones. In this regard, RE80 was found to abrogate the inhibitory effect of progesterone on casein accumulation. Taken together, the results of these studies suggest that several mechanisms may contribute to the chemopreventive and/or therapeutic effects of retinoids in breast cancer, including an inhibition of proliferation, stimulation of cell death, and/or induction of differentiation.

EFFECT OF *IN VIVO* MOBILIZED LIPIDS ON THE EXPRESSION OF ONCOGENE PRODUCTS IN HUMAN COLON TUMOR CELLS. **C. Gercel-Taylor and D.D. Taylor. Dept of Obstetrics and Gynecology, Univ. of Louisville School of Medicine, Louisville, Kentucky**

There are compelling data on the association of nutritional factors, most prominently the levels and types of dietary fats and the incidence of cancers. The actual lipid component and the mechanism of fat-related carcinogenesis is not well understood. In our efforts to understand the particular fatty acids involved in carcinogenesis, we have studied lipids mobilized by cancer patients and their effect on the expression of oncogene products by tumor cells. To investigate the effects of *in vivo* mobilized lipids on certain parameters of tumor cells, lipids were isolated from patients with advanced cancer. The *in vivo* derived lipids were added to cell cultures (human colon tumor cells) at a concentration of 50 ug/ml. The cellular expression of the *ras* oncogene product was enhanced in the Sk-Co-1 and HT29 cells following lipid treatment, while *jun* oncogene product was enhanced in HT29 cells and decreased in Sk-Co-1 cells. No changes were observed in *cerbB-2* expression following lipid treatment of either cell line. p53 expression was decreased following treatment of Sk-Co-1 cells, while treatment of HT29 cells with fats produced a dramatic amplification of the p53 gene product. Immunoprecipitation studies have demonstrated that this enhanced p53 represents the mutated form. Fat treatment of these colon lines also was observed to enhance either the 55kD band or both the 55 and 62 kD bands of *c-fos*. In addition, the effect of these fats on the doubling times of HT29 and Sk-Co-1 was studied. The doubling times were 16.8 and 24.6 hours for HT29 and Sk-Co-1 respectively in controls, while the cultures treated with patient-derived fats have doubling times of 14.4 (HT29) and 22.8(Sk-Co-1) hours. Further analysis of tumor-derived lipids indicated that the diglyceride and monoglyceride fractions were responsible for the effects outlined above.

Phytate Inhibits Colon Carcinogenesis in F344 Rats when Administered for Different Periods of Time after Initiation with Azoxymethane. Pretlow, TP, Hudson, L, O'Riordan, MA, and Pretlow, TG Institute of Pathology, Case Western Reserve University School of Medicine, Cleveland, OH

Previous studies demonstrated that diets supplemented with 2% phytate in the drinking water inhibit colon carcinogenesis in rats. Aberrant crypt foci (ACF) are putative preneoplastic lesions that are readily quantified in whole-mount preparations of colon soon after carcinogen treatment. These studies were designed to determine if different regimens of phytate are effective in preventing colon cancer and if the ACF assay can predict this activity. Six-week-old male F344 rats were given one sc injection of 20 mg/kg azoxymethane; and their drinking water was modified one week later as follows: group A, plain drinking water until sacrifice; B, drinking water with 2% phytate until sacrifice; C, drinking water with 2% phytate for weeks 1-18 and plain water for weeks 18-36; and D, plain water for weeks 1-18 and 2% phytate for weeks 18-36. Groups of rats were killed at 4, 8, 12, 18, and 36 weeks after injection. Tumors were seen only in rats treated with AOM and only those killed 18 or more weeks after injection. The number of tumors per rat in groups B (0.44) and C (0.28) was significantly less ($P \leq 0.04$) than group A (1.25) at 36 weeks; the tumor incidence was reduced (44, 28, 48% of rats) in groups B, C, and D but not significantly different than group A (56%). The average number of crypts per focus and the percentage of ACF with 4 or more crypts were significantly ($P \leq 0.01$) less in group B than group A at all 5 time periods evaluated, while the total number of foci often was not different. These data suggest that the large ACF are more predictive of the future development of tumors than the total number of foci and that phytate inhibits the progression of foci but has little effect on established tumors. Supported by grant 92B28 from the American Institute for Cancer Research and CA48032 from the National Cancer Institute.

INHIBITORY EFFECTS OF COMBINATIONS OF WHEAT BRAN AND PSYLLIUM ON NMU-INDUCED RAT MAMMARY TUMORIGENESIS

L.A. Cohen, E. Zang, A. Rivenson
American Health Foundation, Valhalla, NY 10595

In this study, we compared the effects of high fat (HF) diets containing varying amounts of an insoluble (soft white wheat bran (SWWB)), and a soluble (psyllium) fiber on the development of mammary tumors induced by N-nitrosomethylurea (NMU) in F-344 female rats. Five groups of 30 rats each were fed AIN-76A semipurified diets containing (1) 12% SWWB, (2) 8% SWWB + 2% psyllium, (3) 6% SWWB + 3% psyllium, (4) 4% SWWB + 4% psyllium, and (5) 6% psyllium, starting 3 days post-NMU administration. Corn oil at 20% (wt/wt) was incorporated into all diets. Twenty weeks post-NMU, the incidence, latency and multiplicity of histologically verified mammary tumors in the different experimental groups were compared. The salient results were as follows: Group 4 exhibited a significantly longer latent period, lower incidence (30% vs 60%), and lower total number of mammary adenocarcinoma (15 vs 33) compared to group 1. Overall, the 4%:4% combination of SWWB and psyllium afforded maximum protection against the tumor promoting effects of a high-fat diet, while the 8%:2%, 6%:3% and 6% psyllium treatment groups alone produced intermediate results. No significant differences were found when all mammary tumors (fibroadenomas + adenocarcinomas) were included in the analysis. Since SWWB contains 45% and psyllium 80% total dietary fiber, an insoluble/soluble fiber ratio of approximately 0.5 appears to be the most protective in this breast cancer model. The underlying mechanism by which a 4%:4% combination of SWWB and psyllium suppressed the tumor promoting effects of a HF diet remains to be determined. These results suggest that diets supplemented with specific combinations of insoluble and soluble fiber may provide a means of reducing breast cancer risk in western high risk populations.

Relationships between iron status and mammary cancer development in the DMBA-induced rat

Adria R. Sherman*, Deborah Hrabinski*, Vance Berger**
*Departments of Nutritional Sciences and **Statistics
Rutgers, The State University of New Jersey
PO Box 231
New Brunswick, NJ 08903

Previous experimental and epidemiological research has shown a relationship between iron and cancer incidence. Animal studies from our laboratory and others have shown that iron deficiency is protective against tumorigenesis. High body iron stores have been linked to increased cancer risk in several population groups and clinical diseases. The objective of this study was to determine the dose response relationship between dietary iron and mammary tumor development in the DMBA-induced Sprague Dawley rat. Twenty-one day old rats were fed ad libitum AIN 76 diets containing 5, 10, 15, 35, 50, 500, 1000 mg Fe/kg diet or the 35 mg Fe/kg diet in the amount consumed by rats in the 5 mg Fe/kg group (pair-fed). The iron levels were selected to produce the following iron states: 5= severe anemia; 10= mild anemia, depleted iron stores; 15= depleted iron stores; 35= iron adequacy; 50= 1.4 times the NRC recommendation, similar to intake of iron-rich and iron-fortified foods; 500= 14 times requirement, similar to supplement use; 1000= body iron stores equivalent to those found in iron storage diseases. After feeding the diets for 6 weeks, DMBA (5 mg/100 g body weight) in corn oil was administered intragastrically. Tumor incidence was determined weekly by palpation. Necropsy was performed at 20 weeks post-DMBA. Hemoglobin concentrations and hematocrit values revealed severe anemia in the 5 mg Fe/kg group. Tumor incidence in rats fed 5 mg Fe/kg diet was significantly lower than in all other treatment groups. Tumor onset was delayed in severely iron-deficient rats compared to other groups. Preliminary analyses revealed no consistent linear dose response relationship between dietary iron level and tumor incidence at iron levels greater than 5 mg/kg. Feeding high iron diets for 26 weeks was not tumorigenic compared to feeding less dietary iron. (Supported by a grant from AICR)

Increased Sequence-Specific DNA Binding by p53 Following Treatment of Rat Liver, Balb 3T3 and Human Breast Cells with Food Colorants

Craig Dees[1], Don Henley[1], C. Murray Ardies[2], and Curtis Travis[1]

[1]Health Sciences Research Division, Oak Ridge National Laboratory*, Oak Ridge TN 37831, [2]Center for Exercise and Cardiovascular Research, Northeastern Illinois University, Chicago, IL 60625

Diet has been implicated in the etiology of breast cancer and the incidence of breast cancer in U.S. women is increasing. The reasons for the increased incidence of breast cancer are unknown. Dietary factors thought to be responsible include an increase in meat consumption or high total dietary fat. Exposure to environmental estrogens like DDT (1,1,1-trichloro-2,2-bis (chlorophenyl) ethane) have also been implicated but their role, if any, is controversial.

Recently, it has been suggested that women with diets that contained high levels of the food colorant Food Drug and Cosmetics Red No. 3 (Erythrosine Bluish; CL Food Red) had an increased risk of developing breast cancer. Therefore, we examined the effects of Red No. 3, Blue No. 1, Green No. 3, and Yellow No. 5 for an effect on sequence specific binding of p53 in Rat Liver Epithelial cells (RLE), Balb 3T3 and the T-47 line from a human ductal carcinoma. DDT, methylnitrosoguansine (MNNG), phorbol 12-myristate, 13-acetate (PMA), ethidium bromide, perchloroethylene (PCE), Dibenzo(a,l)pyrene, Dibenzo(a)pyrene, 7,12, Dimethyl-benzo(a)anthracene and phenanthrene were also used to treat cells and increase sequence specific DNA binding.

All the benzopyrenes after treatment with Arochlor 1254-induced S9 liver extract increased p53 DNA sequence specific binding. Phenanthrene, which is not known to be a carcinogen, did not. PMA, DDT, PCE also did not increase p53 sequence specific binding activity. Blue No. 1, Green No. 3 and Yellow No. 5 increased p53 binding after treatment with S9 but not without treatment. Red No. 3 gave the largest increase in sequence specific binding without S9 treatment. S9 treatment of Red No. 3 had no effect.

These results suggest that the food colorants tested are capable of damaging DNA. and are potentially human carcinogens. DDT without S9 treatment does not appear to damage DNA since p53 sequence specific DNA binding is not increased in treated cells, whereas Red No. 3 appears to damage DNA. It may be that increased consumption of colorants in processed foods and meat is responsible for the increased incidence of breast cancer in U.S. women.

RETINOIC ACID RECEPTOR RARß EFFECTS IN CERVICAL TUMOR CELL LINES

Doris M. Benbrook, Kevin Brewer, Coy Heldermon, Evelyn Nunez and Przemko Walisewski
University of Oklahoma Health Sciences Center, Oklahoma City, OK

Two cervical tumor cell lines (HeLa and SiHa) which do not express the retinoic acid receptor RARß were transfected with a eukaryotic expression vector (pXT1, Stratagene) containing the receptor cDNA. Permanent RARß expressing cell lines, were established by selection in G418 and exhibited different morphologies than their parent cell lines. Receptor expression was confirmed by Northern and Western analysis.

The effect of RARß expression on transactivation of RARE's was evaluated using Transient Transfection Assays. Induction of the RARß RARE by retinoic acid was much greater in the HeLa RARß subline than in the parent HeLa line. The CRBPII RARE was not induced by retinoic acid in either cell line.

Gel Retardation Analysis demonstrated specific binding activity to the HPV 16 LCR fp4er sequence in the RARß lines and not in the parent lines. This activity was not competed by consensus glucocorticoid responsive element (GRE) or retinoic acid responsive element (RARE) sequences, but was inhibited with a polyclonal antibody specific for the retinioic acid receptor (RXR). The effects of this binding activity on HPV expression is currently being evaluated.

The HeLa and SiHa parent lines developed into multiple cell layers when grown in organotypic culture. Immunohistochemical analysis of these cultures demonstrated a lack of expression of differentiation markers. This is in contrast to organotypic cultures of a less differentiated cervical tumor cell line, CC-1, and an HPV-immortalized keratinocyte cell line, PE-4, which expressed involucrin focally. Ten micromolar 13-cis-retinoic acid inhibited the upward stratification into cell layers and the expression of Proliferating Cell Nuclear Antigene (PCNA) of all cell lines except HeLa. When grown in organotypic culture, the SiHa RARß subline stratified into cell layers which expressed higher levels of keratin than the parent SiHa line. The HeLa RARß expressing subline did not develop multiple cell layers in organotypic culture. The sensitivities of the parent and sublines, to retinoids and other agents as well as radiation are being compared using organotypic cultures and an MTT Proliferation Assay.

ANTI-NEUROBLASTOMA EFFICACY OF 13-CIS-RETINOIC ACID

CP Reynolds, P Einhorn, P Schindler, JJ Zuo, AA Khan, VI Avramis, JG Villablanca. Division of Hematology-Oncology, Childrens Hospital Los Angeles/University of Southern California, Los Angeles,CA 90027.

To compare the relative efficacy of retinoic acid (trans-RA) to 13-cis-retinoic acid (13-cis-RA) at inducing differentiation and growth arrest of neuroblastoma *in vitro*, we used a panel of 15 human neuroblastoma cell lines, 10 with and 5 without N-*myc* gene amplification. Cells were exposed to 13-cis-RA, trans-RA, or solvent control for 7 days and cell growth over 28 days was determined in a 96 well assay capable of measuring > 3 logs of growth inhibition. Complete growth inhibition over 28 days was seen in 7/15 cell lines treated with 5 μM 13-cis-RA and 8/15 lines treated with 10 μM trans-RA. Average peak serum levels in children for trans-RA have been shown to be < 0.6 μM compared to > 5 μM for 13-cis-RA, so we compared 5 μM 13-cis-RA to 0.5 μM trans-RA and found 5 μM 13-cis-RA was more effective than 0.5 μM trans-RA in 5 of 11 cell lines, and 6 of 11 lines showed no difference in response. As 5 μM 13-cis-RA is achieved in patients using an intermittent schedule (2 weeks of drug/month), we tested this dose schedule *in vitro*. Two such 2 week courses of 13-cis-RA extinguished N-*myc* expression and produced complete growth arrest for 120 days in the SMS-LHN human neuroblastoma cell line. Our phase I trial of 13-cis-RA in 50 neuroblastoma patients following bone marrow transplantation showed the maximal tolerated dose to be 160 mg/m^2, which gave an average peak serum level of 6.7 $\pm$ 5.3 μM, and produced responses in some patients with measurable disease. Thus, 13-cis-RA has lower clinical toxicity and achieves higher serum levels than trans-RA, and produces responses in some patients with neuroblastoma. (Supported by American Institute for Cancer Research grant 92B69)

Incorporation of Exogenous ω-6 and ω-3 Long Chain Fatty Acids into CT-26, a Murine Colon Tumor Cell Line.

Gaposchkin, D.P. and Broitman, S.A. Departments of Pathology and Microbiology, Boston University School of Medicine, Boston, Mass

Previously our lab has shown that marine oils with high levels of eicosapentaenoic acid (C20:5ω-3), an analog of arachidonic acid (C20:4ω-6), inhibit the growth of CT-26, a murine colon carcinoma cell line, when implanted into the colon of male BALB/c mice. To study this phenomenon, we have examined the uptake and incorporation of linoleic (C18:2 ω-6), arachidonic and eicosapentaenoic acids at both 50uM and 100uM complexed to BSA (3 Fatty Acids : 1 BSA) into CT-26 cells in culture as well as examining the lipid profiles of CT-26 grown in four different concentrations of fetal bovine serum, (1%, 2.5%, 5%, and 10%). These studies have shown that the long chain polyunsaturated fatty acids (PUFA) are preferentially incorporated into the phospholipids of these cells as serum concentration increases. Incorporation into phospholipids is rapid in the first six hours and levels off after twelve hours. We have also demonstrated that as greater amounts of these PUFAs are incorporated, the amount of oleic acid in the phospholipid fraction of these cells decreases to between 35% and 50% of control values. CT-26 elongates both arachidonic and eicosapentaenoic acids to docosatetraenoic (C22:4 ω-6) and docosapentaenoic (C22:5 ω-3) acids respectively. However, approximately 50% of incorporated eicosapentaenoic acid is elongated after 24 hours yet only around 30% of the arachidonic acid present is elongated. This suggests that differences seen *in vivo* might be due to the fact that the biochemical pathways of CT-26 incorporate and utilize eicosapentaenoic acid differently than arachidonic acid. This work shows that this is a suitable model system to study by what mechanisms differences seen *in vivo* are occurring.

Dependence of CT-26 Murine Colon Tumor Inhibition On Fatty Acid Supplementation Methodology *In Vitro*: Implications for the *In Vivo* Model.
Kosacolsky Singer C., Broitman S.A. Boston University Department of Pathology, Boston University School of Medicine, Boston, MA 02118.
Previously, we have shown growth inhibition of CT-26 murine colon tumor bowel implants, but not midscapular implants, in mice fed a marine oil diet. We have also shown growth inhibition of CT-26 monolayers by EPA, a major component of marine oil. Further, EPA exerts its effects only when delivered in ethanol, not when supplied complexed to BSA. Phospholipid analysis of cell membranes from cultures supplemented by either method indicate no difference in EPA content. Therefore, inhibition cannot be attributed to varying quantities of EPA in the membrane. Evaluation of peroxidation indicates a dose dependent increase in peroxidation products only with increasing quantities of EPA delivered in ethanol, not as a BSA complex. Neither ethanol nor BSA alone stimulated peroxidation. Because delivery of PUFAs such as EPA by ethanol or BSA complex are both accepted methods in the literature, it is important to note that the two methods are not equivalent. Inhibition of CT-26 *in vitro* by EPA is related to stimulation of peroxidation by ethanol, which implicates a source of peroxidation in bowel tumors that is missing in midscapular tumors. Thus, CT-26 supplemented with EPA delivered in ethanol may be an appropriate model for bowel implants, whereas supplementation with EPA complexed to BSA may be a model for midscapular implants. Work is currently underway to determine the extent of peroxidation detected *in vivo*, and whether it is related to alterations in second messenger systems, or shifts in tumor eicosanoid production resulting in modified immune function and TNF-α production.

INHIBITORY EFFECTS OF CAPSAICIN AND DIALLYL SULFIDE ON DIMETHYLNITROSAMINE-INDUCED MUTAGENESIS

M. Shlyankevich, R. Lee, K. Garden, Y.-C. Lee, and Y.-J. Surh

Department of Epidemiology and Public Health, Yale University
School of Medicine, 60 College Street, New Haven, CT 06520-8034

Capsaicin (8-methyl-*N*-vanillyl-6-nonenamide) is the major pungent principle present in *capsicum* fruits such as hot chili peppers which are commonly used as spices in food. Capsaicin has been found to possess a wide variety of pharmacological and physiological effects such as analgesic, antiinflammatory, and hypolipidemic activities. However, information on the effects of capsaicin on carcinogenesis or mutagenesis has been limited. In the present study, we have found the protective effect of capsaicin on the mutagenicity of dimethylnitrosamine (DMN). Thus, mutagenicity of DMN (83 mM) in *Salmonella typhimurium* TA100 was inhibited 31% and 41% by capsaicin at the concentrations of 0.13 mM and 0.4 mM, respectively. In a parallel experiment, the hydroxylation of *p*-nitrophenol (0.3 mM) catalyzed by rat hepatic cytochrome P-450 2E1 (CYP2E1) activity was inhibted 40% and 52%, respectively by 50 μM and 250 μM of capsaicin. *N*-Demethylation of DMN was also inhibited by capsaicin in a dose-dependent manner. Diallylsulfide (DAS), one of the major sulfur compounds present in garlic, has been reported to be a suicidal inhibitor of CYP2E1. We found that the mutagenic activity of DMN in *Salmonella typhimurium* TA100, in the presence of rat liver postmitochondrial supernatant fortified with NADPH, was markedly inhibited by DAS. These results suggest that capsaicin, like DAS, exerts its protective activity against DMN-induced mutagenesis through inhibition of hepatic CYP2E1 activity which is responsible for the activation of DMN to the mutagenic/carcinogenic species. The effect of capsaicin pretreatment on chemically-induced carcinogenesis is under investigation.
Supported by C.P.R.U. Developmental Grant (CA-42101) and Brystol-Myers Squipp Laboratory Science Training Program.

CHEMOPREVENTION OF MAMMARY PRENEOPLASIA: IN VITRO EFFECTS OF TUMOR INHIBITORS. **Nitin T. TELANG, Strang-Cornell Cancer Research laboratory, Cornell University Medical College, New York**

The RIII strain of mouse expresses the murine mammary tumor virus (MTV) and exhibits a high incidence of mammary hyperproliferation/hyperplasia prior to the appearance of adenocarcinoma. Established mammary epithelial cell lines RIII/MG (origin: non-involved mammary tissues) and RIII/Pr_1 (origin: mammary adenocarcinoma) were utilized in the present study to examine i) relative expression of biochemical and cellular markers in preneoplastic and neoplastic phenotype and ii) inhibition of the perturbed biomarkers in the preneoplastic RIII/MG cells by synthetic retinoids and other naturally-occurring tumor inhibitors. The biomarkers representing quantitative endpoints included: MTV-associated reverse transcriptase (MTV-RT) activity, C16α/C2 hydroxylation of estradiol, and anchorage-independent growth (AIG) in vitro. Both the cell lines exhibited AIG *in vitro* prior to tumorigenicity *in vivo* (Tumor incidence: RIII/MG: 60%, RIII/Pr_1 95% 16 weeks post transplantation). Treatment of RIII/MG cells with the highest non-cytotoxic doses of N-(4-hydroxyphenyl) retinamide (HPR), indole-3-carbinol (I3C) and eicosapentaenoic acid (EPA) induced a 63.3%, 72.1% and 51.1% inhibition in AIG, respectively. This antiproliferative effect was accompanied by corresponding decrease in MTV-RT activity and in C16α/C2 ratio of estradiol hydroxylation. Tumor inhibitor-induced differential down-regulation in the three biomarkers suggests involvement of distinct mechanisms for chemopreventive effects of synthetic and naturally-occurring agents.

ABERRANT HYPERPROLIFERATION IN MOUSE MAMMARY EPITHELIAL CELLS: A BIOMARKER FOR PRENEOPLASIA

Meena S. Katdare, Michael P. Osborne, Nitin T. Telang. Strang-Cornell Cancer Research Laboratory, Cornell University Medical College, New York

Aberrant hyperproliferation (AH) is a late occurring post-initiational event that precedes tumorigenesis in vivo. Experiments on the spontaneously immortalized, non-tumorigenic C57/MG and MMEC cells were designed to validate AH as a cellular marker for preneoplastic transformation in vitro. The cell lines were initiated with either chemical carcinogens or with Ras and myc oncogenes. Initiated cells were tested for AH by determining colony forming efficiency in anchorage-independent growth condition in vitro, and for tumorigenicity by the mammary fat pad transplantation assay in vivo. C57/MG and MMEC cells, upon treatment with chemical carcinogens or transfection with oncogenes, exhibited at least a 60-300 fold increase in AH relative to that seen in appropriate controls. Mammary fat pad transplantation of initiated cells produced rapidly growing tumors in about 4-6 weeks. The tumor-derived $T1/Pr_1$ and myc_3/Pr_1 cells (positive controls) exhibited at least an 800-900 fold increase in AH. Upregulation of AH in initiated mammary epithelial cells in vitro prior to tumorigenesis in vivo therefore provides evidence for AH as an endpoint for induction of mammary preneoplastic transformation. [Support: Indo-US Fulbright Fellowship # 17267; NIH grants R 29 CA 44741 and P 01 CA 29502].

Reduction in the number of azoxymethane induced colonic aberrant crypt foci in rats on high risk western style diet by subsequent dietary administration of β-carotene and Vitamin E

Narayan Shivapurkar, Zhaocheng Tang and Oliver Alabaster
Institute for Disease Prevention, George Washington University Medical Center, Washington D.C. 20037

Aberrant crypt foci (ACF) can be histochemically detected in unembedded segments of colon from rodents exposed to carcinogens and human colonic mucosae from colon cancer patients. Available evidence suggests that ACF are putative precursor lesions. Reduction in the number of colonic ACF in an experimental design where the cancerpreventive agent was administered to rodents before the induction of ACF has been commonly used as the criteria for evaluation of cancerpreventive potential. However, it is recognized that the high risk populations as those in United States are quite likely to already bear the ACF, with significantly higher malignant potential. Vitamin E, β-carotene, folic acid and wheat bran are some of the dietary agents which have been shown in some experimental models for their ability to inhibit or arrest the carcinogenic process that has already been started. Thus we decided to evaluate the effect of dietary administration of high fat, low fiber diet containing Vitamin E, β-carotene, folic acid or wheat bran (experimental diets) on the pre-existing ACF induced in Fischer rats exposed to AOM and administered high fat, low fiber diet (Contro) alone for 10 weeks before the crossover. Groups of rats were sacrificed at 4 weeks and 8 weeks after the crossover and number of ACF of different multiplicities were quantitated in the control and experimental diet. The remaining rats were continued on the respective diets til 30 weeks and effect on the colon tumor development was measured.The results show significant decrease in ACF of different multiplicities only in Vitamin E and β-carotene groups compared to control groups. The results also further showed significant inhibition in the development colon tumors only in Viamin E and β-carotene groups. The results suggest that Viamin E and β-carotene could have a potential cancer preventive value in western population with high risk to develop colon cancer.

Catabolism of Isobutyric Acid by Colonocytes. **Jerzy A. Jaskeiwicz, Yu Zhao, Yoshiharu Shimomura, David W. Crabb, and Robert A. Harris. Departments of Biochemistry & Molecular Biology and Medicine, Indiana University School of Medicine, Indianapolis, IN 46202-5122, USA and Department of Bioscience, Nagoya University of Technology, Nagoya 466, Japan.**

Isobutyric acid, a short-chain fatty acid found in the colon content of omnivores originates from two possible sources - as a product of degradation of valine from dietary or endogenous proteins or as a product of fermentation of dietary fiber. Both the amount and proportion of isobutyric acid originating from these two possible sources are unknown; nor has it been established if isobutyric acid in colon contents can be degraded via the valine catabolism pathway in colon mucosal cells. The latter is particularly important because an intermediate compound of this pathway, methacryl-CoA, is expected to be toxic under conditions that promote its accumulation.

To investigate this problem, colonocytes were isolated from rat colons and incubated in medium enriched with isobutyric acid and other substrates. Isobutyric acid was found to be rapidly catabolized by way of the valine pathway in colonocytes. Normal colonocytes express high amounts of 3-hydroxyisobutyryl-CoA hydrolase, an enzyme of importance in the maintenance of physiological methacryl-CoA concentration during the catabolism of isobutyrate. Colonocytes produce considerable amounts of 3-hydroxyisobutyrate as an end product during isobutyrate catabolism, but they also contain significant amounts of 3-hydroxyisobutyrate dehydrogenase and therefore can use the carbon of isobutyrate for anaplerosis. Butyric acid, another short-chain fatty acid produced in the colon, very effectively inhibits isobutyric acid catabolism by colonocytes.

ALTERATION OF CELLULAR METABOLISM OF A MODEL ENVIRONMENTAL CARCINOGEN BY DIETARY LIPIDS AND LIPIDS *IN VITRO*. Jan Zaleski[a], Patricia A. Richter[a], Gloria Y. Kwei[b] and Frederick C. Kauffman[a]. [a]Lab. for Cell. and Biochem. Toxicol., Rutgers University, Piscataway, NJ 08854, [b]Merck Res. Lab., West Point, PA 19486.

The impact of amount and type of fat consumed on development of certain chemically-induced and spontaneous tumors is well recognized; however, underlying biochemical mechanisms are not fully understood. To obtain some insights into this phenomenon, the influence of intra- and extracellular lipids on hepatic metabolism of benzo[*a*]pyrene (B[*a*]P), an ubiquitous environmental pollutant requiring activation by arylhydrocarbon hydroxylase (AHH) to exert its tumorigenic activity, was studied in freshly isolated rat hepatocytes. The action of this enzyme is necessary to provide precursors for formation of conjugated metabolites (glutathiones, glucuronides and sulfates) which could be excreted in bile and urine. Metabolism of B[*a*]P was quantified by determining formation of total oxygenated products and the three types of conjugates. It was found that: (i) elevation of triacylglycerol content in hepatocytes *in vivo* achieved by maintaining rats on a high-fat diet or *in vitro* by preincubation of cells with long-chain fatty acids decreased the rate of B[*a*]P metabolism by 20-30%, (ii) supplementing the hepatocyte suspension with glyceryl trioleate or choline phospholipids, but not glyceryl tripalmitin, cholesterol or cholesteryl oleate, in amounts normally present in human plasma, attenuated B[*a*]P metabolism by 40-50%, (iii) the inhibitory effect of triacylglycerols and phospholipids on B[*a*]P metabolism was also observed when they were added as natural components of low-density lipoproteins or normal serum. The ability of certain circulatory and intracellular storage lipids to modulate the rate of B[*a*]P metabolism was related to their capacity to sequester this highly lipophilic compound, thereby limiting its access to AHH. Lipid-related attenuation of B[*a*]P metabolism in the liver may not only slow down B[*a*]P elimination from the body, but may also promote delivery of B[*a*]P to extrahepatic tissues susceptible to tumor induction.
(Supported by NIEHS Center Grant ES-05052 and NIH Grant CA-20807)

Maternal dietary fat increases breast cancer risk in female offspring

L. Hilakivi-Clarke, I. Onojafe, E. Cho, R. Clarke, and M.E. Lippman
Lombardi Cancer Research Center, Georgetown University, 3800 Reservoir Rd, NW, Washighton, DC 20007

Our data indicate that an exposure to high levels of dietary fat during fetal life increases the incidence of DMBA-induced mammary tumors in rats. Pregnant rats were exposed to diets containing 46% calories from fat (corn oil: n-6 PUFA) or 12% calories from fat. The diets were made either isocaloric by increasing the fiber content of the high fat diet, or the high fat diet had slightly higher caloric content. In the latter case, animals on the low fat diet ate more and had higher daily caloric intake than the animals on the high fat diet. The low fat diets contained also more carbohydrates than the high fat diets. Consequently, the body weights of offspring were higher among the low fat group. We dissected the 4th abdominal mammary glands from 35- and 75-day-old rats that were exposed to high- or low-fat diets *in utero*. Changes suggesting alterations in the differentiation of the mammary glands were noted. Both in the isocaloric and non-isocaloric groups, the tumor latency was shorter and the proportion of animals with mammary tumors were significantly higher among the rats who were exposed *in utero* via a maternal high-fat diet than in the low-fat group. The size of the DMBA-induced mammary tumors upon first detection and the growth rate of tumors were similar in rats exposed to low fat or high fat diet *in utero*. These endpoints more closely reflect tumor promotion/progression events (*i.e.* rate of growth of initiated cells), and strongly suggest that the effects of high fat diets are more closely associated with influencing early (initiation) events. We are currently investigating the possibility that estrogen or estrogen-regulated growth factors mediate the effects of early dietary fat on the risk to develop breast cancer.

INDEX